Climate Change and Public Health

Climate Change and Public Health

Second Edition

Edited by

BARRY S. LEVY AND JONATHAN A. PATZ

With the assistance of

HEATHER L. McSTOWE

OXFORD
UNIVERSITY PRESS

Oxford University Press is a department of the University of Oxford. It furthers
the University's objective of excellence in research, scholarship, and education
by publishing worldwide. Oxford is a registered trade mark of Oxford University
Press in the UK and certain other countries.

Published in the United States of America by Oxford University Press
198 Madison Avenue, New York, NY 10016, United States of America.

© Oxford University Press 2024

Library of Congress Cataloging-in-Publication Data
Names: Levy, Barry S., editor. | Patz, Jonathan A., editor.
Title: Climate change and public health / Barry S. Levy and Jonathan A. Patz
Description: Second edition. | New York, NY : Oxford University Press, [2024] |
Includes bibliographical references and index. |
Identifiers: LCCN 2023032624 (print) | LCCN 2023032625 (ebook) |
ISBN 9780197683293 (paperback) | ISBN 9780197683316 (epub) |
ISBN 9780197683323
Subjects: LCSH: Public health. | Environmental health.
Classification: LCC RA427 .C482 2024 (print) | LCC RA427 (ebook) |
DDC 613/.1—dc23/eng/20230921
LC record available at https://lccn.loc.gov/2023032624
LC ebook record available at https://lccn.loc.gov/2023032625

DOI: 10.1093/oso/9780197683293.001.0001

Printed by Marquis Book Printing, Canada

To the next generations of scientists, policymakers, and activists who are taking bold steps to address the climate crisis and are leading us to a healthy, equitable, and sustainable future.

Contents[*]

Foreword to the First Edition   xi

Preface   xiii

Acknowledgments   xv

About the Editors   xvii

Contributors   xix

Disclaimer   xxv

PART I: INTRODUCTION

1. **Applying a Public Health Context to Climate Change**   3
   Jonathan A. Patz and Barry S. Levy

   Box 1-1: Disparities   15

   Box 1-2: A Brief History of Organizational Responses to
   Climate Change   20

2. **Applying Climate Science to Climate Change and Extreme
   Weather Events**   30
   Stephen J. Vavrus, Aimee Puz, Samuel Kruse, and Jonathan A. Patz

PART II: HEALTH IMPACTS

3. **Heat-Related Disorders Among Workers**   51
   Tord Kjellstrom, Jeremy Lim, and Jason K.W. Lee

   Box 3-1: Safe Work Depends on Work Intensity Level
   and Temperature   55

4. **Heat-Related Disorders Among Community Populations**   70
   Rupa Basu and Xiangmei (May) Wu

5. **Respiratory Disorders**   86
   *Ioana O. Agache, Vanitha Sampath, Juan Aguilera,
   and Kari C. Nadeau*

[*] Authors of textboxes written by one or more of the chapter's authors are not shown in the Contents.

Box 5-1: Occupational Respiratory Disorders 87
Crystal M. North and David C. Christiani

Box 5-2: Achieving Health and Climate Co-Benefits by Reducing
Household Air Pollution from Solid Cooking Fuels 100
Lisa M. Thompson and Jamesine V. Rogers Gibson

6. **Vectorborne Diseases** 109
Christopher M. Barker and William K. Reisen

7. **Waterborne Diseases** 133
Jennifer R. Bratburd and Sandra L. McLellan

8. **Food Insecurity and Malnutrition** 153
Jessica Fanzo, Kate R. Schneider, and Stanley Wood

Box 8-1: Threats from Plant Pathogens 158
Caitilyn Allen

9. **Mental Health Impacts** 180
Thomas J. Doherty and Amy D. Lykins

Box 9-1: Assessing and Responding to Mental Health Impacts
of Wildfires in Alaska 183
Micah Hahn

Box 9-2: Australia's "Black Summer Bushfire Season" 185

10. **Violence** 205
Barry S. Levy

Box 10-1: Migration Due to Climate Change 215

PART III: DEVELOPING AND IMPLEMENTING
POLICIES FOR MITIGATION

11. **The Public Policymaking Process and the Power of Participation** 225
Kathleen M. Rest

Box 11-1: Examples of Existing Federal Laws, Regulations, and
Other Policies Related to Climate Change 227

Box 11-2: Local Community Takes Action on Deadly Heat 238
Roseann Bongiovanni

Box 11-3: Approaching Climate Action Through the Lens
of Local Needs   239
Shimekia Nichols and James Gignac

Appendix: Glossary of Some Policymaking Terms   248

12. **Energy Policy**   251
Nova M. Tebbe, Nicholas A. Mailloux, and Gregory F. Nemet

Box 12-1: Air Quality and Health Co-Benefits of Mitigation Policies   252

13. **Transportation Policy**   272
Kathryn A. Zyla

Box 13-1: Epidemiological Evidence for the Health Co-Benefits of
Active Transportation   274
Natalie Levine and Maggie L. Grabow

14. **Agriculture Policy**   292
Valerie J. Stull and Jonathan A. Patz

PART IV: DEVELOPING AND IMPLEMENTING ACTIONS FOR ADAPTATION

15. **Developing and Implementing Interventions for
Health Adaptation**   321
Kristie L. Ebi and Peter Berry

Box 15-1: Guiding Principles for Vancouver Coastal Health
and Fraser Health Climate Change and Health Adaptation
Framework   325

Box 15-2: Systems Approach to Climate Services for Health
in Ethiopia   328

Box 15-3: Combining Mitigation and Adaptation:
Climate-Proofing of Healthcare Facilities   331

16. **Planning Healthy and Sustainable Built Environments**   339
Jason Vargo

17. **Promoting Health Through Nature-Based Climate Solutions**   360
Howard Frumkin, Brendan Shane, and Taj Schottland

Box 17-1: Ecosystem Services and Nature's Contributions to People   367

Box 17-2: How Many Trees Can the World Support?   371

PART V: STRENGTHENING PUBLIC AND POLITICAL SUPPORT

18. **Communicating the Health Relevance of Climate Change**   387
    Mona Sarfaty and Edward Maibach

    Box 18-1: Principles of Climate Change Communication   396
        Howard Frumkin and Edward Maibach

19. **Building Movements to Address Climate Change**   404
    Teddie M. Potter, Julia Frost Nerbonne, and Vishnu Laalitha Surapaneni

    Box 19-1: The Climate and Health Movement   410
        Linda Rudolph

    Box 19-2: Greening of the Healthcare Sector   413
        Gary Cohen

    Box 19-3: Three Innovative Climate Change Initiatives   416
        Olivia M. Dyrbye-Wright

20. **Promoting Climate Justice**   422
    Rohini J. Haar and Barry S. Levy

    Box 20-1: Steps for a Just Energy Transition   430

Index   439

Foreword to the First Edition

As a global society, we face a series of threats to the environment and natural resources, including climate change, water scarcity, desertification, deforestation, biodiversity loss, and dependence on dangerous sources of energy. These threats are interconnected and closely linked to increasing population pressure, poverty and socioeconomic inequalities, and the health—and, ultimately, the survival—of humankind.

Our current path is unsustainable. Short-sighted, often narrow, political and economic interests repeatedly supersede common interests and common responsibility. We have a "tragedy of the commons."

We, as a global society, need to recognize our shared responsibility to address these global threats. In order to successfully address these threats, we need a holistic, integrated, and truly cross-sectoral approach.

Public health professionals, working in close collaboration with environmental scientists, social scientists, and other professionals, have important roles to play in addressing climate change.

First, climate change has profound implications for public health, as well described in this book. Second, as health professionals and scientists, we have expertise in the methods and tools to systematically analyze and assess the health threats posed by climate change. Third, we have the capability to contribute to the policies and actions to mitigate climate change and adapt to it—and reduce its health consequences. And finally, we have the communication, leadership, and advocacy skills to help build the popular and political will to successfully address climate change.

As a physician and public health professional and as a political leader, it has been essential for me, over the past several decades, to take initiatives, witness and participate in the evolution of science, policy, and action to address climate change and other global environmental threats. This evolution began in the 1980s, when the World Commission of Environment and Development alerted the global community to the seriousness and interconnectedness of major global environmental threats and the need for a collaborative, interdisciplinary approach to address them. This evolution includes the 1992 Rio Summit, the critical work of the Intergovernmental Panel on Climate Change, and a number of other international, national, and local initiatives to study climate change and its consequences. They are essential to help raise public awareness and foster a shared sense of responsibility and cooperation to address it. Now there is scientific consensus regarding climate change and the major contribution of human activity in causing it.

There is much to be done, including identifying and studying climate-related health problems, raising awareness among and educating the public and policymakers, and developing and implementing policies and actions to adapt to climate change and mitigate its causes. This book will be an extremely valuable resource for informing and empowering health professionals, environmental scientists, social scientists, and others for their vital roles in this work.

Time is running out. We, as a global community, need to take action to mitigate and adapt to climate change—now.

Gro Harlem Brundtland, M.D., M.P.H.
Former Director General, World Health Organization
Former Prime Minister of Norway
Former Deputy Chair of "The Elders"

Preface

Climate change is having profound impacts on the public's health, including more heat-related disorders, respiratory disorders, vectorborne and waterborne diseases, malnutrition, mental health impacts, and violence.

Since the first edition of this book was published in 2015, the climate crisis is worsening. Atmospheric levels of carbon dioxide and methane, the two leading greenhouse gases, are continuing to increase. There have been substantially more reports of severe storms, floods, droughts, and wildfires associated with climate change. And scientific studies and other reports of climate-related health consequences have been steadily increasing. Since 2015, public awareness of climate change and its consequences has risen, responses by governmental agencies and nongovernmental organizations have substantially increased, and all sectors of society have heightened their awareness of and responses to climate change. As a reflection of the increased attention to these issues by the medical and scientific communities, the number of journal articles listed in PubMed in response to a search of "climate change AND public health" identified about 10,800 articles by the end of 2015 and more than 29,800 articles by late August 2023.

This book addresses climate change and its adverse health consequences. It also addresses what public health professionals, environmental scientists, and others can do to mitigate climate change and its consequences, adapt to climate change, and address inequities and promote climate justice.

Although this book is primarily designed for students and mid-career professionals in public health and environmental sciences, students and mid-career professionals in other fields will likely find the book to be informative and useful. We believe that it will also be of value to governmental agencies, nongovernmental organizations, professional associations, and other groups and organizations.

With 55 contributors, most of whom are leading scholars and practitioners, this book aims to be comprehensive, concise, and engaging to readers. Its contents represent a core curriculum that can be used in courses for students in the health professions, environmental sciences, and other fields.

Part I focuses on climate change in the contexts of public health and climate science. Part II describes the adverse health consequences of climate change and specific measures to prevent these consequences. Part III discusses policies for mitigation. Part IV explores actions for adaptation. And Part V presents various means of strengthening public and political support for addressing climate change. (See Figure 1-5 in Chapter 1.)

We developed this book to stimulate and support conversations about what needs to be done to address climate change in the context of public health—what we, as a society, do collectively to assure the conditions in which people can be healthy.

Barry S. Levy and Jonathan A. Patz
Sherborn, Massachusetts
Madison, Wisconsin
September 2023

Acknowledgments

Developing *Climate Change and Public Health* has involved the skills and resources of many people, to whom we are profoundly grateful.

We thank all of the contributors who wrote chapters and textboxes that reflect their expertise and insights. Their commitment to addressing climate change and promoting public health is evident in their work and their contributions to this book.

We express our deep appreciation to Heather McStowe for her excellent work in preparing multiple drafts of the manuscript, obtaining information, confirming citations, and coordinating communication with contributors.

We greatly appreciate the guidance, assistance, and support of Sarah Humphreville, Executive Editor for Public Health, and Emma Hodgdon and Emily Benitez, Project Managers, at Oxford University Press; Suriya Narayanan, Project Manager, at Newgen KnowledgeWorks; and Joann Woy, Copy-editor.

Finally, we express our gratitude and love to Nancy Levy and Jean Patz for their continuing inspiration, encouragement, and support.

Barry S. Levy and Jonathan A. Patz

About the Editors

Barry S. Levy, M.D., M.P.H., is an Adjunct Professor of Public Health at Tufts University School of Medicine and a consultant in environmental and occupational health. He has more than 50 years of experience in public health practice, education, research, policy development, and consultation. He has served as a medical epidemiologist with the Centers for Disease Control, a professor at the University of Massachusetts Medical School, a leader of several international health programs and projects, and president of the American Public Health Association. Dr. Levy has edited 20 previous multicontributor books, 10 of which have focused on environmental and occupational health, and he has written more than 250 journal articles and book chapters. He has taught environmental and occupational health for more than four decades. He is the author of *From Horror to Hope: Recognizing and Preventing the Health Impacts of War* (Oxford University Press, 2022). He has worked in Kenya, Thailand, China, Jamaica, and several other countries. He has received the Sedgwick Memorial Medal of APHA, the Duncan Clark Award of the Association of Prevention Teaching and Research, and other honors.

Jonathan A. Patz, M.D., M.P.H., is the Vilas Distinguished Achievement Professor and John P. Holton Chair of Health and the Environment at the University of Wisconsin-Madison. His faculty appointments are in the Nelson Institute for Environmental Studies and the Department of Population Health Sciences. He served as inaugural director of UW-Madison's Global Health Institute from 2011 to 2022. He has taught and conducted research on the health effects of climate change and global environmental change (planetary health) for 30 years and has written more than 200 scientific publications and edited several books on these subjects. Dr. Patz co-chaired the health report for the first Congressionally mandated US National Assessment on Climate Change and, for 15 years, served as a lead author for the United Nations Intergovernmental Panel on Climate Change, the organization that shared the 2007 Nobel Peace Prize. He has received many awards, including the Aldo Leopold Leadership Fellows Award, a Fulbright Scholarship, and the American Public Health Association's Homer Calver Award. He is an elected member of the National Academy of Medicine.

Dr. Levy and Dr. Patz edited the first edition of this book, which was published by Oxford University Press in 2015 and was selected as the Environmental Health Book of the Year.

Contributors

Ioana O. Agache, M.D., Ph.D.
Professor of Allergy and Clinical
    Immunology
Transylvania University of Brasov
Brasov, Romania
ibrumaru@unitbv.ro

Juan Aguilera, M.D., Ph.D., M.P.H.
Assistant Professor of Health Promotion
    and Behavioral Sciences
Center for Community Health Impact
The University of Texas Health Science
    Center at Houston School of
    Public Health
El Paso, TX
juan.aguilera@uth.tmc.edu

Caitilyn Allen, Ph.D.
Professor
Plant Pathology
University of Wisconsin-Madison
Madison, WI
callen@wisc.edu

Christopher M. Barker, M.S., Ph.D.
Professor
School of Veterinary Medicine
University of California, Davis
Director
Pacific Southwest Center of Excellence in
    Vector-Borne Diseases
Davis, CA
cmbarker@ucdavis.edu

Rupa Basu, Ph.D., M.P.H.
Chief, Air and Climate Epidemiology
    Section
Office of Environmental Health Hazard
    Assessment
California Environmental Protection Agency
Oakland, CA
rupa.basu@oehha.ca.gov

Peter Berry, Ph.D.
Adjunct Assistant Professor, Geography
    and Environmental Management
University of Waterloo
Waterloo, Ontario
Canada
pberry@uwaterloo.ca

Roseann Bongiovanni, M.P.H.
Executive Director
Greenroots
Chelsea, MA
roseannb@greenrootschelsea.org

Jennifer R. Bratburd, Ph.D.
Outreach Program Manager
Center for Sustainability and the Global
    Environment
University of Wisconsin-Madison
Madison, WI
bratburd@wisc.edu

David C. Christiani, M.D., M.P.H.
Professor of Medicine
Harvard Medical School
Elkan Blout Professor of Environmental
    Genetics
Harvard T. H. Chan School of
    Public Health
Boston, MA
dchris@hsph.harvard.edu

Gary Cohen
President and Co-Founder
Health Care Without Harm and Practice
    Greenhealth
Boston, MA
gary.cohen@hcwh.org

Thomas J. Doherty, Psy.D.
Licensed Psychologist
Sustainable Self, LLC
Portland, OR
thomas@selfsustain.com

Olivia M. Dyrbye-Wright
Undergraduate Research Assistant
University of Wisconsin-Madison
Madison, WI
odyrbyewrigh@wisc.edu

Kristie L. Ebi, Ph.D., M.P.H.
Professor
Center for Health and the Global
    Environment (CHanGE)
University of Washington School of
    Public Health
Seattle, WA
krisebi@uw.edu

Jessica Fanzo, Ph.D.
Professor of Climate
Director, Food for Humanity Initiative
Interim Director, International Research
    Institute for Climate and Society (IRI)
The Columbia Climate School
Columbia University
New York, NY
jf671@columbia.edu

Howard Frumkin, M.D., Dr.P.H.
Senior Vice President and Director, Land
    and People Lab
Trust for Public Land
Professor Emeritus, University of
    Washington School of Public Health
Seattle, WA
howard.frumkin@tpl.org

James Gignac, J.D.
Midwest Senior Policy Manager, Climate
    & Energy Program
Union of Concerned Scientists
Chicago, IL
jgignac@ucsusa.org

Maggie L. Grabow, Ph.D., M.P.H.
Primary Care Research Fellow
School of Medicine and Public Health
University of Wisconsin-Madison
Madison, WI
grabow@wisc.edu

Rohini J. Haar, M.D., M.P.H.
Assistant Adjunct Professor, Epidemiology
School of Public Health
University of California, Berkeley
Berkeley, CA
rohinihaar@berkeley.edu

Micah Hahn, Ph.D., M.P.H.
Associate Professor of
    Environmental Health
Institute for Circumpolar Health Studies
University of Alaska Anchorage
Anchorage, AK
mbhahn@alaska.edu

Tord Kjellstrom, Ph.D. (Med), M.M.E.
Director, Health and Environment
    International Trust
Nelson, New Zealand
Visiting Fellow
National Centre for Epidemiology and
    Population Health
Australian National University
Canberra, Australia
kjellstromt@yahoo.com

Samuel Kruse, M.S.
Research Assistant
Global Health Institute
University of Wisconsin-Madison
Madison, WI
samdkruse@gmail.com

Jason K.W. Lee, Ph.D.
Associate Professor, Human Potential
    Translational Research Programme
Director, Heat Resilience and
    Performance Centre
Yong Loo Lin School of Medicine
National University of Singapore
Singapore
phsjlkw@nus.edu.sg

Natalie Levine, M.S.
Researcher
Center for Sustainability and the Global
  Environment
Nelson Institute for Environmental
  Studies
University of Wisconsin-Madison
Madison, WI
natalie.levine@wisc.edu

Barry S. Levy, M.D., M.P.H.
Adjunct Professor of Public Health
Department of Public Health and
  Community Medicine
Tufts University School of Medicine
Sherborn, MA
blevy@igc.org

Jeremy Lim, M.B.B.S., M.P.H.
Associate Professor and Director,
  Leadership Institute for Global Health
  Transformation (LIGHT)
Saw Swee Hock School of Public Health
National University of Singapore
Singapore
jeremyfylim@gmail.com

Amy D. Lykins, Ph.D.
Associate Professor of Clinical Psychology
University of New England
Armidale, New South Wales
Australia
alykins@une.edu.au

Edward Maibach, M.P.H., Ph.D.
Director, Center for Climate Change
  Communication
University Professor, Department of
  Communication
George Mason University
Fairfax, VA
emaibach@gmu.edu

Nicholas A. Mailloux
Ph.D. Student
Center for Sustainability and the Global
  Environment
Nelson Institute for Environmental
  Studies
University of Wisconsin-Madison
Madison, WI
namailloux@wisc.edu

Sandra L. McLellan, Ph.D.
Lynde B. Uihlein Professor of
  Ecosystem Health
School of Freshwater Sciences
University of Wisconsin-Milwaukee
Milwaukee, WI
mclellan@uwm.edu

Kari C. Nadeau, M.D., Ph.D.
John Rock Professor of Climate and
  Population Studies
Chair, Department of
  Environmental Health
Harvard T.H. Chan School of
  Public Health
Boston, MA
knadeau@hsph.harvard.edu

Gregory F. Nemet, Ph.D.
Professor of Public Affairs
La Follette School of Public Affairs
University of Wisconsin-Madison
Center for Sustainability and the Global
  Environment
Madison, WI
nemet@wisc.edu

Julia Frost Nerbonne, Ph.D.
Adjunct Assistant Professor.
Department of Fisheries, Wildlife and
  Conservation Biology
University of Minnesota
Executive Director
Minnesota Interfaith Power & Light
Minnesota, MN
nerbonne@umn.edu

Shimekia Nichols
Executive Director
Soulardarity
Highland Park, MI
shimekia@soulardarity.com

Crystal M. North, M.D., M.P.H.
Assistant Professor
Division of Pulmonary and Critical Care
    Medicine, Department of Medicine
Harvard Medical School
Boston, MA
cnorth@mgh.harvard.edu

Jonathan A. Patz, M.D., M.P.H.
Vilas Distinguished Achievement
    Professor
John P. Holton Chair of Health and the
    Environment
University of Wisconsin-Madison
Madison, WI
patz@wisc.edu

Teddie M. Potter, Ph.D., R.N.
Clinical Professor, School of Nursing
Director, Center for Planetary Health and
    Environmental Justice
University of Minnesota
Minneapolis, MN
tmpotter@umn.edu

Aimee Puz
Undergraduate Research Assistant
University of Wisconsin-Madison
Madison, WI
apuz@wisc.edu

William K. Reisen, Ph.D.
Professor Emeritus, Center for
    Vectorborne Diseases
Department of Pathology, Microbiology
    and Immunology
School of Veterinary Medicine
University of California, Davis
Davis, CA
wkreisen@ucdavis.edu

Kathleen M. Rest, Ph.D., M.P.A.
Senior Fellow
Boston University Institute for Global
    Sustainability
Former Executive Director
Union of Concerned Scientists
Boston, MA
kathleen.rest@gmail.com

Jamesine V. Rogers Gibson, M.P.H.
Senior Advisor-Climate
Bay Area Air Quality Management
    District
San Francisco, CA

Linda Rudolph, M.D., M.P.H.
Senior Policy Advisor, Climate, Health
    and Equity
Public Health Institute
Medical Society Consortium on Climate
    and Health
Oakland, CA
Linda.Rudolph@phi.org

Vanitha Sampath, Ph.D.
Research Scientist
Harvard T.H. Chan School of
    Public Health
Boston, MA
vsampath@hsph.harvard.edu

Mona Sarfaty, M.D., M.P.H.
Executive Director Emeritus and Founder
Medical Society Consortium on Climate
    and Health
Center for Climate Change
    Communication
George Mason University
Fairfax, VA
msarfaty@gmu.edu

Kate R. Schneider, Ph.D., M.P.A.
Research Scholar
Johns Hopkins University, School of
    Advanced International Studies
Washington, DC
kschne29@jhu.edu

Taj Schottland
Associate Director, Climate Program
Trust for Public Land
Montpelier, VT
taj.schottland@tpl.org

Brendan Shane, J.D., M.S.
Director, Climate Program
Trust for Public Land
Washington, DC
brendan.shane@tpl.org

Valerie J. Stull, Ph.D., M.P.H.
Research Scientist
Center for Sustainability and the Global
    Environment
Nelson Institute for Environmental
    Studies
University of Wisconsin-Madison
Madison, WI
vstull@wisc.edu

**Vishnu Laalitha Surapaneni,
M.D., M.P.H.**
Assistant Professor of Medicine, Division
    of Hospital Medicine
University of Minnesota Medical School
Minneapolis, MN
vsurapan@umn.edu

Nova M. Tebbe, M.P.H., M.P.A.
Ph.D. Student
Center for Sustainability and the Global
    Environment
University of Wisconsin-Madison
Madison, WI
ntebbe@wisc.edu

Lisa M. Thompson, R.N., M.S., Ph.D.
Professor
Nell Hodgson Woodruff School of
    Nursing
Gangarosa Department of Environmental
    Health, Rollins School of Public Health
Emory University
Atlanta, GA
lisa.thompson@emory.edu

Jason Vargo, M.C.R.P., M.P.H., Ph.D.
Senior Researcher in Climate Change and
    Health Equity
San Francisco, CA

Stephen J. Vavrus, Ph.D.
Senior Scientist
Nelson Institute Center for Climatic
    Research
University of Wisconsin-Madison
Madison, WI
sjvavrus@wisc.edu

Stanley Wood, Ph.D.
Senior Program Office, Agricultural
    Development
Bill & Melinda Gates Foundation
Affiliate Professor
Evans School of Public Policy and
    Governance
University of Washington
Seattle, WA
stanley.wood@gatesfoundation.org

Xiangmei (May) Wu, Ph.D.
Research Scientist, Air and Climate
    Epidemiology Section
Office of Environmental Health Hazard
    Assessment
California Environmental
    Protection Agency
Oakland, CA
Xiangmei.Wu@oehha.ca.gov

Kathryn A. Zyla, M.E.M., J.D.
Senior Lecturer
Georgetown Law
Executive Director
Georgetown Climate Center
Washington, DC
zyla@law.georgetown.edu

Disclaimer

Authors' statements are independent of the institutions, agencies, or organizations with which they are affiliated. Chapter authors are not responsible for the statements of other authors of textboxes within their chapters, and textbox authors are not responsible for the statements of authors of chapters within which their textboxes appear.

PART I
INTRODUCTION

1

Applying a Public Health Context to Climate Change

Jonathan A. Patz and Barry S. Levy

The northwestern United States and western Canada experience a massive heat wave, in which more than 1,037 people die and many areas have prolonged temperatures over 40°C (104°F). The temperature in one town reaches 49.6°C (121.3°F). Forty-eight hours later, it burns to the ground.

A monsoon season in Pakistan, three times stronger than average, causes extensive flooding, submerging one-third of the country. More than 1,300 people die.

Extreme rainfall occurs in Western Europe, causing extensive floods that result in 200 deaths and considerable damage to infrastructure.

A severe drought and an extreme heat wave occur in China. Falling water levels in major rivers limit shipping, leading to food shortages and factory closures as well as a 50% reduction in hydroelectric power.

Somalia and other countries in the Horn of Africa experience a prolonged drought, which destroys crops, kills livestock, and leads to extreme hunger for 13 million people.

In Cambodia, tropical storms, monsoon rains, and mountain snowmelt cause flooding of the Mekong River and its tributaries, leading to diarrheal diseases in children.

In large cities in South America, people with chronic respiratory disorders experience frequent exacerbations on summer days, when there are high ambient ozone concentrations.

Jonathan A. Patz and Barry S. Levy, *Applying a Public Health Context to Climate Change* In: *Climate Change and Public Health*. Edited by: Barry S. Levy and Jonathan A. Patz, Oxford University Press. © Oxford University Press 2024.
DOI: 10.1093/oso/9780197683293.003.0001

In Norway, tick-borne Lyme disease cases increase as temperatures increase. The annual peak in cases is now 6 weeks earlier than it was 25 years ago.

In Australia, increased community violence occurs on hot summer days.

And throughout the world, people suffer from anxiety, depression, and other mental disorders related to climate change and its adverse consequences.

These are actual occurrences that demonstrate some of the environmental impacts and resultant health consequences of climate change, which is now well established as being caused by human activities.[1] There is scientific consensus that human-caused climate change is increasing the frequency and intensity of extreme weather events and other adverse consequences of climate change (Chapter 2).

This book summarizes what health professionals, environmental scientists, and others know about climate change and its adverse health consequences and what can be done to mitigate its causes and adapt to its consequences.

DEFINITIONS

Climate change has been defined as "a change of climate that is attributed directly or indirectly to human activity that alters the composition of the global atmosphere and that is in addition to natural climate variability observed over comparable time periods."[2] Throughout this book, we use the term *climate change* to mean both climate change and increased climate variability.

The terms *climate* and *weather* are distinct. *Climate* has been defined as "the average course or condition of the weather at a place usually over a period of years as exhibited by temperature, wind velocity, and precipitation"[3] and as "the weather conditions prevailing in an area in general over a long period."[4] In contrast, *weather* has been defined as "the state of the air and atmosphere at a particular time and place: the temperature and other outside conditions (such as rain, cloudiness, etc.) at a particular time and place"[5] and "the state of the atmosphere with respect to heat or cold, wetness or dryness, calm or storm, clearness or cloudiness."[6] It has also been defined as "the state of the atmosphere at a place and time as regards heat, dryness, sunshine, wind, rain, etc."[4] Climate scientists frequently use a period of 30 years to distinguish between *climate* and *weather*.

Public health has been defined as "what we, as a society, do collectively to assure the conditions in which people can be healthy."[7] The three major categories of public health activities are the following:

- <u>Assessment</u>: Collecting, analyzing, and disseminating data, including performing surveillance and monitoring for health effects and antecedent exposures and other risk factors
- <u>Policy development</u>: Developing, implementing, and evaluating policies, in both the health sector and in other sectors, including energy, transportation, and agriculture
- <u>Assurance</u>: Developing, implementing, and evaluating broad approaches, including planning for disaster preparedness and response, building sustainable communities, and developing renewable energy.

All of these activities are relevant in applying a public health context to climate change.

MECHANISMS

Climate change, whether it is caused by natural variability or human activity, depends on the balance between incoming (solar) short-wave radiation and outgoing (infrared) long-wave radiation (Figure 1-1). This balance is affected by the atmosphere in much the same as a blanket affects heat loss from skin. Greenhouse gases (GHGs) act in a similar way as a blanket on the Earth. (See a more detailed discussion on how GHGs cause global warming in Chapter 2.)

Table 1-1 describes key information concerning major GHGs. While carbon dioxide accounts for 79% of GHG emissions in the United States and can remain

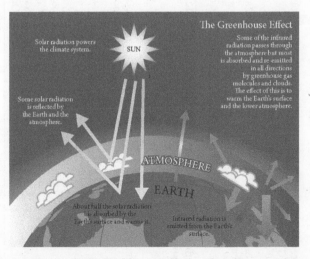

Figure 1-1 A schematic diagram of the greenhouse effect. (*Source*: Adapted from: Solomon S, Qin D, Manning M, et al. [eds.]. Climate Change 2007: The Physical Science Basis. Cambridge, UK: Cambridge University Press, 2007.)

Table 1-1 Key information about major greenhouse gases

Parameter	Carbon dioxide	Methane	Nitrous oxide	Fluorinated gases[a]
Percentage of U.S. GHG emissions in 2020	79%	11%	7%	3%
Global warming potential (100-year)	1	28	298	HFCs: Up to 14,800 PFCs: Up to 12,200 NF_3: 17,200 SF_6: 22,800
Lifetime in atmosphere	Varies (can be up to 120 years)	12 years	114 years	HFCs: Up to 270 years PFCs: 2,600–50,000 years NF_3: 740 years SF_6: 3,200 years
Sources	Transportation: 33% Electric power: 31% Industry: 16% Residential and commercial: 12% Other (non-fossil fuel combustion): 8%	Natural gas and petroleum systems: 32% Enteric fermentation: 27% Landfills: 17% Manure management: 9% Coal mining: 6% Other: 9%	Agricultural soil management: 74% Wastewater treatment: 6% Stationary combustion: 5% Chemical production and other product uses: 5% Manure management: 5% Transportation: 4% Other: 1%	Substitution of ozone depleting substances: 93% Electronics industry: 2% Electrical transmission and distribution: 2% HCFC-22 production: 1% Production and processing of aluminum and magnesium: 1%

[a]Fluorinated gases consist of hydrofluorocarbons (HFCs), perfluorocarbons (PFCs), nitrogen trifluoride (NF_3), and sulfur hexafluoride (SF_6)

Source: U.S. Environmental Protection Agency. Overview of Greenhouse Gases. Available at: https://www.epa.gov/ghgemissions/overview-greenhouse-gases. Accessed March 13, 2023.

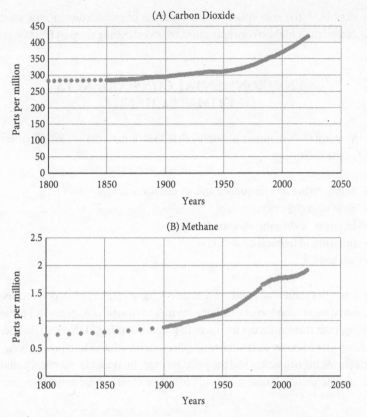

Figure 1-2 Atmospheric levels of carbon dioxide (A) and methane (B) in parts per million by volume, averaged over each calendar year, 1800-2021. For carbon dioxide (CO_2), levels up to 1958 are from ice core data, and since 1958 from the Mauna Loa Observatory in Hawaii. For methane (CH_4), levels before 1984 are from ice core data, and since 1984 from the Mauna Loa Observatory. (*Sources*: CO_2 data for 1800-1850: https://www1.ncdc.noaa.gov/pub/data/paleo/icecore/antarctica/law/law_co2.txt; CO_2 data for 1850-1957: http://data.giss.nasa.gov/modelforce/ghgases/Fig1A.ext.txt; and CO_2 data for 1958 to the present: https://www.esrl.noaa.gov/gmd/webdata/ccgg/trends/co2/co2_annmean_mlo.txt. CH_4 data for 1800-1983: https://cdiac.ess-dive.lbl.gov/ftp/trends/atm_meth/EthCH498B.txt; and CH_4 data for 1984 to the present: https://www.esrl.noaa.gov/gmd/webdata/ccgg/trends/ch4/ch4_annmean_gl.txt.)

in the atmosphere for decades, it is less potent than methane and far less potent than nitrous oxide and fluorinated gases.

Atmospheric concentrations of GHGs, such as carbon dioxide and methane, have increased substantially over the past several decades (Figure 1-2). Since the beginning of the Industrial Era in the 1800s, atmospheric concentrations of carbon dioxide have increased by more than 50% and those of methane have more than doubled.[8] The atmospheric concentrations of carbon dioxide and methane now exceed their highest concentrations recorded during the past 800,000 years.[9]

ENVIRONMENTAL CONSEQUENCES OF CLIMATE CHANGE

Environmental phenomena possibly related to climate change include increases in all of the following:

- Temperature and frequency and/or duration of heat waves
- Heavy precipitation events
- Intensity and/or duration of droughts
- Intensity of tropical cyclones
- Sea level.[10]

The Intergovernmental Panel on Climate Change (IPCC) has determined the likelihood for attribution of weather extremes to climate change and for projected changes with three different increases in global warming (Table 1-2). In addition to those listed above, environmental phenomena related to climate change include shrinking of glaciers and the polar ice caps, increases in chemical pollutants and aeroallergens in ambient air, and ecosystem changes that reduce biodiversity.

Increased Temperature

Between 1880 and 2020, the average surface temperature of the Earth increased 1.09°C (1.96°F) (Figure 1-3).[1] From 1990 to 2018, more than 30% of heat-related deaths from 43 countries have been attributed to increased temperature.[11] In the United States, the frequency of daily temperatures over 38.0°C (100.4°F) is expected to increase substantially; temperature levels that now occur once in 20 years could, in the future, occur every 2 to 4 years.[12]

Extreme heat events are projected to become longer, more severe, and more frequent (Chapters 3 and 4).[13,14] Substantially increased temperatures have been observed since 2012. The 2016-2020 period was hotter than any other 5-year period in the instrumental record by that time (since approximately 1850).[1] In some areas, prolonged periods of record high temperatures associated with droughts contribute to dry conditions, which increase the risk of wildfires (Chapter 5).[14,15]

Changes in Precipitation

Globally, in the second half of this century, El Niño-related variability of precipitation will intensify. However, in many dry regions, mean precipitation will

Table 1-2 Observed changes in extremes, their attribution since 1950 (except where stated otherwise), and projected changes at +1.5°, +2.0°, and +4.0°C of global warming, on global and continental scales

Change in Indicator	Observed Since 1950	Attributed Since 1950	Projected at global warming level		
			+1.5°C	+2.0°C	+4.0°C
Warm/hot extremes: Frequency or intensity	↑***	✓*** Main driver	↑***	↑***	↑***
Cold extremes: Frequency or intensity	↓***	✓*** Main driver	↓***	↓***	↓***
Heavy precipitation events: Frequency, intensity, and/or amount	↑* Over majority of land regions with good observational coverage	✓* Main driver of the observed intensification of heavy precipitation in land regions	↑* In most land regions	↑* In most land regions	↑** In most land regions
Agricultural and ecological droughts: Intensity and/or frequency	↑ In some regions	✓ In some regions	↑* In more regions compared to observed changes	↑* In more regions compared to 1.5°C of global warming	↑** In more regions compared to 2.0°C of global warming
Precipitation associated with tropical cyclones	↑	✓*	↑* Rate +11%	↑* Rate +14%	↑* Rate +28%
Tropical cyclones: Proportion of intense cyclones	↑*	✓	↑* +10%	↑* +13%	↑* +20%
Compound events: Co-occurrent heat waves and droughts	↑* Frequency	✓* Frequency	↑* Frequency and intensity increases with warming	↑* Frequency and intensity increases with warming	↑* Frequency and intensity increases with warming

(*continued*)

Table 1-2 Continued

Change in Indicator	Observed Since 1950	Attributed Since 1950	Projected at global warming level		
			+1.5°C	+2.0°C	+4.0°C
Marine heat waves: Intensity and frequency	↑* Since 1900	✓** Since 2006	↑* Strongest in tropical and Arctic Ocean	↑* Strongest in tropical and Arctic Ocean	↑* Strongest in tropical and Arctic Ocean
Extreme sea levels: Frequency	↑* Since 1960	✓	↑** Scenario-based assessment for 21st century	↑** Scenario-based assessment for 21st century	↑** Scenario-based assessment for 21st century

***Virtually certain, **Very likely, *Likely/high confidence, No asterisks = Medium confidence

Source: Arias PA, Bellouin N, Coppola E, et al. Technical Summary. In V Masson-Delmotte, P Zhai, A Pirani, et al. [eds.]. Climate Change 2021: The Physical Science Basis. Contribution of Working Group I to the Sixth Assessment Report of the Intergovernmental Panel on Climate Change. Cambridge, UK and New York: Cambridge University Press, 2021, p. 67. Available at: https://www.ipcc.ch/report/ar6/wg1/downloads/report/IPCC_AR6_WGI_TS.pdf. Accessed March 13, 2023.

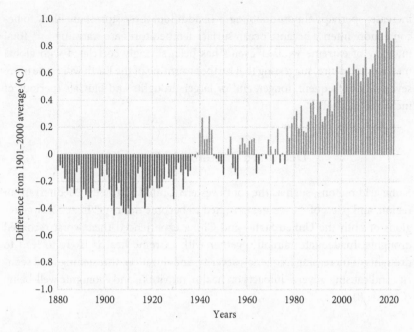

Figure 1-3 Yearly surface temperature from 1880 to 2022, compared to the 20th-century average. Bars below 0 indicate cooler-than-average years, and those above 0 indicate warmer-than-average years. (*Source*: National Oceanic and Atmospheric Administration Climate. gov graph, based on data from the National Centers for Environmental Information.)

decrease.[8] In parts of the United States, precipitation is projected to become less frequent,[16] but episodes of heavy rainfall are projected to become more frequent because warmer air holds more water vapor,[17–19] continuing a trend over the past several decades.[10] Flooding from heavy rainfall events causes injuries and, because of sewage contamination of drinking water, cases of infectious water-borne diseases (Chapter 7). In many regions, heavy rainfall will lead to increased flooding from surface runoff that is not adequately transported into and through underground drainage systems.[20] In 2022, about 58 million people globally were adversely affected by floods, about 50% more than the average of the previous 5 years.[21] Floods are expected to increase globally, especially in Asia, Africa, South America, and Central America.[9]

Extreme Weather Events

While it is not clear if the frequency of all hurricanes and cyclones will increase, the frequency of strong (Category 4 and Category 5) hurricanes will likely increase—perhaps almost doubling.[22] In addition, these storms are likely to become more intense because ocean surface temperatures are warming.[23,24] Total intensity of extreme weather events has been strongly correlated with global mean temperature, suggesting that further warming of the Earth will cause more severe, more frequent, longer, and/or larger droughts and pluvials (periods of increased rainfall).[25]

Decreased Rainfall and Drought

Some arid regions, such as the southwestern United States, the Mediterranean region, and parts of Africa, are expected to become drier.[10,16] In 2022, large regions of both the United States and China experienced their worst recorded droughts. Inadequate rainfall together with extreme heat is likely to lead to droughts of increasing frequency, severity, and duration, threatening food security and causing adverse impacts on health, nutrition, and economic well-being (Chapter 8).

Wildfires

Increased temperatures and decreased rainfall are projected to increase the occurrence of wildfires.[26] Wildfires adversely affect health directly and indirectly by emitting particulate matter, toxic gases, and other air pollutants. Short-term exposure to fine particulate matter related to wildfires has been associated with all-cause, cardiovascular, and respiratory mortality (Chapter 5).[27] And wildfires are associated with severe mental health impacts (Chapter 9).

Changes in Land Use

Additive and possibly synergistic effects can occur due to the combined impacts of climate change and changes in land use, as illustrated by the following examples:

- As residential areas extend into flood plains and vulnerable coastal areas, the consequences of floods and coastal storm surges will likely increase.
- In deforested areas, heavy rainfall is likely to increase landslides. In the past, strong hurricanes have caused thousands of deaths in Central America and elsewhere due to mudslides from deforested hillsides.[28]
- In 2005, Hurricane Katrina and resultant flooding killed more than 1,800 people (mainly in Louisiana), injured many more, and displaced thousands. The severity of its consequences was probably increased by

receding coastal wetlands, which provided reduced buffering from storm surges.[29]

• In cities, increased frequency, intensity, and duration of heat waves together with dark surfaces, buildings, and industrial activities can create the *urban heat island effect* (Chapter 4). And cities with few trees receive less cooling from evapotranspiration. (See Chapters 16 and 17.)

Sea-Level Rise

Global mean sea level has increased approximately 20 centimeters (8 inches) during the past century—far more than in the previous 2,000 years.[9] Sea-level rise will exacerbate storm surges, worsen coastal erosion, and inundate low-lying areas, including Small Island Developing States. It will also cause salinization of aquifers, exacerbating hypertension and other diseases in people living in coastal areas. (See Chapter 2.)

Ocean Acidification

Since the start of the Industrial Era, acidification of oceans has been occurring due to absorption of increasing atmospheric concentrations of carbon dioxide.[8] Ocean acidification is projected to increase throughout this century, threatening shell-forming organisms, such as corals and fish species that feed on them and, ultimately, individuals and communities that depend on healthy fisheries.

FRAMING APPROACHES TO CLIMATE CHANGE AND HEALTH

Climate change is caused by human activity, primarily from the combustion of fossil fuels and deforestation.[30-32] However, there are many remaining questions concerning (a) the risks and consequences of climate change; (b) vulnerabilities due to socioeconomic, demographic, geographic, and other factors; and (c) the effectiveness and feasibility of various measures to mitigate and adapt to climate change.

ADVERSE HEALTH CONSEQUENCES

As detailed in Part II of this book, there is a wide range of adverse health consequences—both direct and indirect—that can occur as a result of climate change. These health consequences include:

- Heat-related disorders (Chapters 3 and 4)
- Respiratory disorders (Chapter 5)
- Vectorborne and waterborne infectious diseases (Chapters 6 and 7)
- Malnutrition (Chapter 8)
- Mental health impacts (Chapter 9)
- Violence (Chapter 10).

Mental health consequences have often been overlooked. Increased temperatures, extreme weather events, sea-level rise, and other consequences of climate change adversely affect mental health. One analysis estimated that, for every 1.0°C (1.8°F) increase in ambient temperature, there will be a 2% increase in deaths due to mental disorders, especially those related to substance abuse.[33]

Vulnerable Populations

Vulnerability to these consequences is determined by several factors, including level of exposure to climate risk factors, sensitivity to these factors, and capacity to adapt to them.[34,35] Poverty, minority health status, female gender, young or old age, diseases, and disabilities place people at higher risk of these consequences. Geography influences vulnerability to these consequences; for example, Arctic peoples, such as the Inuit, are suffering from major consequences due to exceptional warming in the Arctic (Box 9-1 in Chapter 9).[36]

Workers in many occupations are also at increased risk for the health consequences of climate change. Like "canaries in a coal mine," outdoor laborers who are exposed to both hot ambient temperatures and internal body heat generated by work exertion may suffer the earliest impacts of climate change—thereby providing early warning of the health consequences of climate change (Chapter 3).[37]

Pregnant women represent another vulnerable population. For example, pregnant women in low-lying coastal areas who consume groundwater with increased salt concentration as a result of sea-level rise have an increased risk of developing preeclampsia and gestational hypertension, which can be life-threatening.[38] And pregnant women exposed to increased ambient temperature have an increased risk of delivering a pre-term infant.[39]

Inequities

The health consequences of climate change are generally highest in marginalized populations, including people with low income, members of racial and ethnic minority groups, Indigenous Peoples, and other disadvantaged groups. These populations have generally had few resources to adapt to climate change and limited input into policies addressing climate change. (See Box 1-1.)

Globally, there are great inequalities among countries with respect to both (a) amounts of GHG emissions and (b) magnitude and severity of the health

---

Box 1-1 Disparities

Barry S. Levy

Disparities in the health consequences of climate change are determined by (a) differences in the nature and magnitude of climate-related exposures, such as heat and humidity, floods and droughts, airborne contaminants, disease-carrying vectors, and extreme weather events; (b) socioeconomic status, geographic location, racial and ethnic discrimination, and immigration status; (c) sensitivity to climate-related factors, such as acclimatization to heat, respiratory allergies, nutritional status, and access to healthcare; and (d) capability to protect against and respond to climate-related health threats (resilience).[1,2] Some health consequences of climate change are clustered in urban areas because of population density, inadequate greenspace, and the urban heat island effect.[3] Others are clustered in rural areas because of wildfires, droughts, floods, and coastal erosion.[4]

Poverty is associated with vulnerability to climate-related hazards. Many low- and middle-income countries (LMICs) are projected to have losses in agricultural productivity of 30% or more due to climate change—which will worsen poverty and food security. More than 780 million people are exposed to the combined risks of poverty and major flooding, mainly in LMICs. Many LMICs are substantially poorer today than they would have been without climate change—a trend that might result in 80% or more lost income in many LMICs by 2100. Within countries, low-income people suffer more from the consequences of climate change than do high-income people.[5]

In the United States, Blacks and other people of color are at higher risk of climate-related health impacts than Whites, exacerbating racial disparities in health status. In urban areas, racial and socioeconomic groups that experience the most heat exposure are those with the poorest preexisting health status; for example, in the Boston area on a hot summer day, there can be a 3.3°C (6°F) difference in surface temperature between neighborhoods with the highest and lowest socioeconomic status.[6,7] In the United Kingdom, people who are members of racial or ethnic minority groups, immigrants, and others who face discrimination are more likely to suffer the health consequences of climate change and less likely to have the resilience, capability, and resources to respond to these consequences.[8]

The adverse effects of climate change are generally greater in LMICs than in high-income countries (HICs)—although LMICs have only minimally contributed to climate change. This gross disparity has resulted from global policies that undermine equity, preexisting disparities in health status between LMICs and HICs, exposure to extreme heat, high prevalence of vectorborne diseases, frequent human rights violations, and the presence of armed conflict in LMICs. Children in LMICs are more vulnerable to the health consequences of climate change than those in HICs because of limited structural and economic resources, higher prevalence of malnutrition and communicable diseases, poor sanitation, and less access to public health services and medical care.[9] In addition, LMICs have scarce resources to adapt to the most severe and long-term consequences of climate change[10]; for example, studies in New Delhi and elsewhere have found that socially vulnerable groups, such as children and agricultural workers, are inequitably exposed to urban heat without the mitigating effects of greenspace.[11] Small Island Developing States (SIDS) are vulnerable to sea-level rise and storm surges; they have limited resources to prepare for and respond to these challenges, and, because most SIDS are small, they cannot evacuate their populations inland.[12] And climate change will force millions of people to migrate (Box 10-1 in Chapter 10).

Old age, preexisting medical conditions, and social deprivation make people vulnerable to the adverse health impacts of climate change. Health education and public preparedness measures that consider differences in exposure, sensitivity, and adaptive capacity among groups help address health and social inequalities related to climate change. But adaptation strategies based on individual preparedness, action, and behavior change may worsen health and social inequalities due to their selective uptake unless they are implemented with broad public information campaigns and financial support.[8]

Climate change is adversely affecting housing by increasing costs for maintenance and repair after extreme weather events (and increasing insurance rates), by causing floods that lead to indoor moisture and mold, and by increasing threats from climate-related wildfires and storms. People of color, low-income individuals, older people, and individuals who are medically vulnerable are disproportionately affected by climate impacts on housing. And they face major barriers to preventing, responding to, and mitigating these adverse impacts.

Historically, Indigenous Peoples throughout the Global North have often been forcibly displaced to undesirable locations, such as flood-prone areas, unfarmable lands, and regions that could not support their populations.

Climate-related disasters have frequently impacted Indigenous Peoples because of these past policies and lack of reparation.[13]

Box References

1. Ebi KL, Hess JJ. Health risks due to climate change: Inequity in causes and consequences. Health Affairs (Millwood) 2020; 39: 2056–2062.

2. Smith GS, Anjum E, Francis C, et al. Climate change, environmental disasters, and health inequities: The underlying role of structural inequalities. Current Environmental Health Reports 2022; 9: 80–89.

3. Friel S, Hancock T, Kjellstrom T, et al. Urban health inequities and the added pressure of climate change: An action-oriented research agenda. Journal of Urban Health 2011; 88: 886. doi: 10.1007/s11524-011-9607-0.

4. Walker J. Rural health inequities and the impact of climate change. Australian Journal of Rural Health 2019; 27: 583–584.

5. Chancel L, Bothe P, Voituriez T. Climate Inequality Report 2023. World Inequality Lab Study 2023/1. Available at: https://wid.world/document/climate-inequality-report-2023/. Accessed February 22, 2023.

6. Renteria R. Social disparities in neighborhood heat in the Northeast United States. Environmental Research 2022; 111805. https://doi.org/10.1016/j.envres.2021.111805

7. Plumer B, Popovich N. How decades of racist housing policy left neighborhoods sweltering. New York Times, August 24, 2020.

8. Paavola J. Health impacts of climate change and health and social inequalities in the UK. Environmental Health 2017; 16(Suppl 1): 113. doi: 10.1186/s12940-017-0328-z.

9. Arpin E, Gauffin K, Kerr M, et al. Climate change and child health inequality: A review of reviews. International Journal of Environmental Research and Public Health 2021; 18: 10896. doi: 10.3390/ijerph182010896.

10. Chapman AR, Ahmed AK. Climate justice, humans rights, and the case for reparations. Health and Human Rights 2021; 23: 81–94.

11. Mitchell BC, Chakraborty J, Basu P. Social inequities in urban heat and greenspace analyzing climate justice in Delhi, India. International Journal of Environmental Research and Public Health 2021; 18: 4800. https://doi.org/10.3390/ijerph18094800.

12. Shultz JM, Kossin JP, Ettman C, et al. The 2017 perfect storm season, climate change, and environmental injustice. Lancet Planet Health 2018; 2: e370–e371.

13. Jones R. Climate change and Indigenous health promotion. Global Health Promotion 2019; 26(3 Suppl): 73–81.

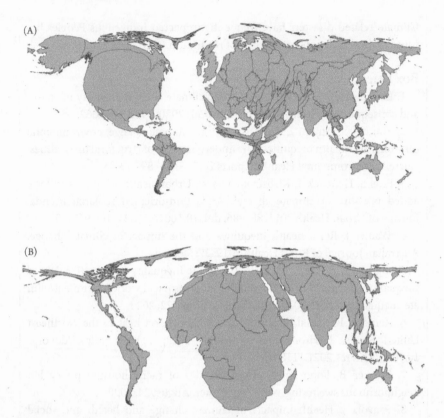

Figure 1-4 Cartogram maps demonstrating (A) relative proportions of GHG emissions, by country, and (B) magnitude and severity of the consequences of climate change, by country. (*Source*: Patz JA, Gibbs HK, Foley JA, et al. Climate change and global health: Quantifying a growing ethical crisis. EcoHealth 2007; doi.10.1007/s10393-007-0141-1.)

consequences of climate change. In general, those countries that have emitted relatively low amounts of GHGs face the greatest health consequences of climate change (Figure 1-4).

EFFECTS ON BIOLOGICAL SYSTEMS

The adverse effects of climate change on biological systems include:

- Changes in the lifecycles of vectors, reservoirs, and pathogens
- Diseases of wildlife and plants
- Disruptions of interactions among species
- Destruction of habitats and ecosystems.

Extreme weather and climate variability have widespread effects on microorganisms that cause disease in humans, animals, and plants. For example, a systematic review found that 218 (58%) of 375 infectious diseases affecting humans globally, such as vectorborne diseases like malaria and waterborne diseases like cholera, have at some point been aggravated by climate variability.[40]

EFFECTS ON SOCIAL SYSTEMS

Climate change causes a variety of adverse consequences on social systems. For example, droughts, floods, and sea-level rise have often forced people to flee their homes and communities in search of safer places with better economic opportunities. And climate-induced food insecurity and resultant food price shocks have often been associated with violence (Chapters 8 and 10).

Box 1-2 provides a brief history of organizational responses to climate change, including the establishment and ongoing activities of the IPCC, which performs repeated international assessments on climate science and the societal and biological impacts as well as the economic and policy implications of climate change. These assessments inform the United Nations Framework Convention on Climate Change (UNFCCC), which convenes Conference of the Parties (COP) meetings almost annually.

PREVENTIVE MEASURES

The two broad categories of preventive measures to address climate change are the following:

- Mitigation (a form of primary prevention): Measures to reduce the production of GHGs, which cause climate change (Chapters 12–14)
- Adaptation (a form of secondary prevention): Measures to reduce the impacts of climate change (Chapters 15–17).

Mitigation and adaptation measures are often implemented by governmental agencies. However, nongovernmental agencies (NGOs), the private sector, and individuals play critically important roles.

Box 1-2   A Brief History of Organizational Responses to Climate Change

---

Jonathan A. Patz and Barry S. Levy

In 1987, the United Nations World Commission on Environment and Development (WCED) published the book *Our Common Future* (the Brundtland Report), which envisioned environment and development as inextricably connected and placed environmental issues on the political agenda. In doing so, it created the concept of *sustainable development*.[1] In 1988, the World Meteorological Organization and the United Nations Environment Program established the United Nations Intergovernmental Panel on Climate Change (IPCC), which has performed international scientific assessments on climate science, vulnerabilities and adaptation, and climate mitigation policies and financing. IPCC members include leading scientists from various sectors and from major educational and research institutions.[2]

In 1989, the IPCC published its First Assessment Report, which asserted that climate change is a challenge requiring international cooperation. The scientific evidence in the report ultimately led to the creation of the United Nations Framework Convention on Climate Change (UNFCCC), the key international treaty for mitigating climate change and adapting to its consequences, which entered into force in 1994.[3]

In 1992, the first United Nations Conference on Environment and Development (UNCED)—the "Rio Earth Summit"—took place in Rio de Janeiro. Fulfilling the main goal of the conference, the United Nations, multilateral organizations, and national governments voted to adopt an action plan on sustainable development (Agenda 21). This plan was revised and commitments to its implementation were made at subsequent conferences of the parties (COPs) to the UNFCCC, which have been held almost annually.

In 1995, the IPCC Assessment Report emphasized the linkages between climate change and human health. In 1997 at COP3, this report provided the basis for the adoption of the Kyoto Protocol, an international treaty that set binding obligations on industrialized countries to reduce emissions of greenhouse gases (GHGs). COP6 in The Hague and COP13 in Bali, significantly advanced the Climate Justice Movement (Chapter 20).

The landmark Paris Agreement in 2015, at COP21, created a common accounting framework of nationally determined contributions to cut GHG emissions to prevent temperatures from rising more than 2.0°C (3.6°F) above pre-industrial levels. As of February 2023, 194 countries and the European Union had ratified or acceded to the Paris Agreement.

By 2021, the IPCC had produced six comprehensive scientific reports about climate change, the most recent of which was released during that year, and had supplemented these reports with methodological and technical papers. It had also produced special reports. For example, in 2018, it produced a report in which it concluded that limiting global warming to 1.5°C (2.7°F) above pre-industrial levels would require, compared to 2010, a 45% reduction in GHG emissions by 2030 and net-zero GHG emissions by 2050.[4]

Other important initiatives have included

- The C40 Cities Climate Leadership Group, which consists of representatives of megacities who cooperate on reducing GHG emissions
- The Green Climate Fund, formally adopted in 2017 at COP17 in Durban, South Africa, which assists "developing countries" in mitigation and adaptation measures to address climate change by providing support to reduce their GHG emissions
- Reducing Emissions from Deforestation and Forest Degradation in Developing Countries (REDD+), which is a framework created by the UNFCCC to guide activities to reduce GHG emissions from deforestation and forest degradation and promote sustainable management and conservation of forests in developing countries (See Chapter 17.)
- A 2021 proposal by the World Health Organization in partnership with the European Parliament and Vanuatu and Tuvalu (two Small-Island Developing States) for the establishment of a "fossil fuel nonproliferation treaty"[5]
- The Grand Challenge on Climate Change, Human Health and Equity, a multi-year initiative of the U.S. National Academy of Medicine to improve and protect human health, well-being, and equity by working to transform systems that contribute to and are impacted by climate change.[6]

In recent years, there has been substantially increased engagement of health professionals in addressing climate change. Boxes 19-1 and 19-2 in Chapter 19 describe the Climate and Health Movement, and Chapter 20 describes the Climate Justice Movement.

Box References

1. Brundtland GH. Report of the World Commission on Environment and Development: Our Common Future. Geneva: United Nations, 1987. Available at: https://www.are.admin.ch/are/en/home/media/publications/sustainable-development/brundtland-report.html. Accessed March 19, 2023.

2. IPCC. History of the IPCC. Available at: https://www.ipcc.ch/about/history/. Accessed March, 2023.

3. United Nations Framework Convention on Climate Change. Full text of the Convention Available at: https://unfccc.int/resource/docs/convkp/conv eng.pdf. Accessed February 21, 2023.

4. Myles A, Babiker M, Chen Y, et al. Summary for Policymakers. In V Masson-Delmotte, P Zhai, H-O Pörtner, et al. Special Report: Global Warming of 1.5°C. Cambridge: Cambridge University Press, 2018, pp. 3–24. Available at: https://www.ipcc.ch/sr15/. Accessed March 19, 2023.

5. The Fossil Fuel Non-Proliferation Treaty. Health Professionals Call for a Fossil Fuel Treaty. Available at: https://fossilfueltreaty.org/health-letter. Accessed February 21, 2023.

6. Dzau VJ, Levine R, Barrett G, Witty A. Decarbonizing the U.S. health sector — A call to action. New England Journal of Medicine 2021; 385: 2117–2119. doi: 10.1056/NEJMp2115675.

Mitigation

Reducing the emission of GHGs is achieved mainly by implementation of policies and use of technologies. Understanding the governmental policymaking process is important for individuals and organizations that are developing and advocating for policies to address climate change (Chapter 11). Policies to promote mitigation span several sectors, including the following:

- Energy policies to promote increased use of renewable energy, decreased use of fossil fuels, and reduced energy demand (Chapter 12)
- Transportation policies to promote walking and bicycling (*active transportation*) and use of fuel-efficient vehicles (Chapter 13)
- Agriculture policies to promote reduced meat production and consumption, appropriate development of biofuels, and decreased methane emissions (Chapter 14).

Reducing population growth, which could substantially reduce GHG emissions, could be facilitated by improving access to education, especially for girls; reducing infant and child mortality; and increasing access to family planning.[41]

There are many opportunities to implement mitigation measures that both reduce emissions of GHGs and improve human health. Health co-benefits of mitigation measures can result from improved air quality, increased physical activity, improved nutrition and food security, reduced risk of infectious diseases, reduced exposure to environmental hazards, improved water quality, reduced risks to mental health, and other factors.[42]

Indoor and outdoor air pollution causes an estimated 7 million premature deaths annually.[43] Burning fossil fuels causes much air pollution. Analyses have projected of the following:

- Phasing out fossil fuels could prevent 5.1 million deaths annually[44] (Box 12-1 in Chapter 12).
- Improved physical fitness achieved by active transportation could save 5 million lives annually[45] (Box 13-1 in Chapter 13).
- The "Universal Healthy Reference Diet" could prevent an estimated 11 million premature deaths annually (Chapter 14).[46]
- If nine major countries abided by their Paris Agreement commitments to reduce GHGs, as a result, by 2040 (a) improved air quality would prevent 1.2 million deaths annually, (b) increased physical activity would prevent 1.2 million deaths annually, and (c) healthier diets would prevent 5.9 million deaths annually.[47]

Adaptation

Adaptation measures can reduce the impact of climate change on public health and social systems. For example, improved multisectoral preparedness for extreme weather events can improve emergency responses and reduce morbidity and mortality. Vulnerability and adaptation assessment can be used to identify likely events, populations at increased risk, and opportunities to reduce harm.[48–51] Health adaptation actions can reduce the impacts of climate change on health systems (Chapter 15). Design of healthy and sustainable "built environments" can facilitate adaptation to climate change, improve health status, and contribute to mitigation of GHG emissions (Chapter 16). Nature-based climate solutions can promote both adaptation and mitigation of GHG emissions (Chapter 17).

Adaptation and Mitigation Indicators

The Lancet Countdown: Tracking Progress on Health and Climate Change monitors 43 global indicators of climate-related health hazards, exposures, and impacts; climate adaptation, planning, and resilience for health; climate mitigation actions related to health co-benefits; relevant economics and finance; and public and political engagement in health and climate change.[52] Some of its findings of great concern have included the following:

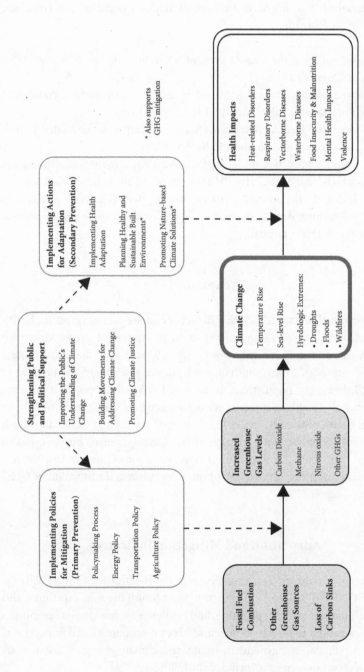

Figure 1-5 A conceptual framework of climate change and its health impacts, which reflects the overall organization and individual chapters of this book. (Diagram by Jonathan A. Patz and Barry S. Levy.)

- During the 2021-2022 period, there were 3.7 billion more heat-wave days than in 1986–2005 period.
- During the 2017-2021 period, heat-wave mortality was 68% higher than in the 2000-2004 period.
- Months in which malaria transmission was possible increased 32% in the highlands of Africa and 14% in the highlands of the Americas between the 1951-1960 and 2012-2021 periods.
- By 2020, 98 million more people annually reported moderate to severe food insecurity than during the 1981-2010 period.
- In 2021, only 27% of urban centers contained greenspace.

In 2022, the Lancet Countdown concluded that, if most countries failed to abide by their nationally determined commitments to reduce GHG emissions as pledged in the 2015 Paris Agreement, today's children could see, during their lifetimes, global temperature increases of 2.4°-3.5°C (4.3°-6.3°F) above pre-industrial levels.[52]

Strengthening Public and Political Support for Mitigation and Adaptation

Strengthening public and political support is essential for implementing policies for mitigation and actions for adaptation. Among the numerous ways of strengthening this support are communicating the health relevance of climate change (Chapter 18), building movements for addressing climate change (Chapter 19), and promoting climate justice (Chapter 20).

CONCEPTUAL OVERVIEW AND ORGANIZATION OF THIS BOOK

Figure 1-5 provides a conceptual overview of climate change and its health impacts and illustrates the organization of this book. It shows the path by which fossil fuel combustion, other GHG sources, and loss of carbon sinks leads to increased GHG levels, which, in turn, leads to climate change and its health impacts (Part II of this book). It shows how implementing policies for mitigation (primary prevention) can prevent further increases in (and reduce) GHG levels (Part III). It shows how implementing actions for adaptation (secondary prevention) can prevent climate change from causing health impacts (Part IV). And it shows that strengthening public and political support can reinforce policies for mitigation and actions for adaptation (Part V).

SUMMARY POINTS

- Human-caused climate change is creating, both directly and indirectly, profound and widespread environmental and health consequences.
- The health risks from climate change disproportionately affect vulnerable populations and exacerbate disparities within and between countries.
- Mitigation policies in the energy, transportation, and agriculture sectors can reduce GHG emissions, while simultaneously providing health co-benefits.
- Adaptation actions can reduce the health consequences of climate change, and some adaptation actions can also mitigate GHG emissions.
- Some strategies, such as improving the public's understanding of climate change, building movements, and promoting climate justice can strengthen public and political support to address climate change.

References

1. Arias PA, Bellouin N, Jones RG, et al. Technical Summary. In V Masson-Delmotte, P Zhai, A Pirani et al. (eds.). Climate Change 2021: The Physical Science Basis. Contribution of Working Group I to the Sixth Assessment Report of the Intergovernmental Panel on Climate Change. Cambridge: Cambridge University Press, 2021, pp. 33–144. doi: 10.1017/9781009157896.002.
2. United Nations Framework Convention on Climate Change. Full text of the Convention. Available at: https://unfccc.int/resource/docs/convkp/conveng.pdf. Accessed February 21, 2023.
3. Merriam-Webster Dictionary. Climate. Available at: http://www.merriam-webster.com/dictionary/climate. Accessed February 21, 2023.
4. Stevenson A, Lindberg CA (eds.). New Oxford American Dictionary (3rd Edition). New York: Oxford University Press, 2010.
5. Brittanica.com. Weather. Available at: https://www.britannica.com/dictionary/weather. Accessed February 21, 2023.
6. Merriam-Webster Dictionary. Weather. Available at: https://www.merriam-webster.com/dictionary/weather. Accessed February 21, 2023.
7. Institute of Medicine. The Future of Public Health. Washington, DC: National Academy Press, 1988.
8. National Oceanic and Atmospheric Administration. Carbon dioxide now more than 50% higher than pre-industrial levels. June 3, 2022. Available at: https://www.noaa.gov/news-release/carbon-dioxide-now-more-than-50-higher-than-pre-industrial-levels. Accessed February 21, 2023.
9. Stocker TF, Qin D, Plattner GK, et al. (eds.). Climate Change 2013: The Physical Science Basis. Working Group I Contribution to the Fifth Assessment Report of the Intergovernmental Panel on Climate Change. Cambridge, UK and New York: Cambridge University Press, 2013.
10. Field CB, Barros VR, Dokken DJ, et al. (eds.). Climate Change 2014: Impacts, Adaptation, and Vulnerability. Contribution of Working Group II to the Fifth

Assessment Report of the Intergovernmental Panel on Climate Change. Cambridge, UK: Cambridge University Press, 2014.

11. Vicedo-Cabrera AM, Scovronick N, Sera F, et al. The burden of heat-related mortality attributable to recent human-induced climate change. Nature Climate Change 2021; 11: 492–500.

12. Karl TR, Melillo JM, Peterson TC (eds.). Global Climate Change Impacts in the United States. New York: Cambridge University Press, 2009.

13. Kunkel KE, Easterling DR, Redmond K, Hubbard K. Temporal variations of extreme precipitation events in the United States: 1895–2000. Geophysical Research Letters 2003; 30: 1900. Doi: 10.1029/2003GL018052.

14. Centers for Disease Control and Prevention. Heat-related deaths after an extreme heat event-four states, 2012, and United States, 1999–2009. Morbidity and Mortality Weekly Report 2013; 62: 433–436.

15. Trenberth KE, Fasullo JT, Mackaro J. Atmospheric moisture transports from ocean to land and global energy flows in reanalyses. Journal of Climate 2011; 24: 4907–4924.

16. Wuebbles DJ, Fahey DW, Hibbard KA, et al. (eds.). Climate Science Special Report: Fourth National Climate Assessment, Volume I. Washington, DC: U.S. Global Change Research Program, 2017.

17. Fowler AM, Hennessey KJ. Potential impacts of global warming on the frequency and magnitude of heavy precipitation. Natural Hazards 1995; 11: 283–303.

18. Mearns LO, Giorgi F, McDaniel L, Shields C. Analysis of daily variability of precipitation in a nested regional climate model: Comparison with observations and doubled CO2 results. Global Planetary Change 1995; 10: 55–78.

19. Trenberth KE. Conceptual framework for changes of extremes of the hydrologic cycle with climate change. Climatic Change 1999; 42: 327–339.

20. Palla A, Colli M, Candela A, et al. Pluvial flooding in urban areas: The role of surface drainage efficiency. Journal of Flood Risk Management 2018; 11(Suppl 2): S663–S676. https://doi.org/10.1111/jfr3.12246.

21. Ritchie H, Rosado P, Roser M. Natural Disasters. Our World In Data. Available at: https://ourworldindata.org/natural-disasters. Accessed October 20, 2023.

22. Bender MA, Knutson TR, Tuleya RE, et al. Modeled impact of anthropogenic warming on the frequency of intense Atlantic hurricanes. Science 2010; 327: 454–458.

23. Webster PJ, Holland GJ, Curry JA, Chang HR. Changes in tropical cyclone number, duration, and intensity in a warming environment. Science 2005; 309: 1844–1846.

24. Emanuel K, Sundararajan R, Williams J. Hurricanes and global warming: Results from downscaling IPCC AR4 simulations. Bulletin of the American Meterological Society 2008; 89: 347–367.

25. Rodell M, Li B. Changing intensity of hydroclimatic extreme events revealed by GRACE and GRACE-FO. Nature Water 2023; 1: 241–248.

26. Handmer J, Honda Y, Kundzewicz Z, et al. Changes in Impacts of Climate Extremes: Human Systems and Ecosystems. New York: Cambridge University Press, 2012, pp. 231–290.

27. Chen G, Guo Y, Yue X, et al. Mortality risk attributable to wildfire-related $PM_{2.5}$ pollution: A global time series study in 749 locations. Lancet Planetary Health 2021; 5: e579–e587.

28. National Oceanic and Atmospheric Administration (NOAA), National Hurricane Center, and Central Pacific Hurricane Center. Hurricanes in History (Mitch 1998).

1998. Available at: https://www.nhc.noaa.gov/outreach/history/#mitch. Accessed March 17, 2023.

29. Nelson SA. Why New Orleans is vulnerable to hurricanes: Geologic and historical factors. 2012. Available at: http://www.tulane.edu/~sanelson/New_Orleans_and_ Hurricanes/New_Orleans_Vulnerability.htm. Accessed February 21, 2023.

30. Molina M, McCarthy J, Wall D, et al. What We Know: The Reality, Risks, and Responses to Climate Change. Washington, DC: American Association for the Advancement of Science, 2014.

31. Anderegg WR, Prall JW, Harold J, Schneider SH. Expert credibility in climate change. Proceedings of the National Academy of Sciences USA 2010; 107: 12107–12109.

32. Cook J, Nuccitelli D, Green SA, et al. Quantifying the consensus on anthropogenic global warming in the scientific literature. Environmental Research Letters 2013; 8: 024024. doi: 10.1088/1748-9326/8/2/024024.

33. Liu J, Varghese BM, Hansen A, et al. Is there an association between hot weather and poor mental health outcomes? A systematic review and meta-analysis. Environment International 2021; 153: 106533. https://doi.org/10.1016/j.env int.2021.106533.

34. Samson J, Berteaux D, McGill BJ, Humphries MM. Geographic disparities and moral hazards in the predicted impacts of climate change on human populations. Global Ecology and Biogeography 2011; 20: 532–544.

35. Hess JJ, Malilay JN, Parkinson AJ. Climate change: The importance of place. American Journal of Preventive Medicine 2008; 35: 468–478.

36. Ford JD. Vulnerability of Inuit food systems to food insecurity as a consequence of climate change: A case study from Igloolik, Nunavut. Regional Environmental Change 2009; 9: 83–100.

37. Roelofs C, Wegman D. Workers: The climate canaries. American Journal of Public Health 2014; 104: 1799–1801.

38. Khan AE, Scheelbeek PFD, Shilpi AB, et al. Salinity in drinking water and the risk of (pre)eclampsia and gestational hypertension in coastal Bangladesh: A case-control study. PloS One 2014; 9: e108715. doi: 10.1371/journal.pone.0108715.

39. Chersich MF, Pham MD, Area A, et al. Associations between high temperatures in pregnancy and risk of preterm birth, low birth weight, and stillbirths: Systematic review and meta-analysis. British Medical Journal 2020; 371: m3811. https://www.bmj. com/content/371/bmj.m3811.

40. Mora C, McKenzie T, Gaw IM, et al. Over half of known human pathogenic diseases can be aggravated by climate change. Nature Climate Change 2022; 12: 869–875. https://doi.org/10.1038/s41558-022-01426-1.

41. Porter E. Reducing carbon by curbing population. New York Times, August 5, 2014.

42. Mailloux NA, Henegan CP, Lsoto D, et al. Climate solutions double as health interventions. International Journal of Environmental Research and Public Health 2021; 18: 13339. https://doi.org/10.3390/ijerph182413339.

43. Landrigan PJ, Fuller F, Acosta NJR, et al. The Lancet Commission on Pollution and Health. Lancet 2018; 391: 462–512.

44. Lelieveld J, Haines A, Burnett R, et al. Air pollution deaths attributable to fossil fuels: Observational and modelling study. British Medical Journal 2023; 383: e077784. doi: 10.1136/bmj-2023-077784.

45. Kohl HW, Craig CW, Lambert EV, et al. The pandemic of physical inactivity: Global action for public health. Lancet 2012; 380: 294–305.

46. Willett W, Rockström J, Loken B, et al. Food in the anthropocene: The EAT-Lancet Commission on healthy diets from sustainable food systems. Lancet 2019; 393: 447–492.
47. Hamilton I, Kennard H, McGushin A, et al. The public health implications of the Paris Agreement: A modelling study. Lancet 2021; 5: E74–E83. doi: 10.1016/S2542-5196(20)30249-7.
48. Ebi KL, Schmier JK. A stitch in time: Improving public health early warning systems for extreme weather events. Epidemiologic Reviews 2005; 27: 115–121.
49. Burton I, Malone E, Huq S. Adaptation Policy Frameworks for Climate Change: Developing Strategies, Policies, and Measures. Cambridge: University of Cambridge Press, 2004.
50. Adger WN, Arnell NW, Tompkins EL. Successful adaptation to climate change across scales. Global Environmental Change 2005; 15: 77–86.
51. Field CB, Barros V, Stocker TF, et al. (eds.). Managing the Risks of Extreme Events and Disasters to Advance Climate Change Adaptation: Special Report of the Intergovernmental Panel on Climate Change. Cambridge: Cambridge University Press, 2012. Available at: http://www.ipcc-wg2.gov/SREX/images/uploads/SREX-All_FINAL.pdf. Accessed February 21, 2023.
52. Romanello M, Di Napoli C, Drummond P, et al. The 2022 report of the Lancet Countdown on health and climate change. Lancet 2022; 400: 1619–1654. https://doi.org/10.1016/S0140-6736(22)01540-9.

Further Reading

Sarofim MC, Saha S, Hawkins MD, et al. Temperature-related Health, Death, and Illness. In A Crimmins, J Balbus, JL Gamble, et al. (eds.). The Impacts of Climate Change on Human Health in the United States: A Scientific Assessment. Washington, DC: U.S. Global Change Research Program, 2016, pp. 43–68. Available at: https://health2016.globalchange.gov/low/ClimateHealth2016_02_Temperature_small.pdf. Accessed February 21, 2023.

Cissé G, McLeman R, Adams H, et al. Health, Wellbeing, and the Changing Structure of Communities. In HO Pörtner, DC Roberts, M Tignor, et al. (eds.). Climate Change 2022: Impacts, Adaptation, and Vulnerability. Contribution of Working Group II to the Sixth Assessment Report of the Intergovernmental Panel on Climate Change. Cambridge: Cambridge University Press, 2022, pp. 1041–1170. Available at: https://www.ipcc.ch/report/ar6/wg2/downloads/report/IPCC_AR6_WGII_Chapter07.pdf.   Accessed February 21, 2023.

The above two publications of the USGCRP and the IPCC provide extensive discussion about climate change, its adverse health consequences, and mitigation and adaptation measures.

Romanello M, Di Napoli C, Drummond P, et al. The 2022 report of the Lancet countdown on health and climate change. Lancet 2022; 400: 1619–1654. https://doi.org/10.1016/S0140-6736(22)01540-9.

This is one of a series of annual publications which track quantitative indicators related to climate change and human health. Established indicators address impacts; adaptation and resilience; mitigation and health co-benefits; economics and finance; and public and political engagement in health and climate change.

2

Applying Climate Science to Climate Change and Extreme Weather Events

Stephen J. Vavrus, Aimee Puz, Samuel Kruse, and Jonathan A. Patz

The scientific study of climate is interdisciplinary, including aspects of oceanography, biogeochemistry, atmospheric science, and physical geography.[1] It utilizes records and observations, both from the relatively recent past, such as temperature records, weather observations, patterns of plant distribution, and crop yields, and from the more distant past, such as data from ice cores and sediment deposits (Figure 2-1). By synthesizing this information, climate scientists can learn much about past climate.

Climate scientists can also project how climate may change in the future—a challenging task because of the long time periods involved, complex climate processes, and the uncertainties of knowing exactly what the future influences on climate will be. However, by using climate indices, modeling techniques, satellite observations, historical records, computer analytics, and other methodologies and tools, climate scientists have greatly enhanced their ability to interpret current climate variations and to project future climate change. This chapter introduces key concepts in climate science that are central to understanding why climate change is occurring now and how the climate is expected to change in the future.

CLIMATE PATTERNS

Natural variability of the climate system predominantly occurs with preferred spatial patterns and timescales, through atmospheric circulation and through interactions with land and ocean surfaces. These patterns, which are often termed *regimes*, *modes*, or *teleconnections*, include the Northern Annular Mode (NAM, also known as the Arctic Oscillation) and the Southern Annular Mode (SAM, also known as the Antarctic Oscillation), which result from opposing variations in air pressure between polar regions and middle latitudes (about 30°–60°). The NAM and the SAM tend to align in predictable locations.

Stephen J. Vavrus, Aimee Puz, Samuel Kruse, and Jonathan A. Patz, *Applying Climate Science to Climate Change and Extreme Weather Events* In: *Climate Change and Public Health*. Edited by: Barry S. Levy and Jonathan A. Patz, Oxford University Press. © Oxford University Press 2024. DOI: 10.1093/oso/9780197683293.003.0002

Figure 2-1 Glaciologists install a global positioning system in front of a nearby supraglacial lake on the surface of the Greenland Ice Sheet to measure year-round surface ice velocities and the response of the ice to supraglacial lake drainage events. (Photograph by Laura A. Stevens.)

Both of these modes affect the strength and location of the jet stream in middle latitudes. The Pacific-North American Pattern is characterized by oscillations in the transport of air masses from the Canadian Arctic and Pacific Ocean into North America. The El Niño-Southern Oscillation (ENSO) is characterized by year-to-year fluctuations in tropical Pacific Ocean temperatures and associated atmospheric circulation anomalies. The Atlantic Multidecadal Oscillation (AMO) is characterized by long-term fluctuations in temperature differences between the North Atlantic Ocean and the South Atlantic Ocean.

HOW GREENHOUSE GASES CAUSE GLOBAL WARMING

Climate change, resulting from natural variability or human activity, depends on the balance between incoming (solar) shortwave radiation and outgoing (infrared) longwave radiation. This balance is affected by the atmosphere in much the same way as a blanket affects heat loss from our skin. Greenhouse gases (GHGs) act similarly as the Earth's blanket. After solar (light) energy heats the surface of the Earth, the molecular structure of GHGs enables atmospheric absorption of longer-wavelength (infrared) radiation from the surface, impeding the escape of this heat energy to the cooler upper atmosphere, thereby warming the Earth's surface. This phenomenon is known as the *greenhouse effect*, some of which is beneficial, enabling the environment to be warm enough for survival of living organisms. However, an overabundance of GHGs, as described below, can heat the Earth beyond optimal temperatures for survival (Figure 1-1 in Chapter 1).

The composition of the atmosphere has changed tremendously over the past several centuries, with concentrations of GHGs, such as carbon dioxide and methane, having increased dramatically during this period (Figure 1-2 in Chapter 1). Historical levels of GHGs have been determined from various types of analyses, such as studies of air bubbles trapped in Antarctic ice cores.[2,3] For example, the concentration of carbon dioxide, the major GHG, has increased from about 280 parts per million (ppm) in 1800 to about 420 ppm now, an increase of approximately 50%—primarily due to the combustion of fossil fuels and deforestation. Methane has risen from 751 parts per billion (ppb) in 1800 to 1,912 ppb in 2022. And nitrous oxide has risen from 273 ppb in 1800 to 336 ppb in 2022. The increases in methane and nitrous oxide have both been primarily due to agricultural activities.

Increased GHG concentrations have led to global warming—an effect called *positive radiative forcing*—by increasing the absorption and containment of infrared radiation in the lower atmosphere. Figure 2-2 demonstrates the relative effect on radiative forcing by emissions of major GHGs and drivers.

Human influence on climate, known as *anthropogenic climate change* or *climate disruption*, can be analyzed as a signal that deviates from the background of natural climate variability or climate patterns. To interpret this signal, historical climate data are needed to estimate natural variability. Because instrument records, such as measurements of temperature and precipitation, are available globally only for approximately the past 150 years, previous climates must be deduced from records of paleoclimates (climates prevalent at particular times in the geological past), including data based on tree rings, the presence and prevalence of fauna and flora in deep-sea cores, isotope analyses of coral and ice cores, and other documentary evidence.

CLIMATE MODELING

A *climate model* is "a numerical representation of the climate system based on the physical, chemical and biological properties of its components, their interactions

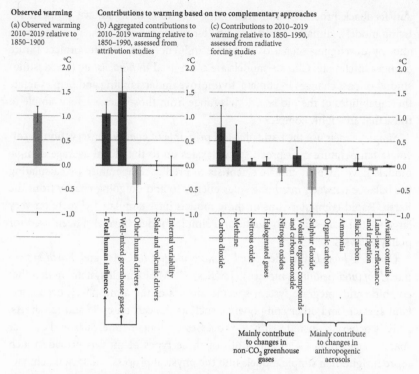

Figure 2-2 Assessed contributions to observed warming in 2010-2019 relative to 1850-1900. <u>Panel (a)</u>: Observed global warming (increase in global surface temperature). Whiskers show the *very likely* range. <u>Panel (b)</u>: Evidence from attribution studies, which synthesize information from climate models and observations. Temperature change attributed to total human influence; changes in well-mixed GHG concentrations; other human drivers due to aerosols, ozone, and land-use change (land-use reflectance); solar and volcanic drivers; and internal climate variability. Whiskers show *likely* ranges. <u>Panel (c)</u>: Evidence from the assessment of radiative forcing and climate sensitivity. Temperature changes from individual components of human influence: emissions of GHGs, aerosols, and their precursors; land-use changes (land-use reflectance and irrigation); and aviation contrails. Whiskers show *very likely* ranges. Estimates account for both direct emissions into the atmosphere and their effect, if any, on other climate drivers. For aerosols, both direct effects (through radiation) and indirect effects (through interactions with clouds) are considered. (*Source*: Allan RP, Arias PA, Berger S, et al. Summary for Policymakers. In V Masson-Delmotte, P Zhai, A Pirani, et al. [eds.]. Climate Change 2021: The Physical Science Basis. Contribution of Working Group I to the Sixth Assessment Report of the Intergovernmental Panel on Climate Change. Cambridge, UK, and New York: Cambridge University Press, 2021, pp. 3-32. doi: 10.1017/9781009157896.001.)

and feedback processes, and accounting for some of its known properties."[4] Using models, climate scientists can better understand past changes in climate and, by developing scenarios of future influences on climate, project future changes in climate. Climate models are often used in *hindcasting* (backtesting) studies of past changes in climate, to improve understanding and help evaluate the capabilities of the model. Models range from those that are very simple to those that are highly complex.

Simple models use the basic physical principle of energy conservation to calculate temperature over broad spatial scales up to the global average. Simple models have enabled climate scientists to predict temperature by calculating the balance between incoming solar energy to and outgoing energy from the Earth. Rapid computer runs of these models have simulated climate on very large spatial scales far into the future. Simple models have also been used for policy analyses.

Complex models include *global climate models* (GCMs) and *Earth-system models of intermediate complexity* (EMICs). GCMs are multisystem models that combine and integrate system-specific models of the atmosphere, ocean, ice, land surface, and other components, such as the carbon cycle and ice sheets. EMICs generally include climate processes on longer timescales and coarser spatial scales than GCMs do. Both of these types of models provide much more insight than simple models into the physical processes that cause climate change.

Many simulations cited in the 2021 report of the Intergovernmental Panel on Climate Change (IPCC)[5] are based on an international collection of coupled ocean-land-atmosphere models, called the *Coupled Model Intercomparison Project Phase 6* (CMIP6). It is composed of a set of coordinated climate model experiments that provide a context for (a) assessing mechanisms responsible for model differences in poorly understood feedbacks, including those associated with the carbon cycle and with clouds; (b) examining climate predictability and exploring the ability of models to predict climate processes and resulting climate change on multidecadal timescales; and (c) determining why similarly forced models produce a range of responses. Compared to previous CMIP phases, CMIP6 includes newer representations of physical, chemical, and biological processes with higher resolution, thereby improving the state of large-scale indicators of climate change.[5]

CMIP6 model outputs also serve as the building blocks for IPCC's future projections. For example, these model outputs are driven by IPCC's Shared Socioeconomic Pathway (SSP) scenarios, such as SSP1-2.6 and SSP2-4.5, to show the range of future climate depending on which SSP scenario is followed (Table 2-1).

Table 2-1 Best estimates for when various global warming levels will be reached, relative to 1850-1900 baseline temperature, according to IPCC's Shared Socioeconomic Pathway (SSP) scenarios

	SSP1-2.6	SSP2-4.5	SSP3-7.0	SSP5-8.5
1.5°C	2028	2028	2028	2026
2.0°C	2064	2046	2043	2039
3.0°C	NA	2094	2069	2060
4.0°C	NA	NA	2091	2078
5.0°C	NA	NA	NA	2094

The numbers, such as 2.6 after SSP1, refer to the stratospheric-adjusted radiative forcing by 2100, in watts per square meter (W/m^2). "NA" means not reaching that warming level by 2100.

Source: Tebaldi C, Debeire K, Eyring V, et al. Climate model projections from the Scenario Model Intercomparison Project [ScenarioMIP] of CMIP6. Earth System Dynamics 2021; 12: 253–293.

DOWNSCALING

The large spatial scale of many projections makes it difficult to infer how global climate systems will affect climate locally. For example, GCMs have typically been run at a spatial resolution of about 161 kilometers (about 100 miles) or coarser and thus have not accurately simulated local climate conditions and interactions. Therefore, climate scientists use *downscaling methods* to simulate local conditions by using the output from GCMs, either as inputs for a *dynamical regional-scale model* with higher resolution, known as *dynamical downscaling*, or in combination with known statistical relationships between large-scale weather patterns and local weather conditions, known as *statistical downscaling*. Downscaling is especially important for analyzing the impacts of climate on the health of local populations and individuals, such as heat stress intensified by urban heat islands.

COMPUTATION IN CLIMATE SCIENCE

Downscaling methods help simulate local climate conditions in the absence of fine-grained spatial resolution, but the coarseness of the input climate models still contributes to large uncertainties in projections. To decrease uncertainty in climate projections, models must better account for small-scale physical processes and local atmospheric conditions. The large computational power and huge amount of storage required for these high-resolution models have been major challenges for climate scientists. For example, the total size of the output of all 100 CMIP6 models run at a spatial resolution of approximately 50 kilometers (about 31 miles) is close to 80 petabytes[6]—four times the amount of data in the Library of Congress.

Advances in computer science have greatly improved capabilities to process the enormous amounts of data required in climate analysis, forecasting, and prediction. High-resolution spatial scales of less than 50 kilometers were once impractical or even impossible to include in models due to the heavy computational overhead. But the exponential increase in computer power in the past few decades has facilitated much progress. *High-performance computer systems* (HPC systems) are groups of computing resources that use specialized hardware and software optimized to increase the speed of computation. These systems, which have become more accessible to academics, provide the computational power necessary to perform simulations at higher spatial and temporal resolution[7] and to make more accurate predictions.[8]

CONSEQUENCES OF CLIMATE CHANGE

Global Heating

The rate of global warming is faster now than in any period in the past 1,000 years.[4] Surface temperatures in the mid to late 20th century were considerably warmer in most regions than they had been in the previous 600 years—and in some regions the previous several thousand years.[9] From 1880 to 2020, the average global temperature rose by 1.09°C (1.96°F) (Figure 1-3 in Chapter 1). By 2100, average global temperature is projected to increase by another 1.0°-3.7°C (1.8°-6.7°F).[10]

Sea-Level Rise

Average global sea level rose annually by approximately 1.9 millimeters (0.07 inches) between 1971 and 2006, and 3.7 millimeters (0.15 inches) between 2006 and 2018, due to thermal expansion of ocean water and ocean discharge of ice from melting glaciers, the Greenland and West Antarctica ice sheets and the Arctic ice cap.[10]

Since 1900, global mean sea level has risen faster than it did during the previous 3,000 years.[10] By 2100, sea level is projected to rise by 40-100 centimeters (15.7-39.4 inches) over its level in the 1986–2005 period. In the next 30 years, sea level at the contiguous U.S. coastline is projected to rise approximately

25-30 centimeters (10-12 inches)—the same as sea-level rise over the past 100 years.[11]

Sea-level rise and land subsidence (land sinking) have important implications for coastal management and climate action. Land subsidence is caused by compaction of sediments in deltas, frequently impacting cities near coasts. Human activity is often the driver of rapid land subsidence because of extraction of groundwater, oil, and gas.[12] For example, in Jakarta, Indonesia, land subsidence by 2050 is projected to increase the area potentially affected by sea-level rise to 110.5 square kilometers (42.7 square miles). In 2019, influenced by the socioeconomic impacts of land subsidence, Indonesia's president announced the moving of the nation's capital from Jakarta to East Kalimantan (on Borneo).[13]

Effects on Physical and Biological Systems

Regional changes in climate, especially increases in temperature, have already affected diverse physical and biological systems in many parts of the world. For example, river and lake ice is breaking up in the spring earlier than before and animal ranges are moving to higher altitudes. Alpine species, such as certain wildflowers, have been displaced to higher altitudes; when they have no further terrain to which to migrate, some could become extinct.

The extent of perennial sea ice in the Arctic has decreased by about 13% per decade since 1979, while global snow cover and glacier volume have decreased substantially since the 1970s (Figure 2-3).[14,15] As Arctic sea ice has thinned since 1979, the proportion of thick ice older than 5 years has decreased by approximately 90% (Figure 2-4).[16]

Widespread and potentially irreversible changes may occur in Earth systems, such as slowing of the ocean circulation that transports warm water to the North Atlantic from low latitudes, large-scale melting of the ice sheets in Greenland and West Antarctica, and accelerated global warming due to positive feedbacks in the carbon cycle, such as by the release of methane from thawing Arctic tundra. The probability of these events is unknown, but it is likely to be affected by how rapidly climate change evolves and how long it lasts.

Extreme Weather Events

The overall frequency of extreme weather events will likely increase due to climate change. Floods and droughts, which accompany extreme weather, will also likely increase. Higher temperature causes water in soil to evaporate more quickly, increasing the probability of drought; however, warm air holds more

Figure 2-3 Agassiz Glacier in Glacier National Park, Montana, 1913 (top) and 2007 (bottom). (Top photograph by W.C. Alden. Bottom photograph by D. Fagre. Both photographs courtesy of the United States Geological Survey.)

moisture than cool air, increasing the probability of heavy precipitation. The concentration of water vapor in the atmosphere has increased, leading to increased frequency of intense rainfall.[17–20]

Even a relatively small change in average climatic conditions, such as a modest rise in the mean temperature, can lead to a large increase in the frequency and intensity of extreme weather events, as shown in Figure 2-5(a). Greater temperature variance (variability) can also lead to more extreme weather events even without a change in mean temperature, as shown in Figure 2-5(b). In recent decades, extreme heat has become more intense and widespread due to both higher mean temperature and greater temperature variance, as shown in Figure 2-5(c).

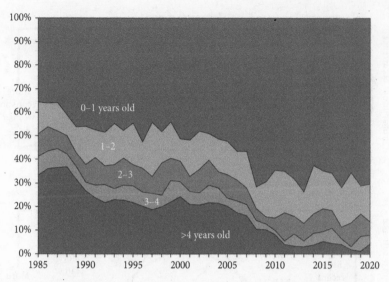

Figure 2-4 Sea ice age percentage within the Arctic Ocean for the week of March 11-18 from 1985 through 2020. (*Source*: Perovich D, Meier W, Tschudi M, et al. Arctic report card: Update for 2020. Available at: https://arctic.noaa.gov/Report-Card/Rep ort-Card-2020/ArtMID/7975/ArticleID/891/Sea-Ice. Accessed January 16, 2023.)

Attribution of Extreme Weather Events to Climate Change

Extreme event attribution is an emerging field of climate science in which the role of climate change in causing or influencing extreme weather events is studied. Recently, there has been rapid progress in the ability to determine the causes of some extreme weather events, especially for extreme weather events related to temperature, those with long historical records of observations, those with adequate simulations in climate models, and those not strongly influenced by infrastructure or resource management.[21] But it has not been possible to determine if climate change caused a *specific* extreme weather event.[22] To improve capabilities and generate more reliable results, the National Academy of Sciences made the following recommendations.

- Resolution of models needs to be improved.
- Physical mechanisms that lead to extreme weather events need to be better understood.
- Historical observational records need to be extended.
- Remote sensing devices need to be improved.[21]

The World Weather Attribution (WWA) initiative uses statistical methods to estimate the role of climate change in extreme weather events such as droughts,

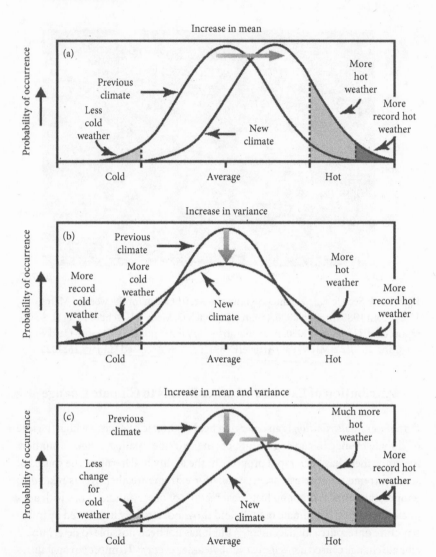

Figure 2-5 Probability of occurrence of hot and cold weather in relation to (a) increase in mean air temperature, (b) increase in variance of air temperature, and (c) increase in both mean and variance of air temperature. (*Source*: Folland, CK, Karl TR, Christy JR, et al. Observed Climate Variability and Change. In JT Houghton, Y Ding, DJ Griggs, et al. [eds.] Climate Change 2001: The Scientific Basis. Contribution of Working Group I to the Third Assessment Report of the Intergovernmental Panel on Climate Change. Cambridge, UK, and New York: Cambridge University Press, 2021, p. 155.)

floods, and heat waves. During late June and early July 2021, a massive heat wave occurred in the Pacific Northwest, in which many areas experienced temperatures above 40°C (104°F) for more than 4 days—far above previous high temperatures for that time of year. More than 900 people died.[23] In Lytton, British Columbia, the temperature reached 49.6°C (121°F), the hottest temperature ever recorded in Canada; then, within 48 hours, the town was completely burned to the ground in a wildfire.[23] The WWA estimated that this heat wave was 150 times more likely because of climate change.

In 2022, a devastating heat wave occurred in India and Pakistan, killing 90 people, causing flooding of a glacial lake in northern Pakistan and forest fires in India.[24] The WWA determined that human-induced climate change made this heat wave more likely and more severe and made the probability of a similar event about 30 times more likely in the future.[25]

In July 2021, extreme rainfall occurred in several European countries, leading to extreme flooding that resulted in approximately 200 deaths and more than $12 billion in damage.[26] Climate change has increased the intensity of daily summer rainfalls by 3-19%.[27]

However, climate change does not play a role in all extreme weather events. For example, the WWA found that it played little role in the devastating 2020 floods in Vietnam.

FEEDBACK PROCESSES

There are many feedback processes operating simultaneously, making future climate change difficult to predict. These feedback processes can either amplify a change in the system, such as a warming caused by increased GHG concentrations, or reduce such a change. Some processes can enhance or amplify local weather conditions but are not climate feedback processes. For example, the *urban heat island effect*, in which buildings, dark surfaces, human activity, and other factors in cities enhance temperature increases, illustrates how a weather condition can be influenced by surface features. However, the heating from the urban heat island effect is localized (Chapter 4).

A *positive feedback process* is one that amplifies an initial change, which is then further amplified, creating a *reinforcing cycle*. For example, in the *ice-albedo feedback mechanism*, a small amount of warming can cause a small amount of sea ice to melt, and, in turn, the resultant loss of bright ice cover can expose a dark ocean surface, which can more effectively absorb more sunlight and cause even more ice to melt. This feedback process is one reason why the Arctic region is warming so quickly and is projected to warm more than any other region in the future. Another important feedback process is the *water vapor feedback*, in which an initial warming triggers more surface evaporation, which, in turn, puts more heat-trapping water vapor into the atmosphere and therefore causes additional warming. Positive feedback processes explain why climate has varied so much over millions of years, ranging from ice-free global conditions to ice ages, in which glaciers covered much of the land.

The climate system also has *negative feedback processes*, which help to keep temperatures in check—similar to how an automobile's braking system offsets the power of the accelerator. An important negative feedback process is the *Planck feedback*, in which a rise in surface temperature causes more long-wave radiation to be emitted by the Earth's surface, thereby venting heat and mitigating additional warming.

Some processes, such as the response of clouds to climate change, are not clearly understood to be either positive or negative feedback processes because of differences in the results from different climate models. As the climate warms due to GHGs, the resultant changes in clouds may act as a positive feedback process to promote additional warming in some regions while acting as a negative feedback process to inhibit further warming in other locations.

PREDICTIVE CAPABILITY

Climate model projections are derived from GCMs that are run on supercomputers, the computing power of which has increased dramatically in recent years and enabled more detailed simulations. Despite this improved capability, GCMs are most accurate when used to make projections over long timescales and large spatial scales, such as how *global* average temperature will change *over the course of this century*. GCMs are much less accurate in projecting the exact trajectory of these long-term changes due to natural variations in the climate system, which occur on short timescales and can temporarily counter or amplify the long-term gradual trends. These natural short-term variations include the major modes of atmospheric circulation variability, such as ENSO and AMO, which can complicate detection of a long-term climate trend. For example, the very cold 2013-2014 winter in much of eastern North America occurred during the long-term trend of global warming.

Even on decades-long timescales, the climate system can deviate from its projected long-term response, as demonstrated by the *global warming hiatus* (period of slower rate of increase of global mean surface air temperature) from 1998 to 2013. However, when many climate-model simulations are run for the same future time period, representing a range of possible scenarios, they demonstrate that such warming pauses are probable.[28]

Despite the growing sophistication of climate models and extent of scientific consensus, there are some aspects of climate change that remain especially uncertain, such as the effect on clouds cited above. It is also unclear what changes will occur in circulation patterns, such as ENSO. And the stability of ice sheets remains uncertain—a knowledge gap that affects estimates of future sea-level rise.

KNOWNS, PROBABLES, AND UNKNOWNS

Projections of future climate change and its adverse environmental and health consequences are generally stated in terms of probability and level of confidence. For example, the confidence levels of key assessments in the 2021 IPCC report were based on evaluations of scientific understanding. These were expressed both qualitatively and quantitatively across the full range from "exceptionally unlikely" (0–1% probability) to "virtually certain" (99–100% probability). Confidence in the correctness of an assessment was based on "the type, amount, quality, and consistency of evidence" and "the degree of agreement in the scientific studies considered."[10]

Climate scientists have concluded that several climate changes already occurring will likely continue, given human activity. For example, the mean surface temperature during each of the past four decades has been successively warmer than any preceding decade since 1850, and global surface temperature has increased faster since 1970 than in any other 50-year period over at least the past 2,000 years.[10] Looking to the future, climate scientists project, with high confidence, that surface warming will continue—although the rate of temperature increase will probably not be steady.

In addition, climate scientists have concluded that, for at least the past three decades:

- The Greenland and West Antarctic ice sheets have lost mass.
- Most glaciers have continued to shrink.
- Arctic sea ice and spring snow cover in the Northern Hemisphere have continued to decrease.

During the rest of this century, it is very likely that global warming will cause Arctic sea ice cover to further diminish and spring snow cover in the Northern Hemisphere to further decline. It is also very likely that before 2050 the Arctic Ocean will be virtually ice-free during at least one summer season.[10]

These trends in ice and snow cover are directly attributable to the rapid warming of the Arctic, where temperatures have been rising two to four times faster than in the rest of the world. This phenomenon, known as *Arctic amplification*, may also promote extreme weather in temperate regions because of changes in the polar jet stream, the strength and waviness of which are largely determined by the temperature difference between the Arctic and middle latitudes (Figure 2-6). When this temperature difference is large, the westerly winds of the jet stream tend to be strong and blow relatively straight eastward across the hemisphere. But when Arctic amplification reduces this temperature difference, these westerly winds weaken and the jet stream tends to meander more in north-south waves.[29]

These jet-stream waves create the weather that we experience. The larger the waves, the more slowly they tend to travel eastward. Slower and more meandering waves promote longer-lasting weather conditions, which can become extremely hot—or cold—events. Although exactly how the jet stream will respond to future climate change is not known, Arctic amplification is expected to continue, possibly leading to long-term extreme weather events and related health consequences.

Climate scientists believe, with high confidence, that since the beginning of the Industrial Era, oceans have absorbed about 30% of the carbon dioxide emitted by human activity. As a result, the acidity of ocean water has increased.

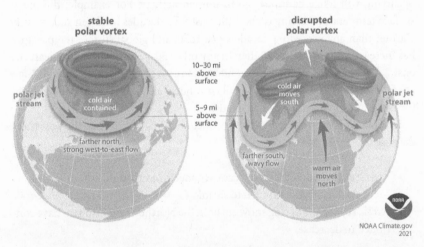

Figure 2-6 Arctic amplification and the polar vortex. When the Arctic polar vortex is especially strong and stable (left globe), it encourages the polar jet stream, down in the troposphere, to shift northward. The coldest polar air stays in the Arctic. When the vortex weakens, shifts, or splits (right globe), the polar jet stream often becomes extremely wavy, allowing warm air to flood into the Arctic and polar air to sink down into the mid-latitudes. (*Source*: Climate.gov. Understanding the Arctic Polar Vortex. Available at: https://www.climate.gov/news-features/understanding-clim ate/understanding-arctic-polar-vortex. Accessed January 16, 2023.)

Further uptake of carbon dioxide by ocean water will cause continued ocean acidification. And climate scientists have concluded that there will generally be more extreme weather, including more extremes in precipitation (heavy rainfalls and droughts), stronger tropical storms, and more extreme heat, but less extreme cold.[10]

SUMMARY POINTS

- GHG emissions, primarily from burning fossil fuels, have increased global surface temperature at a rate faster than in the past 2,000 years.
- Global warming was associated with a 3.7-millimeter rise in sea level per year between 2006 and 2018, and a 13% decline in Arctic sea ice per decade since 1979.
- Climate modeling has advanced dramatically due to improved understanding of climatic processes and the increasing processing speed and data-memory capacity of computers.
- Globally, extreme weather events are becoming more intense and more frequent.
- Positive and negative feedback processes, such as the *ice-albedo feedback mechanism*, which accelerates warming as bright reflective sea ice disappears, have profound influences on climate change.

References

1. Le Treut HR, Somerville U, Cubasch Y, et al. Historical Overview of Climate Change. In SD Solomon, D Qin, M Manning, et al. (eds.) Climate Change 2007: The Physical Science Basis. Contribution of Working Group I to the Fourth Assessment Report of the Intergovernmental Panel on Climate Change. Cambridge, UK, and New York, Cambridge University Press, 2007, pp. 93–127.
2. Etheridge DM, Steele LP, Francey RJ, Langenfelds RL. Atmospheric methane between 1000 AD and present: Evidence of anthropogenic emissions and climatic variability. Journal of Geophysical Research 1998; 103: 15979–15993.
3. Güllük T, Slemr F, Stauffer B. Simultaneous measurements of CO_2, CH_4, and N_2O in air extracted by sublimation from Antarctica ice cores: Confirmation of the data obtained using other extraction techniques. Journal of Geophysical Research 1998; 103: 15971–15978.
4. Stocker TF, Qin D, Plattner GK, et al. (eds.). Climate Change 2013: The Physical Science Basis. Working Group I Contribution to the Fifth Assessment Report of the Intergovernmental Panel on Climate Change. Cambridge, UK; New York: Cambridge University Press, 2013.
5. Allan RP, Arias PA, Berger S, et al. Summary for Policymakers. In V Masson-Delmotte, P Zhai, A Pirani, et al. (eds.) Climate Change 2021: The Physical Science

Basis. Contribution of Working Group I to the Sixth Assessment Report of the Intergovernmental Panel on Climate Change. Available at: https://www.ipcc.ch/report/ar6/wg1/. Accessed November 29, 2022.

6. Schär C, Fuhrer O, Arteaga A, et al. Kilometer-scale climate models: Prospects and challenges. Bulletin of the American Meteorological Society 2020; 101: E567–E587. Available at: https://journals.ametsoc.org/view/journals/bams/101/5/bams-d-18-0167.1.xml

7. Haarsma RJ, Roberts MJ, Vidale PL, et al. High Resolution Model Intercomparison Project (HighResMIP v1.0) for CMIP6. Geoscientific Model Development 2016; 9: 4185–4208.

8. Jiang T, Evans K, Branstetter M, et al. Northern Hemisphere blocking in ~25-km-resolution E3SM v0.3 atmosphere-land simulations. Journal of Geophysical Research: Atmospheres 2019; 124: 2465–2482.

9. Nicholls N, Gruza GV, Jouzel J, et al. Observed Climate Variability and Change. In JT Houton, LG Meiro Filho, BA Callander, et al. (eds.). Climate Change 1995: The Science of Climate Change. Cambridge, UK: Cambridge University Press, 1996, pp. 133–192.

10. Arias PA, Bellouin N. Jones RG, et al. Technical Summary. In V Masson-Delmotte, P Zhai, A Pirani, et al. (eds). Climate Change 2021: The Physical Science Basis. Contribution of Working Group I to the Sixth Assessment Report of the Intergovernmental Panel on Climate Change. Cambridge, UK, and New York: Cambridge University Press, pp. 33–144.

11. Sweet WV, Hamlington DB, Kopp RE, et al. Global and Regional Sea Level Rise Scenarios for the United States: Updated Mean Projections and Extreme Water Level Probabilities Along U.S. Coastlines. NOAA Technical Report NOS 01. Silver Spring, MD: National Oceanic and Atmospheric Administration, National Ocean Service. Available at: https://aambpublicoceanservice.blob.core.windows.net/oceanserviceprod/hazards/sealevelrise/noaa-nos-techrpt01-global-regional-SLR-scenarios-US.pdf. Accessed January 16, 2023.

12. Nicholls RJ, Lincke D, Hinkel J, et al. A global analysis of subsidence, relative sea-level change and coastal flood exposure. Nature Climate Change 2021; 11: 338–342.

13. Cao A, Esteban M, Valenzuela VPB, et al. Future of Asian deltaic megacities under sea level rise and land subsidence: Current adaptation pathways for Tokyo, Jakarta, Manila, and Ho Chi Minh City. Current Opinion in Environmental Sustainability 2021; 50: 87–97.

14. NASA Global Climate Change. Vital Signs. Available at: https://climate.nasa.gov/vital-signs/arctic-sea-ice/. Accessed November 29, 2022.

15. Perovich D, Meier W, Tschudi M, et al. Arctic Report Card 2020: Sea Ice. Available at: https://repository.library.noaa.gov/view/noaa/27904. Accessed November 29, 2022.

16. Meredith M, Sommerkorn M, Cassotta S, et al. Polar Regions. In H-O Pörtner, DC Roberts, V Masson-Delmotte, et al. (eds.). IPCC Special Report on the Ocean and Cryosphere in a Changing Climate. Cambridge, UK and New York: Cambridge University Press, 2022, pp. 203–320. Available at: https://doi.org/10.1017/9781009157964.005. Accessed November 29, 2022.

17. Dai A. Recent climatology, variability, and trends in global surface humidity. Journal of Climate 2006; 19: 3589–3606.

18. Simmons AJ, Willett KM, Jones PD, et al. Low-frequency variations in surface atmospheric humidity, temperature, and precipitation: Inferences from reanalyses

and monthly gridded observational data sets. Journal of Geophysical Research 2010; 115: D01110. doi:10.1029/2009JD012442.

19. Willett KM, Jones PD, Gillett NP, Thorne PW. Recent changes in surface humidity: Development of the HadCRUH dataset. Journal of Climate 2008; 21: 5364–5383.

20. U.S. Global Change Research Program. National Climate Assessment. Washington, DC, 2014. Available at: https://nca2014.globalchange.gov/. Accessed November 29, 2022.

21. National Academies of Sciences, Engineering, and Medicine. Attribution of Extreme Weather Events in the Context of Climate Change. Washington, DC: The National Academies Press, 2016. Available at: https://nap.nationalacademies.org/catalog/21852/attribution-of-extreme-weather-events-in-the-context-of-climate-change. Accessed November 29, 2022.

22. Zhang X, Hegerl G, Seneviratne S, et al. WCRP Grand Challenge: Understanding and Predicting Weather and Climate Extremes. Geneva: World Climate Research Programme, 2014. Available at: https://www.wcrp-climate.org/images/documents/grand_challenges/GC_Extremes_v2.pdf. Accessed November 29, 2022.

23. White R, Anderson S, Booth J, et al. The unprecedented Pacific Northwest heatwave of June 2021. Nature Portfolio 2022; in review.

24. Rashid IU, Abid MA, Almazroui M, et al. Early summer surface air temperature variability over Pakistan and the role of El Niño-Southern Oscillation teleconnections. International Journal of Climatology 2022; 42: 5768–5784.

25. Otto FEL, Zachariah M, Saeed F, et al. Climate change likely increased extreme monsoon rainfall, flooding highly vulnerable communities in Pakistan. World Weather Attribution, 2022. Available at: https://www.worldweatherattribution.org/wp-content/uploads/Pakistan-floods-scientific-report.pdf. Accessed September 5, 2023.

26. Business Insurance. Recent floods caused nearly $12 billion damage in Belgium. July 23, 2021. Available at: https://www.businessinsurance.com/article/00010101/STORY/912343432/Recent-floods-cause-nearly-$12-billion-damage-in-Belgium. Accessed November 20, 2022.

27. Kreienkamp F, Philip SY, Tradowsky JS, et al. Rapid Attribution of Heavy Rainfall Events Leading to the Severe Flooding in Western Europe During July 2021. World Weather Attribution, 2021. Available at: https://www.worldweatherattribution.org/wp-content/uploads/Scientific-report-Western-Europe-floods-2021-attribution.pdf. Accessed September 5, 2023.

28. Easterling DR, Wehner MF. Is the climate warming or cooling? Geophysical Research Letters 2009; 36: L08706. doi:10.1029/2009GL037810.

29. Francis JA, Vavrus SJ. Evidence linking Arctic amplification to extreme weather in mid-latitudes. Geophysical Research Letters 2012; 39. doi:10.1029/2012GL051000.

Further Reading

Masson-Delmotte V, Zhai P, Pirani A, et al. (eds.). Climate Change 2021: The Physical Science Basis. Contribution of Working Group I to the Sixth Assessment Report of the Intergovernmental Panel on Climate Change. Cambridge, UK, and New York: Cambridge University Press, 2021, Available at: https://www.ipcc.ch/report/ar6/wg1/. Accessed November 29, 2022.

This comprehensive assessment of the physical aspects of climate change focuses on elements to help understand past climate change, document current climate change, and project future climate change. It considers evidence based on observations of the climate system, paleoclimate archives, theoretical studies of climate processes, and simulations using climate models.

NOAA. National Centers for Environmental Information. Available at: https://www.ncei.noaa.gov/. Accessed November 29, 2022.

The National Centers for Environmental Prediction maintains the world's largest climate data archive and provides data and climatological services to all sectors in the United States and users throughout the world. Its records include paleoclimatology data, centuries-old journals, and very recent data. It monitors and assesses the state of climate throughout the world in near real time, providing data and other information on climate trends and variability, including comparisons of climate today with climate in the past.

Intergovernmental Panel on Climate Change. Global Warming of 1.5°C: IPCC Special Report on Impacts of Global Warming of 1.5°C Above Pre-industrial Levels in Context of Strengthening Response to Climate Change, Sustainable Development, and Efforts to Eradicate Poverty. Cambridge, UK, and New York: Cambridge University Press, 2022. Available at: https://doi.org/10.1017/9781009157940. Accessed November 29, 2022.

This report provides important insights into how much fossil fuel emissions must be cut—and by when—for limiting warming of the Earth to 1.5°C (2.7°F) above preindustrial temperatures. This key report provides important information for policymakers in choosing interventions to address climate change, promote sustainable development, and eradicate poverty.

Alley RB. The Two-Mile Time Machine: Ice Cores, Abrupt Climate Change, and Our Future. Princeton, NJ: Princeton University Press, 2015. Available at: https://www.degruyter.com/document/doi/10.1515/9781400852246/html?lang=en#contents. Accessed November 29, 2022.

This is a fascinating history of global climate changes, as revealed by reading the annual rings of ice from cores drilled in Greenland.

PART II

HEALTH IMPACTS

3

Heat-Related Disorders Among Workers

Tord Kjellstrom, Jeremy Lim, and Jason K.W. Lee

An immigrant farmworker in the Central Valley of California is harvesting table grapes. He is paid on a piece-work basis—by the amount of grapes that he harvests. His workday starts early and, as the day evolves, he is bothered by the heat from the sun. He has to work fast and hard in order to keep up with the daily output target and to ensure he makes enough money to send to his mother in Mexico. She relies very much on his financial support. The farmworker has brought several liters of drinking water and drinks frequently to compensate for the body liquid lost due to his profuse sweating in the heat. He takes regular short rests, but by 11 AM, he is tired and finding it difficult to get back to work. He feels dizzy and wants to stop working due to the extreme heat. But he cannot afford stopping as the supervisor would be angry and he would not have enough money to send to his mother. By 12 noon, he loses consciousness. He is revived by his co-workers, who pour water on him. He is so sick that he is taken to a local clinic for treatment. For several days that follow, he needs rest to recover.

This incident was reported to the California Occupational Safety and Health Administration (CalOSHA), which told company managers that they had to improve working conditions and procedures to ensure the health and safety of workers by providing

- Regular rest breaks in shaded places
- Buddy-monitoring for symptoms of heat strain in fellow workers
- Unlimited supply of cool drinking water at worksites
- Assurance that hourly worker income would not be reduced by high ambient temperature.

Although the mandated rest breaks might reduce worker productivity on hot days, these breaks are necessary to avoid excessive heat stress—unless the work

Tord Kjellstrom, Jeremy Lim, and Jason K.W. Lee, *Heat-Related Disorders Among Workers* In: *Climate Change and Public Health*. Edited by: Barry S. Levy and Jonathan A. Patz, Oxford University Press. © Oxford University Press 2024.
DOI: 10.1093/oso/9780197683293.003.0003

could be done using air-conditioned tractors. Climate change, accompanied by increasing ambient temperature, will increase occupational heat-related disorders.

NATURE AND MAGNITUDE OF PROBLEMS

The daily lives of most people living in areas with very hot seasons will be adversely affected by heat—not only during heat waves. In many work situations, labor productivity will be reduced, with important economic consequences for businesses, workers and their family members, and communities (Figure 3-1).

One solution is installing air conditioners in workplaces. However, millions of people in hot countries cannot be protected by air conditioning while working, so other approaches are needed. In addition, the excessive use of air conditioning is unsustainable.

Increasing temperature, *heat-level rise*, is the most obvious manifestation of climate change in much of the world.[1] It is also the most commonly modeled impact of increased concentrations of greenhouse gases (GHGs) in the atmosphere.[1] Ambient environmental heat is recorded primarily by air temperature. But, from a physiological perspective, heat is a function not only of air

Figure 3-1 Heat-stressed farmworkers harvesting lettuce. (Photograph by Earl Dotter.)

temperature, but also of humidity, air movement (wind speed), and heat radiation (outdoors, mainly from the sun).[2]

Heat exposure, resulting from the combination of air temperature, humidity, air movement, and heat radiation, represents an environmental health hazard that occurs when the body cannot eliminate excess heat. *Heat stress* is also influenced by clothing and the metabolic rate (resulting from muscle movement in relation to physical activity). *Heat strain* refers to the combined effects of heat on human physiology and health, including heat-related symptoms and their effect on work performance.

The effects of heat on human health have a physiological basis as heat exposure leads to heat stress and possibly heat strain (Figure 3-2). Humid ambient environments with air temperatures above 35°C (95°F) are a threat to physiological balance, both in terms of core body temperature and body hydration—unless fluid lost by profuse sweating is replaced by drinking water. The risk of reduced human performance and development of clinical disorders is especially great among children, older people, and people performing physically demanding activities.[2] In addition, factors such as individual degree of heat acclimatization can lead to inter-individual variability in risk profiles and prevention procedures.[3] Details concerning physiology-based health risks are described in a report of the World Health Organization (WHO) and the World Meteorological Organization (WMO).[4]

Heat exposure, heat stress, and heat strain can be described with more than 160 different *heat stress indices*.[5] The most widely used ones include the Wet Bulb Globe Temperature (WBGT), Heat Index, Corrected Effective Temperature, Humidex, and Universal Thermal Climate Index. Each of the indices has advantages and disadvantages. However, within the range of heat levels in the shade in hot countries, these indices are closely correlated with each other. This chapter focuses on WBGT, which was designed for workplace heat assessment but can be interpreted in relation to physical activity levels and other aspects of daily life, including metabolic rate and clothing.[2] WBGT is included in many international and national standards.[6,7] A WBGT of 28.0°C (82.4°F) limits human performance and work capacity, even in moderate-intensity (300 W) jobs (Box 3-1). As a reflection of climate change, the annual number of days at or above this WBGT level in Hong Kong increased from approximately 10 in the 1980s to more than 40 in recent years (Figure 3-4), and the duration of the annual very hot period has increased by 1 month since the 1980s.

Potential acute health effects of high heat exposure include heat rash (prickly heat), heat syncope (heat fainting), heat exhaustion, and heat stroke (potentially fatal, with very serious clinical symptoms and adverse effects on the central nervous system).[8] Potential chronic effects of high heat exposure include mental health problems due to difficulty coping with heat and its effects on daily life, such as suicide among farmers who have experienced economic

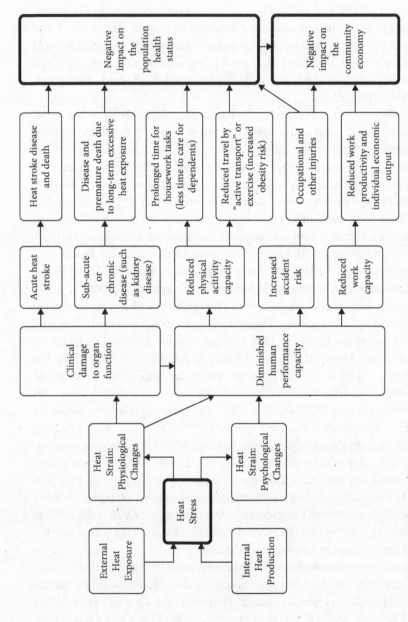

Figure 3-2 Schematic diagram showing the complex relationship between heat stress and its negative impacts on population health status and community economy. (Diagram by Tord Kjellstrom, Jeremy Lim, and Jason K. W. Lee.)

Box 3-1 Safe Work Depends on Work Intensity Level and Temperature

Tord Kjellstrom

Wet Bulb Globe Temperature (WBGT) is used to provide guidelines to reduce heat-related morbidity and mortality among workers. Threshold levels are based on research from the National Institute for Occupational Safety and Health, the American Conference of Governmental Industrial Hygienists, and other organizations.[1,2] To determine safe temperatures for workers, both internal body heat (generated by contracting muscles) and external WBGT are considered together. For internal body heat, metabolic rates have been determined by the varying levels of physical activity required to perform work activities, such as sitting at rest (metabolic rate of 115 watts), moderate walking or lifting (300 watts), and very heavy work (520 watts).

Figure 3-3 shows the Threshold Limit Value (TLV) related to heat hazard based on work level (metabolic rate) and WBGT. The solid line is the TLV temperature level, above which heat hazard is present for an acclimatized worker. The dotted line is the Action Limit, above which heat hazard is present for a non-acclimatized worker. Heat acclimatization programs involve a 7-14-day period in which non-acclimatized workers are incrementally exposed to heat stress while conducting their normal work activities. Once these limits are reached, when the

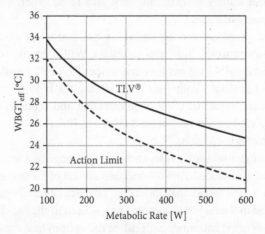

Figure 3-3 Threshold Limit Value (TLV®) and Action Limit for heat hazard, based on worker's metabolic rate (in watts) and effective WBGT (measured WBGT plus clothing adjustment factor). (*Source*: OSHA Technical Manual, Section III: Chapter 4. Heat Stress. Available at: https://www.osha.gov/otm/section-3-health-hazards/chapter-4. Accessed March 27, 2023.)

environmental factors (WBGT) are combined with a laborer's workload level, it is necessary to implement measures to prevent heat-related illness, such as work/rest regimens, shade, cooling, and hydration. The slopes of the lines indicate that heavy work (such as above 500 watts) can only safely be done below a WBGT of 26.0°C (78.8°F)for an acclimatized worker and below a WBGT of 22.0°C (71.6°F) or a non-acclimatized worker. In hot conditions (at or above 30.0°C [86.0°F] WBGT), only light work (below 200 watts) can safely be performed.

Box References

1. Jacklitsch B, Williams WJ, Musolin K, et al. NIOSH criteria for a recommended standard: Occupational exposure to heat and hot environments (DHHS [NIOSH] Publication 2016-106). Cincinnati, OH: National Institute for Occupational Safety and Health, 2016.

2. American Conference of Governmental Industrial Hygienists. Heat Stress and Strain: TLV® Physical Agents, 7th Edition Documentation. TLVs and BEIs with 7th Edition Documentation. Cincinnati, OH: ACGIH, 2017.

stress due to heat (Chapter 9); chronic kidney disease due to repeated dehydration [9]; and congenital malformations of the brain and heart in the offspring of women in jobs requiring high physical exertion because of repeated body temperature increases during early pregnancy.[10,11] In addition, high heat exposure can cause, both acutely and chronically, exacerbation of preexisting heart, lung, and kidney disease; diabetes; and some other specific chronic noncommunicable diseases.[8]

Heat-related problems are increasing in many tropical countries due to increasing temperatures.[12] In several countries, WBGT levels are already high enough to substantially limit outdoor and indoor work. When effective cooling systems are not available, daily heat levels during the hottest seasons in most tropical countries make daily life uncomfortable and reduce productivity at work.[13] The recent impacts of heat on daily life in India and Pakistan highlight emerging problems for workers.[14]

When climate change leads to more hot days and longer hot seasons, work and leisure activities are reduced. An economic assessment of climate change impacts in the United States found that lost labor productivity and coastal flooding damage resulted in the largest economic losses related to climate change.[15]

Crop workers are at extremely high risk of dying from heat exposure at work. From 1992 to 2006 in the United States, 423 crop workers were reported as having died due to heat exposure at work (28 per year, on average). The heat-related death rate of crop workers was almost 20 times higher than the rate for all U.S. civilian workers.[16] (In the past 5 years, the annual number of crop worker

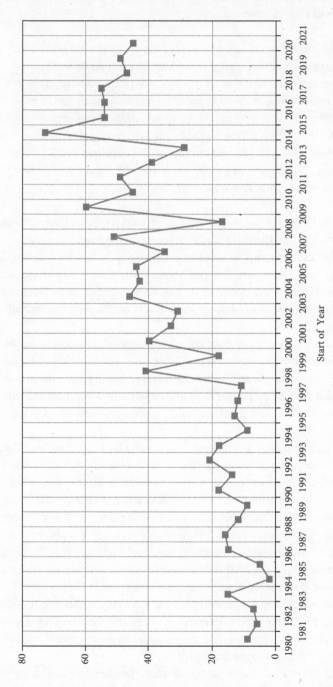

Figure 3-4 Annual number of days when maximum WBGT indoors was 29.0°C (84.2°F) or higher, Hong Kong International Airport, 1980–2021. (Graph by Tord Kjellstrom, Jeremy Lim, and Jason K. W. Lee, using HothapsSoft database, https://climatechip.org/.)

deaths due to heat exposure at work has decreased to an average of 22 per year.) The following was a typical case:

> A 56-year-old worker from Mexico who was harvesting tobacco leaves in North Carolina from early morning, started to work much more slowly and became confused after lunch. But he did not recover. He was taken to a hospital emergency department, where his body temperature was found to be 42°C (about 108°F). Despite treatment, he died. The maximum ambient temperature that day was 34°C (about 93°F), relative humidity was 44%, and the Heat Index was 36°–44°C (about 97°–112°F) in mid-afternoon. The calculated WBGT from these data was 27°C (about 81°F) in the shade and 30°C (about 86°F) in the sun—the bottom of the heat strain risk range.[7,17]

From 2011 to 2019, on average, approximately 38 workers in the United States were reported to have died annually due to heat exposure at work.[18]

Heat can also cause serious illnesses in indoor work.[4] Anecdotal data and observations at factories making low-cost shoes, clothes, and other consumer goods in tropical low-income countries demonstrate reduction in productivity by heat exposure in these work settings without air conditioning or other efficient cooling systems.

All workplaces in hot areas without efficient air conditioning require implementation of heat protective measures on all hot days. The Global Heat Health Information Network coordinates information and action to reduce the impact of heat on health globally by partnerships and capacity building, heat vulnerability and impact science, heat prediction and services, interventions, and communication and outreach.[19]

Epidemiological and experimental studies in India have demonstrated that extreme heat creates major health problems for workers.[20–22] Large cohort studies in Thailand, based on interviews with more than 40,000 young adults, have shown that, in jobs with high heat exposure, self-reported general health status is decreased,[23] prevalence of kidney disease is increased,[24] and history of occupational injuries is increased.[25] Heat-related mortality has also been reported among workers (mainly performing outdoor work) in Japan,[26] among working-age men during the 2003 heat wave in France,[27] and among migrant construction workers from Nepal working in hot conditions in Doha, Qatar.[28]

Older people—including older workers—are especially vulnerable to heat exposure because sweating ability decreases with age and their sensing of heat stress is also reduced.[29] A study of older people in Japan demonstrated a reduction in their physical activity at high temperatures.[30]

People living and working in hot places can adapt their behavior to heat; however, resultant limits in daily activities will undermine livability in these places.[4]

Anthropological studies in India found that workers avoided heat strain by reducing their work hours and intensity of work, appropriate to the thermal conditions.[31]

DIRECT CAUSES

At rest, core body temperature needs to be kept within a narrow range around 37.0°C (98.6°F).[2] The environment immediately around the body exchanges heat with the body. When the body is warmer than the ambient air, heat energy is lost from the body; when the body is cooler, heat energy is added to the body.

Basal metabolism and added metabolic activities create substantial heat inside the body, which, in hot ambient environments, is difficult to release.[2] In these environments, the role of sweat evaporation is essential for maintaining a safe core body temperature. When sweat liquid (mainly water) evaporates from the skin, a substantial amount of body heat is released. However, the evaporation rate depends on the humidity in the surrounding air—the higher the humidity, the slower the evaporation. At 100% relative humidity, there is no heat loss from evaporation; however, sweating continues and this may lead to substantial dehydration unless the fluids lost via sweat are replaced by drinking water.[2] Physiologically intolerable conditions have occurred, or have threatened to occur, in South Asia, the coastal Middle East, and coastal southwest North America—all of which are densely populated regions.[32] Without intervention, many people worldwide will not be able to remove surplus heat via sweat evaporation.

For the sweat evaporation mechanism to function effectively, the blood flow to the skin must increase.[2] Physiological acclimatization to high temperature therefore involves improving one's ability to dissipate body heat by widening skin capillaries and increasing cardiac output. People who cannot rapidly increase cardiac output, such as many older people, are at increased risk of heat stress.[2] (See also Chapter 4.)

Hot and humid ambient environments, with air temperatures above 37.0°C (98.6°F), threaten physiological balance, especially for workers in common protective clothing or overalls. Several factors influence the impact of heat on health and human performance. These examples include high metabolic rates associated with intense work activities, active transportation (walking or bicycling), and participation in sports. Insulating clothing can reduce both heat losses from and heat additions to the body. However, when a person sweats, clothing can prevent cooling by inhibiting evaporation[33]; for example, workers required to use protective clothing or thick overalls while working in hot environments will not achieve the full benefit of sweat evaporation. As another example, during the COVID-19 pandemic, healthcare workers in India and Singapore reported thirst, excessive sweating, and exhaustion while wearing personal protective equipment as well as headaches, dizziness, breathing difficulties, and dehydration, which are precursors to heat-related illnesses.[34]

Cultural and economic factors contribute to the adverse impacts of heat, including the following:

Work hours: If, by tradition, no work is performed during the hottest part of the day and workers take a "siesta," then heat-related problems are reduced—but fewer hours are available for work each day. Heat problems can also be reduced by shifting work and other strenuous daily activities to cooler hours of each day and/or to cooler seasons of the year. However, this shifting may decrease work productivity.[35]

Clothing: Clothing customs also influence the impact of heat. In India, for example, women are expected to wear specific types of traditional clothing even when they need different protective clothing at work. Job tasks there that create substantial internal heat production can become extremely hazardous for women.[36]

Manual labor: Especially in low- and middle-income countries, access to mechanized machinery at work is limited, and, even in food production, a high percentage of energy comes from manual labor.[37]

FUTURE PROJECTIONS

When heat exposure increases and hot periods become longer, millions more people will live and work in areas where the climate is "unlivable" for long periods each year.[38] If WBGTs were to increase another 3.0°C (5.4°F), productivity would be severely reduced while risk of heat-related disorders would greatly increase.[39] The most affected areas will be in tropical and subtropical areas, and low-income communities will be especially affected.[13,40] During severe heat waves, even cooler areas will experience temporary decreases in productivity.[12]

Many other parts of the world, including much of the U.S. South, Latin America, Pacific islands, Australia, China, India, Southeast Asia, and much of Africa, are experiencing—and will continue to experience—similar impacts of heat on human performance and daily life.[13] Local data for grid cells on estimated current and future heat levels, expressed as basic climate variables or heat stress indices based on WBGT, the Universal Thermal Climate Index, and the Heat Index, can be found on the website www.ClimateCHIP.org (Figure 3-5). This website also shows projected monthly estimates of these parameters for each decade until the end of this century.

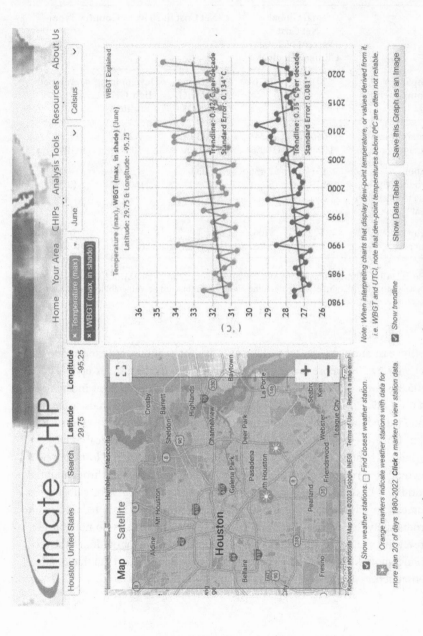

Figure 3-5 Trends of mean maximum temperature and maximum WBGT in June (in the shade), Houston area, 1980–2022. (*Source*: ClimateCHIP.org at https://www.ClimateCHIP.org.)

Table 3-1 Projected economic impacts of climate change in 2030, in billions of U.S. dollars

Impact Component	Total Global Net Cost (Percent of Total Climate Change)		Net Cost in 2030, by Country Type		
	2010	2030	Developing, Low GHG Emitters	Developing, High GHG Emitters	Developed
Total climate change	$609 (100%)	$4,345 (100%)	1,730 (100%)	2,292 (100%)	179 (100%)
Labor productivity loss due to workplace heat	$311 (51%)	$2,436 (56%)	1,035 (60%)	1,364 (60%)	48 (27%)
Clinical health impacts	$23 (3.7%)	$106 (2.4%)	84 (4.9%)	21 (0.9%)	0.002 (0.001%)

Source: DARA. Climate Vulnerability Monitor 2012. Madrid: Fundacion DARA Internacional, 2012. Available at: http://daraint.org/climate-vulnerability-monitor/climate-vulnerability-monitor-2012/. Accessed August 2, 2022.

The east and west coastal regions of India, for example, are expected to experience by 2099 more than 60 days annually with the heat index up to 41.0°C (105.8°F)—moderate risk—and more than 40 days with the heat index up to 54.0°C (129.2°F)—high risk. Under such conditions, sunstroke, muscle cramps, and heat exhaustion are likely, and more severe heat-related illnesses, such as heat stroke, are possible with prolonged exposure or intense physical activity.[41]

Table 3-1 shows estimates of economic losses due to projected lost work productivity as a result of climate change for 21 geographic regions throughout the world.[42,43] Losses for 2030 are estimated at $4.3 trillion (3.1% of global gross domestic product). More than half of these losses accrue from reduced work productivity due to workplace heat. Although these estimates do not consider preventive adaptation measures, except reducing heavy labor outdoors, the economic impact of lost work productivity may be much greater than that of other consequences of climate change.[15]

PREVENTIVE MEASURES

Heat-warning systems and preparedness are key preventive measures, not only for workers but also for community residents who can benefit from information on local experiences and heat-related risks. Accessing local historical and projected data can be a first step. In local studies, low-cost data loggers for temperature and humidity can be used to monitor local workplace conditions.[44] In addition, wearable devices, such as smartwatches,[45] can be used to predict thermal discomfort and the onset of heat-related conditions[46] based on physiological indicators of heat strain, such as high heart rate and increased average skin temperature.[47] Specific physiological solutions, such as aerobic fitness conditioning, heat acclimatization, work-rest cycles, cooling, and hydration, can also be effective and sustainable in reducing heat strain in workers.[2]

Addressing the problem of human heat exposure in workplaces, residences, or other locations cannot simply be done by increasing air conditioning, because air conditioning generally uses electricity produced from non-renewable energy sources and therefore increases GHG emissions. For example, electricity consumption for refrigeration and cooling accounted for more than 20% of China's total electricity consumption (7,225 billion kilowatt-hours) in 2019,[48] and 27% of its GHG emissions during that year.[49] In Bangkok, Thailand, the additional need for air conditioning for every 1.0°C (1.8°F) increase of external heat was approximately 2,000 megawatts—the output from a modern major electricity power station.[50] And hundreds of millions of people living and working in low-income countries cannot afford air conditioning.

Other methods to cool indoor spaces may be effective in certain conditions, such as the *wet surface evaporation approach* (using evaporation of water from large wet blankets hanging in a workplace to reduce temperature), which may work well in places with low relative humidity and sufficient air movement. However, this method increases relative humidity, thereby increasing heat exposure. Fans directed at individual workers or groups create a feeling of cooler air movement. However in many indoor workplaces, fans can create high noise levels.

Locating buildings near shade-providing trees or structures is another approach to reducing heat exposure outdoors. Urban design is another important approach, especially in rapidly urbanizing tropical and subtropical countries. (See Chapter 16.)

Architectural methods to avoid indoor heat loads can create cooler buildings and reduce the need for air conditioning.[51] Examples include insulating walls and ceilings, using double-glazed windows, and, in very sunny locations, using external panels over roofs or along walls facing the sun in the morning and evening. The type of roof material used in homes and workplaces is also important.

In urban areas, black-painted or black-stained sheets of corrugated iron or asbestos cement on roofs create heat traps for sunlight and transfer that heat to the indoor space below. As a result of this urban heat island effect, inner-city residents are often at increased risk of heat stress (Chapter 4). The added heat absorption from solar heat radiation from black or very dark-colored roof materials can be a problem in urban areas. Traditional building practices in tropical hot areas have recognized that wind movement can cool buildings. Modern building design can create air channels that enable cooler air from underground or lower levels to rise to higher levels in the building. Both ongoing cooling approaches and building design approaches need to consider their sustainability.

Preparedness and heat-warning systems based on weather forecasts and better development and enforcement of health and safety regulations are key elements of programs to reduce the impact of climate change. Special needs must be addressed for high-risk workers, such as immigrant farmworkers, construction workers, and workers in factories and workshops without effective cooling systems.

Preventive measures can often have multiple benefits in reducing heat exposure and addressing climate change. For example, some air conditioning systems use solar heat radiation as the energy source for the cooling process, thereby providing cooler indoor spaces and reducing GHG emissions.[52]

SUMMARY POINTS

- High heat exposure in workplaces can cause heat stroke and other heat-related disorders as well as severe complications of chronic noncommunicable diseases.
- Chronic exposure to high heat at work can cause mental health problems and, due to repeated dehydration, chronic kidney disease.
- Increasing heat and humidity are having profound adverse effects on worker productivity.
- Increasing temperature and humidity due to climate change will make it increasingly difficult for strenuous work to be performed in many parts of the world.
- Preventive measures include heat-warning systems and preparedness, providing air conditioning and other means to ensure cool indoor spaces, and designing and siting work buildings to reduce indoor heat.

ACKNOWLEDGMENT

The authors acknowledge the assistance of Ms. Marianne Lim in collating some of the research used in preparation of this chapter.

References

1. Intergovernmental Panel on Climate Change (IPCC). Climate Change 2021: The Physical Science Basis. Working Group 1 report. IPCC Sixth Assessment. Geneva: IPCC: 2021. Available at: https://www.ipcc.ch/report/sixth-assessment-report-working-group-i/. Accessed June 10, 2022.
2. Parsons K. Human Thermal Environments. The Effects of Hot, Moderate, and Cold Temperatures on Human Health, Comfort, and Performance (3rd ed.). London: Taylor & Francis, 2014.
3. University of California, Merced. Heat illness prevention procedures. Available at: https://ehs.ucmerced.edu/sites/ehs.ucmerced.edu/files/documents/field-safety/heat_illness_and_prevention_program.pdf. Accessed May 11, 2022.
4. World Health Organization/World Meteorological Organization. Climate Change and Occupational Heat Stress: Technical report by WHO and WMO. Geneva: WHO, 2023.
5. De Freitas CR, Grigorieva EA. A comprehensive catalogue and classification of human thermal climate indices. International Journal of Biometeorology 2015; 59: 109–120.
6. International Standards Organization. Hot environments - Estimation of the heat stress on working man, based on the WBGT-index (wet bulb globe temperature). ISO Standard 7243. SO, 1989. Available at: https://www.iso.org/standard/13895.html. Accessed August 3, 2023.
7. American Conference of Government Industrial Hygienists. TLVs for Chemical Substances and Physical Agents & Biological Exposure Indicators. Cincinnati: ACGIH, 2009.
8. Jacklitsch B, Williams WJ, Musolin K, et al. NIOSH Criteria for a Recommended Standard: Occupational Exposure to Heat and Hot Environments (DHHS [NIOSH] Publication 2016-106). Cincinnati, OH: National Institute for Occupational Safety and Health, 2016.
9. Sanchez Polo V, Garcia-Trabanino R, Rodriguez G, Madero M. Mesoamerican nephropathy (MeN): What we know so far. International Journal of Nephrology and Renovascular Disease 2020; 13: 261–272.
10. Edwards MJ, Shiota K, Smith MSR, Walsh DA. Hyperthermia and birth defects. Reproductive Toxicology 1995; 9: 411–425.
11. Graham JM Jr. Update on the gestational effects of maternal hyperthermia. Birth Defects Research 2020; 112: 943–952.
12. Kjellstrom T, Maître N, Saget C, et al. Working on a Warmer Planet: The Impact of Heat Stress on Labour Productivity and Decent Work. Geneva: International Labor Organization, 2019. Available at: https://www.ilo.org/wcmsp5/groups/public/---dgreports/---dcomm/---publ/documents/publication/wcms_711919.pdf. Accessed August 2, 2022.
13. Kjellstrom T, Otto M, Lemke B, et al. Climate Change and Labour: Impacts of Heat in the Workplace. United Nations Development Programme, 2016. Available at: https://

www.ilo.org/wcmsp5/groups/public/---dgreports/---dcomm/documents/genericd ocument/wcms_476051.pdf. Accessed August 2, 2022.

14. Jain Y, Jain R. India and Pakistan emerge as early victims of extreme heat conditions due to climate injustice. British Medical Journal 2022. doi: 10.1136/bmj.o1207.

15. Kopp R, Hsiang S, Muir-Wood R, et al. American Climate Prospectus. Economic Risks in the United States. New York: Rhodium Group, 2014. Available at: https:// gspp.berkeley.edu/assets/uploads/research/pdf/American_Climate_Prospectus.pdf. Accessed August 2, 2022.

16. Centers for Disease Control and Prevention (CDC). Heat-related deaths among crop workers-United States, 1992-2006. Morbidity and Mortality Weekly Report 2008; 57: 649–653.

17. Lemke B, Kjellstrom T. Calculating workplace WBGT from meteorological data. Industrial Health 2012; 50: 267–278.

18. U.S. Bureau of Labor Statistics. TED: The Economics Daily. 43 work-related deaths due to environmental heat exposure in 2019. Available at: https://www.bls.gov/opub/ ted/2021/43-work-related-deaths-due-to-environmental-heat-exposure-in-2019. htm. Accessed June 10, 2022.

19. Global Heat Health Information Network. Heat & Health. Available at: https://ghhin. org/heat-and-health/. Accessed May 11, 2022.

20. Sahu S, Sett M, Kjellstrom T. Heat exposure, cardiovascular stress and work productivity in rice harvesters in India: Implications for a climate change future. Industrial Health 2013; 51: 424–431.

21. Nag PK, Nag A, Sekhar P, Pandit S. Vulnerability to Heat Stress: Scenario of Western India. WHO report APW No. SO 08 AMS 6157206. Delhi: World Health Organization Regional Office, 2009. Available at: https://kipdf.com/vulnerability-to-heat-stress-scenario-in-western-india_5ad988777f8b9aeb898b4597.html. Accessed August 2, 2022.

22. Venugopal V, Chinnadurai J, Viswanathan V, et al. The social impacts of occupational heat stress on migrant workers engaged in public construction: A case study from Southern India. The International Journal of the Constructed Environment 2016; 7: 25–36.

23. Tawatsupa B, Lim L, Kjellstrom T, et al. The association between overall health, psychological stress and occupational heat stress among a large national cohort of 40,913 Thai workers. Global Health Action 2010; 3: 10.3402/gha.v3i0.5034.

24. Tawatsupa B, Lim LL-Y, Kjellstrom T, et al. Association between occupational heat stress and kidney disease among 37,816 workers in the Thai Cohort Study (TCS). Journal of Epidemiology 2012. doi:10.2188/jea.JE20110082.

25. Tawatsupa B, Yiengprugsawan V, Kjellstrom T, et al. Association between heat stress and occupational injury among Thai workers: Findings of the Thai Cohort Study. Industrial Health 2013; 51: 34–46.

26. Sawada S. Recent occupational heat related problems and national policies for occupational heat stress prevention in Japan (Abstract SS097-4). 30th International Congress on Occupational Health, Cancun, Mexico, 2012. Available at: https://icoh. confex.com/icoh/2012/webprogram/Paper7069.html. Accessed August 2, 2022.

27. Hemon D, Jougla E. Excess mortality associated with heat wave in August 2003 (Surmortalité liée à la canicule d'août 2003 – Rapport d'étape). Technical report to the Minister of Health, Family and Disabled Persons (Rapport remis au Ministre

de la Santé, de la Famille et des Personnes Handicapées), 25 September 2003. Paris: INSERM, 2003 (in French).

28. Pradhan B, Kjellstrom T, Atar D, et al. Heat stress impacts on cardiac mortality in Nepali migrant workers in Qatar. Cardiology 2019; 143: 37–48.

29. Kenney WL, DeGroot DW, Holowatz LA. Extremes of human heat tolerance: Life at the precipice of thermoregulatory failure. Journal of Thermal Biology 2004; 29: 479–485.

30. Togo F, Watanabe E, Park H, et al. Meteorology and the physical activity of the elderly: The Nakanojo Study. International Journal of Biometeorology 2005; 50: 83–89.

31. Planalp JM. Heat Stress and Culture in North India (special technical report). Washington, DC: U.S. Army Medical Research and Development Command, 1971.

32. Raymond C, Matthews T, Horton RM. The emergence of heat and humidity too severe for human tolerance. Science Advances 2020; doi: 10.1126/sciadv.aaw1838.

33. Kuklane K, Lundgren K, Gao C, et al. Ebola: Improving the design of protective clothing for emergency workers allows them to better cope with heat stress and help to contain the epidemic. Annals of Occupational Hygiene 2015; 59: 258–261.

34. Lee J, Venugopal V, Latha PK, et al. Heat stress and thermal perception amongst healthcare workers during the COVID-19 pandemic in India and Singapore. International Journal of Environmental Research and Public Health 2020; doi: 10.3390/ijerph17218100.

35. Kjellstrom T, Holmer I, Lemke B. Workplace heat stress, health and productivity: An increasing challenge for low and middle income countries during climate change. Global Health Action 2009; doi: 10.3402/gha.v2i0.2047.

36. Kuklane K, Lundgren K, Kjellstrom T, et al. Insulation of traditional Indian clothing: Estimation of climate change impact on productivity from PHS (Predicted Heat Strain) model. In: Proceedings of CIB W099 International Conference; Achieving Sustainable Construction Health and Safety. Lund, Sweden, June, 2014. Available at: https://dspace.lboro.ac.uk/dspace-jspui/bitstream/2134/15024/3/2014%20Kuklane%20et%20al%20insulation%20of%20traditional%20indian%20clothing%20.pdf. Accessed August 2, 2022.

37. Pimentel D. Energy inputs in food crop production in developing and developed nations. Energies 2009; 2: 1–24.

38. Intergovernmental Panel on Climate Change (IPCC). Climate change 2022: Impacts, adaptation, and vulnerability. Working Group II Contribution to the IPCC Sixth Assessment Report. IPCC, 2022. Available at: https://www.ipcc.ch/report/ar6/wg2/downloads/report/IPCC_AR6_WGII_FinalDraft_FullReport.pdf. Accessed May 13, 2022.

39. Kjellstrom T, Lemke B, Otto M. Mapping occupational heat exposure and effects in South-East Asia: Ongoing time trends 1980–2009 and future estimates to 2050. Industrial Health 2013; 51: 56–67.

40. Kjellstrom T, Freyberg, C, Lemke B, et al. Estimating population heat exposure and impacts on working people in conjunction with climate change. International Journal of Biometeorology 2018, 62: 291–306.

41. Koteswara Rao K, Lakshmi Kumar TV, Kulkarni A, et al. Projections of heat stress and associated work performance over India in response to global warming. Scientific Reports 2020; 10: 16675. doi: 10.1038/s41598-020-73245-3.

42. Kjellstrom T, Kovats S, Lloyd SJ, et al. The direct impact of climate change on regional labour productivity. International Archives of Environmental & Occupational Health 2009; 64: 217–227.

43. DARA. Climate Vulnerability Monitor 2012. Madrid: Fundacion DARA Internacional, 2012. Available at: http://daraint.org/climate-vulnerability-monitor/climate-vulnerability-monitor-2012/. Accessed August 2, 2022.

44. ClimateCHIP.org. Basic local heat monitoring and occupational exposure assessment. Available at: http://climatechip.org/Local_Monitoring. Accessed August 2, 2022.

45. Brown D. Fitbit's new line-up features a smartwatch that might pick up fever, stress symptoms. USA Today, August 25, 2020. Available at: https://www.usatoday.com/story/tech/2020/08/25/fitbits-new-smartwatch-may-pick-up-fever-stress-symptoms/3429380001/. Accessed May 11, 2022d.

46. Nazarian N, Liu S, Kohler M, et al. Project Coolbit: Can your watch predict heat stress and thermal comfort sensation? Environmental Research Letters 2021; 16: 034031.

47. Garzón-Villalba XP, Wu Y, Ashley CD, et al. Ability to discriminate between sustainable and unsustainable heat stress exposures – part 2: Physiological indicators. Annals of Work Exposures and Health 2017; 61: 621–632.

48. Shen B, An F, Chen J, et al. People's Republic of China: Developing a climate-friendly cooling sector through market and financing innovation. Asian Development Bank, 2021. Available at: https://www.adb.org/sites/default/files/project-documents/52249/52249-003-tacr-en.pdf. Accessed May 12, 2022.

49. Larsen K, Pitt H, Grant M, et al. China's greenhouse gas emissions exceeded the developed world for the first time in 2019. Rhodium Group, 2021. Available at: https://rhg.com/research/chinas-emissions-surpass-developed-countries/. Accessed May 12, 2022.

50. Lundgren K, Kjellstrom T. Sustainability challenges from climate change and air conditioning use in urban areas. Sustainability 2013; 5: 3116–3128; doi:10.3390/su5073116

51. Asian Development Bank (ADB). Beating the Heat: Investing in Pro-poor Solutions for Urban Resilience. Mandaluyong City: ADB, 2022. Available at: https://www.adb.org/publications/beating-heat-pro-poor-solutions-urban-resilience. Accessed March 6, 2023.

52. Desideri U, Proietti S, Sdringola P. Solar powered cooling systems: Technical and economic analysis on industrial refrigeration and air conditioning applications. Applied Energy 2009; 86: 1376–1386.

Further Reading

International Labor Organization. Working in a Warming World. Geneva: ILO, 2019.
This report provides country-level data on heat effects on working people, based on internationally accepted climate modeling data and impact models. It highlights the links between heat effects on work capacity and the local economy.

McKinnon M, Lissner T, Romanello M, et al. (eds.). Climate Vulnerability Monitor (3rd edition): A planet on fire. Geneva: Climate Vulnerability Monitor, 2022. Available at: https://www.v-20.org/climatevulnerabilitymonitor. Accessed March 7, 2023.

This report highlights the particular effects of increasing heat on working people in low- and middle-income tropical countries.

El Khayat M, Halwani DA, Hneiny L, et al. Impacts of climate change and heat stress on farmworkers' health: A scoping review. Frontiers in Public Health 2022; 10: 782811. doi: 10.3389/fpubh.2022.782811.

Based on 92 published articles, this review focused primarily on heat-related illnesses and kidney disease among farmworkers. It described in detail risk factors and preventive measures.

Acharya P, Boggess B, Zhang K. Assessing heat stress and health among construction workers in a changing climate: A review. International Journal of Environmental Research and Public Health 2018; 15:247. doi: 10.3390/ijerpph15020247.

This publication reviewed the severity with which construction workers are affected by heat, risk factors and co-morbidity associated with heat-related illnesses in the construction industry, vulnerable populations, and implementation of preventive measures.

Flouris AD, Dinas PC, Ioannou LG, et al. Workers' health and productivity under occupational heat strain: A systematic review and meta-analysis. Lancet Planet Health 2018; 2: e521–e531.

This meta-analysis, based on 110 studies in 30 countries, provides valuable information on heat strain, kidney disease, and productivity losses.

Morrissey MC, Brewer GJ, Williams WJ, et al. Impact of occupational heat stress on worker productivity and economic cost. American Journal of Industrial Medicine 2021; 64: 981–988.

This review highlights the physiological, physical, psychological, and financial harms of heat stress on worker productivity and proposes strategies to quantify heat-related productivity losses.

4

Heat-Related Disorders Among Community Populations

Rupa Basu and Xiangmei (May) Wu

On a hot day in August, a young couple went hiking on an 8-mile loop trail in the Sierra Nevada Mountains in Northern California with their 1-year-old infant and their dog. They had only 2.5 liters of water. The temperature was below 26.7°C (80.1°F) in the morning but rose to 42.8°C (109.0°F) in the afternoon. Two days later, their bodies were found on the trail. No cause of death was immediately identified. An investigation, partly based on their cell phone data, determined the cause of their deaths were hyperthermia and dehydration.[1]

Heat-related disorders, which are often fatal, have profound effects on public health.[2] Morbidity and mortality due to heat-related disorders will likely continue to increase as a result of increased temperatures not limited to heat waves. Most heat-related disorders and exacerbations of medical conditions due to heat can be prevented with effective heat-warning systems and preparedness and response plans.

NATURE AND MAGNITUDE

According to the latest report from the Intergovernmental Panel on Climate Change (IPCC), the rate of global warming since 1981 has been 0.18°C (0.32°F) per decade.[3] The warmest 7 years in recorded history have all occurred since 2015. Heat-related health impacts are generally underreported, because only heat-related illnesses coded in the International Classification of Diseases (ICDs) are traditionally considered when evaluating health impacts. For example, the United Nations has estimated that there were more than 166,000 deaths due to extreme heat between 1998 and 2017.[4] In contrast, a group of climate scientists

Rupa Basu and Xiangmei (May) Wu, *Heat-Related Disorders Among Community Populations* In: *Climate Change and Public Health*. Edited by: Barry S. Levy and Jonathan A. Patz, Oxford University Press. © Oxford University Press 2024.
DOI: 10.1093/oso/9780197683293.003.0004

estimated that 356,000 deaths due to extreme heat occurred in 2019 alone.[5] The Centers for Disease Control and Prevention has reported an annual average of 67,512 emergency department visits, 9,235 in-patient hospital admissions, and 702 deaths in the United States due to heat-related illnesses.[6] However, the U.S. Global Change Research Program has estimated that there are more than 1,300 heat-related deaths in the United States annually, not only from heat-related disorders, but also from exacerbations of coronary artery disease, cerebrovascular disease, diabetes, and other preexisting chronic noncommunicable diseases, which have usually not been counted as heat-related deaths.[7] Heat-related morbidity and mortality are also underreported because they tend to be recognized and reported mainly during or shortly after heat waves (surveillance bias) and when no other cause of death or illness is suspected.[8-11]

HEAT WAVES

There are various definitions of a *heat wave*. The World Meteorological Organization defines a heat wave as a period of 5 or more days when ambient temperature exceeds the average maximum from the 1961–1990 period by 5°C (9°F). Some definitions of a heat wave are based not only on temperature but also on relative humidity. The definitions are based on specific geographical regional or local climate history. In places with high relative humidity, the U.S. National Weather Service (NWS) issues heat warnings and heat advisories based on the *heat index* (also known as apparent temperature), which combines air temperature and relative humidity (Chapter 3). In the western United States, where relative humidity is generally low, heat warnings are based on a prototype index called "HeatRisk."[12] In the northeastern United States, a heat wave is defined as 3 consecutive days during which the heat index reaches or exceeds 32.2°C (90.0°F). In contrast, the threshold heat index for a heat wave could be as high as 40.6°–43.3°C (105°–110°F) in Texas. The thresholds used by NWS for defining heat waves are at the resolution of *forecast zones*, the size of which are between city and county scale, and therefore, can vary within the same state. It is also important to consider both daytime and nighttime thresholds when defining a heat wave in a specific location.

The nature and frequency of heat waves vary between and within geographic areas. In places where people infrequently experience high temperatures, such as Russia and the United Kingdom, temperatures in the mid-80s can cause adverse health effects. In contrast, in tropical regions, where people are more acclimatized to heat, the temperatures that can cause heat-related health effects in cooler regions are often considered normal. Such geographical differences can occur between neighboring areas because of microclimates with different temperature, relative humidity, and other factors. Duration, intensity, and timing of heat waves are important factors to consider because they correlate with mortality.[13] Heat waves that occur earlier in the warm season, not necessarily with very high absolute temperature, may cause more morbidity and mortality because people may be unprepared and not yet adapted to hot weather.

During heat waves in the United States, mortality increases almost 4%—somewhat more for the first heat wave of summer.[14] The most well-studied U.S. heat wave is the 1995 Chicago heat wave, which lasted 5 days and was reported to cause more than 700 excess deaths.[15,16] Research, which identified social isolation, older age, living on higher stories of apartment buildings, and other risk factors, guided the development of citywide heat preparedness and response plans to reduce mortality from heat waves. Because of this research, the death toll attributed to a Chicago heat wave in 1999 was substantially lower—80 deaths, mainly among people under 65 years of age, suggesting that heat preparedness and response plans overlooked some vulnerable subgroups of the population.[17]

In 2006, California experienced a 12-day heat wave with both high temperature and high relative humidity, during which approximately 140 recorded deaths and 16,000 excess hospitalizations occurred due to heat-related illnesses. Research later found that the actual number of deaths associated with this heat wave was almost four times greater than the 140 recorded.[18-20] Deaths in this heat wave, which occurred primarily among middle-aged outdoor workers, including farmworkers, were largely attributed to high nighttime temperatures.[21] This pattern of increased nighttime temperatures and high relative humidity is expected to continue in future California heat waves.

Globally, some major heat waves during the past three decades have included:

- In 1995, a heat wave in Greater London was associated with a 2.6% increase in hospital admissions and a 10.8% increase in mortality.[22] An estimated 619 excess deaths occurred, mostly among women and mostly from respiratory and cerebrovascular diseases.[23]
- In 2003, a protracted heat wave in Europe caused more than 70,000 deaths in 12 countries, primarily among older people living in urban areas.[24-26] Approximately 15,000 of these deaths occurred in France, where there was no heat-response plan and the healthcare system was not prepared for extreme heat.
- In 2007, during two successive heat waves in Athens, temperatures reached 46.2°C (115.2°F) and resulted in massive wildfires and hundreds of cases of heat exhaustion and heat stroke.[27]
- In 2010, a 44-day heat wave in Russia caused approximately 56,000 deaths (11,000 in Moscow) as well as wildfires and drought throughout the country.[28]

- From December 2014 to January 2015, Argentina experienced a 22-day heat wave, which caused at least 1,877 deaths.
- From April to June 2015, heat waves caused more than 2,200 deaths in India and 2,500 in Pakistan.
- In the spring and summer of 2018, much of Europe experienced above-average temperatures and drought, a combination that resulted in wildfires. The European Union estimated that there were 104,000 heat-related deaths among older people.[29,30]
- In 2018, a heat wave in Japan caused more than 22,000 hospitalizations and 80 deaths.
- In 2019, during a 32-day heat wave in India and Pakistan with a near-record high temperature of 50.8°C (123.4°F), hundreds of deaths were recorded, but the actual number of deaths is unknown. Temperatures were so high that the asphalt on road surfaces melted and airports had to be temporarily closed.
- In 2019, a series of heat waves in northwestern Europe caused more than 2,500 deaths.[31] Proactive preparedness measures likely reduced the death toll.
- In 2021, western Canada and the northwestern United States experienced an extreme heat wave during which Canada recorded its highest temperature ever, 49.6°C (121.3°F) (Figure 4-1). More than 1,037 excess deaths occurred.[32]
- In 2022, a heat wave, which occurred in India and Pakistan from mid-March through April—unusually early, killed at least 90 persons.

BEYOND HEAT WAVES

Widespread adverse health effects occur due to increased mean daily temperature. A study in 213 U.S. counties between 1999 and 2008 reported a 4.3% increase in emergency hospitalizations for respiratory diseases among older people per 5.6° C (10.0°F) increase in same-day temperature.[33] Another study, using data from more than 1.2 million emergency department visits in California during the warm seasons from 2005 to 2008, found that the same increment (5.6°C or 10.0°F) in mean apparent temperature was associated with increased incidence of a wide spectrum of medical conditions on the same day, including 1.7% in ischemic heart disease, 2.8% in ischemic stroke, 2.8% in cardiac dysrhythmia, 16% in acute renal failure, 26% in dehydration, and 393% in heat-related illnesses; the study also found that these heat-related medical conditions were higher among older people, those who had preexisting medical conditions,

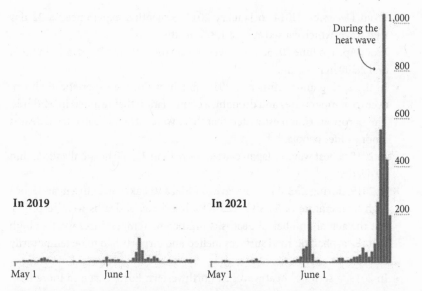

Figure 4-1 Number of emergency department visits for heat-related illness in U.S. Department of Health and Human Services Region 10 (Alaska, Idaho, Oregon, and Washington) and nationwide (excluding Region 10), May 1-June 30 for the years 2019 and 2021, based on data from the National Syndromic Surveillance Program. (*Source*: Schramm PJ, Vaidyanathan A, Radhakrishnan L, et al. Heat-related emergency department visits during the northwestern heat wave—United States, June 2021. Morbidity and Mortality Weekly Report 2021; 70: 1020–1021.)

underweight individuals, and members of minority groups.[34] A large-scale study during the warm seasons from 2002 to 2008 in the United States found that a 2.8°C (5.0°F) increase in temperature during the week before delivery was associated with a 12–16% higher occurrence of premature births.[35]

In addition to the studies described above, there is some evidence that each of the following is associated—but not necessarily causally associated—with heat waves or increased ambient heat:

- Cerebrovascular diseases, such as ischemic stroke[36,37]
- Cardiovascular diseases, such as ischemic heart disease (including myocardial infarction), congestive heart failure, atrial fibrillation, and abnormal heart rate variability[38–42]
- Respiratory diseases, such as chronic obstructive pulmonary disease and respiratory tract infections[38]
- Gastrointestinal diseases among children[37]
- Aggravation of diabetes[36,37]
- Kidney and liver disorders, including renal failure[43]

- Adverse birth outcomes, such as low birthweight, pre-term birth, and stillbirth[44]
- Violence, including homicide and suicide, and mental disorders (Chapter 10).[45]

In addition, heat can increase stress, adversely affect brain function, disrupt sleep, and trigger anxiety, thereby causing impaired cognitive function, insomnia, mood swings, depression, and commission of violent acts.[46]

In response to increasing ambient temperature, body mechanisms dissipate heat by perspiration and dilatation of blood vessels in the skin (vasodilation), which brings more blood to the skin surface to dissipate heat by radiation and sweating.[47] The body's capacity to respond to heat stress is limited by its capability to increase cardiac output to increase blood flow to the skin. However, the body may not be able to quickly adapt to heat using these mechanisms. During heat exposure, blood viscosity and serum cholesterol increase, possibly contributing to the incidence of myocardial infarction and ischemic stroke.

During pregnancy, heat exposure can divert blood from the uterus and inhibit fetal growth.[48] Dehydration from heat exposure may also decrease uterine blood flow and may increase pituitary secretion of antidiuretic hormone and oxytocin, which can prematurely induce labor.

Heat can cause morbidity or mortality indirectly. For example, higher temperatures can increase concentrations of ozone and other outdoor air pollutants,[49] causing or worsening asthma and respiratory allergies[50] and increasing respiratory symptoms and impaired lung function,[51] lost school days due to respiratory illnesses,[52] emergency department visits,[53,54] hospitalizations,[55] and deaths.[56] Ground-level ozone concentrations, which are highly correlated with ambient temperature, exceed federal and state standards in much of the United States, especially in heavily populated urban areas. (See Chapter 5.)

RISK FACTORS AND VULNERABLE GROUPS

Heat-related disorders could be mediated or exacerbated by both intrinsic and extrinsic risk factors. Intrinsic risk factors include age, educational level, socioeconomic status, and health conditions. Extrinsic risk factors include time spent outdoors, access to air conditioning, and levels of heat-related ambient air pollutants (such as ozone, nitrogen dioxide, and fine particulate matter).[57] Certain medications, such as diuretics, anticholinergic agents, beta-blockers, and antidepressants, may impair sweating and increase risk of heat-related illnesses and deaths.[58]

Population subgroups with increased risk of heat-related morbidity and mortality include older people, infants and young children, pregnant women, people with chronic medical conditions (such as diabetes, depression, cardiovascular disease, and cerebrovascular disease), those with mental disorders, and those who live in cooler climate zones.[59] These populations may have a higher threshold for sweating and reduced ability to dissipate heat. People with diseases that impair awareness, mobility, and behavior are also at increased risk. Deaths of older people and people with pre-existing chronic diseases during heat waves cannot be simply considered "mortality displacement," in which terminally ill people die a few days earlier with heat exposure than they would have died if the exposure had not occurred, as the contribution by heat in such situations has been confirmed by research.[60,61]

Vulnerability of a population to climate change is a function of exposure, sensitivity, adaptive capacity, population density, and other factors.[62] The health impacts of heat waves are generally greater in low-income countries and among people of low socioeconomic status because they usually have more heat exposure and less access to air conditioning, cooling centers, transportation, medical care, and health education. Athletes, hikers, and others who exercise during heat waves, especially outdoors, are susceptible to the adverse impacts of heat, but they are often not adequately protected because they do not perceive themselves as vulnerable.[63]

When intrinsic and extrinsic factors are combined, vulnerable subgroups may become more vulnerable to heat exposure. For example, an epidemiologic study found that mothers who were Black or Hispanic, smoked or consumed alcohol during pregnancy, had preexisting hypertension or diabetes, or had gestational hypertension or preeclampsia had greater risks of preterm delivery associated with ambient heat exposure.[64]

People living in urban areas may experience greater health consequences of heat exposure because of the *urban heat island effect*—the difference in temperature measured within a city compared to the neighboring areas (Figure 4-2). The daytime temperatures in urban areas are 0.6°–3.9°C (1.0°–7.0°F) higher than the surrounding areas, and nighttime temperatures are 1.1°–2.8°C (2.0°–5.0°F) higher.[65] In urban areas, temperature increases due to climate change are magnified by concrete and blacktop roads, low ventilation of "canyons" created by tall buildings, and "point-source" heat emitted from vehicles and air conditioners.[58] Cities typically have fewer trees, which exert a cooling effect because of *transpiration* (the process of water movement through a plant and its evaporation from aerial parts), release of oxygen, and absorption of carbon dioxide. Dark surfaces, such as black asphalt on roads, parking lots, and roofs, have low reflectivity and therefore absorb and retain heat readily. In addition, these surfaces radiate heat at night, increasing nighttime temperatures.

Even within urban areas, there may be marked differences in heat exposure between different socioeconomic groups. For example, a study found that within the Boston area on a hot summer day, there was a 3.3°C (6.0°F) difference in surface temperature between the neighborhoods with the highest and lowest

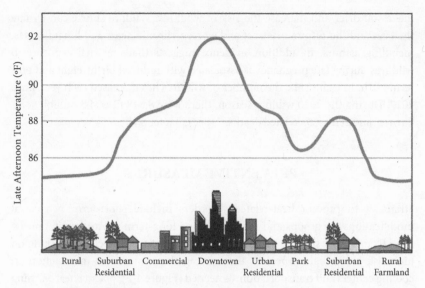

Figure 4-2 Schematic representation of the urban heat island effect. (*Source*: U.S. Environmental Protection Agency.)

residential security, a marker for socioeconomic status, paved surfaces, and the presence of trees.[66]

FUTURE PROJECTIONS

As ambient temperatures continue to rise, heat waves will become more frequent, more intense, and longer in duration, and heat-related morbidity and mortality are likely to increase substantially.[62] Studies have predicted the following:

- In Chicago, from 2081 to 2100, heat-related mortality will account for between 166 and 2,217 excess deaths per year.[67]
- Under a high-emission scenario, heat mortality will increase per decade by about 7.9% in Spain, 1.7% in France, and 0.9% in Germany.[68]
- In China, heat-related excess mortality will increase 2.4% in the 2030s and 5.5% in the 2090s.[69]
- Heat-related hospital costs in Australia will increase by approximately 60% from 2026 to 2032, compared to 20 years earlier, and by about 130% by 2050.[70]
- Annual average temperatures in California will increase up to 2.8°C (5.0°F) by the 2030s and up to 5.6°C (10.0°F) by 2100 or sooner.[71]

An overall drier climate is predicted in Southern California, leading to decreased water supply, water quality, and food production and increased waterborne disease.[72] Temperature increase, drought, and snow melt earlier in the spring will make soil drier and increase the risk of wildfires. Wildfire smoke can irritate the eyes and the respiratory tract and worsen chronic heart and lung diseases, including asthma. In addition, evidence suggests that maternal exposure to wildfires during late pregnancy is associated with reduced birthweight and pre-term birth. In California, the number of wildfires has increased each year since 2017. During the 2020 wildfire season, there were 8,648 fires, 4.3 million acres burned, and at least 33 related deaths.[73] (See Chapter 5.)

PREVENTIVE MEASURES

Strategies to prevent heat-related disorders include short-term behavioral modifications for groups and individuals and long-term environmental measures. Examples of short-term strategies during heat waves include checking on high-risk individuals, especially those living alone; providing information on cooling centers and transportation, as needed (Figure 4-3); placing heat warning signs in outdoor recreation areas, such as parks and hiking trails; increasing frequency of water breaks for outdoor workers and athletes; and conserving energy to avoid power outages. Long-term strategies include the following:

- Preparedness and response planning, such as establishing heat-warning systems to alert people about imminent heat waves

Figure 4-3 Cooling center in Portland, Oregon, during severe heat wave in 2022. (Sipa via AP Images.)

- Making cool environments available, such as by providing air conditioning and shading residences, using renewable resources to the extent possible
- Educating the public, especially vulnerable populations
- Planting trees
- Using building materials that reduce heat build-up. (See Chapters 15, 16, and 17.)

To prepare for catastrophic heat waves, many countries and cities have developed heat preparedness and response plans to prevent heat-related illnesses and deaths. These plans should address the needs of high-risk populations and include long-term surveillance and evaluation. Ideally, heat-response plans are specific for a city or region because exposures and risk characteristics of susceptible populations, such as demographic factors, socioeconomic status, and the ability to acclimatize, vary by geographic location.

Providing education and resources to high-risk populations is a critical part of heat-response plans. Challenges to implementing these measures include difficulties in identifying high-risk populations. In addition, those who are at high risk often do not know where to seek assistance, such as the locations of cooling centers, or may have mental health issues, decreased mobility, language barriers, or inability to stop working if necessary.

A challenge in developing and implementing heat-response plans is determining at what point heat waves become sufficiently hazardous to health to issue public health alerts. If alerts are issued too early, they may be ignored by the public; if they are issued too late, some people will suffer unnecessarily.

The NWS has been issuing heat warnings whenever needed. However, in retrospect, several regional heat waves in California from 1999 to 2009, which caused hospital visits for many disorders, did not trigger heat alerts.[8] The current NWS heat-warning system has been evolving, becoming more timely and localized. An alternate approach for health-based early warning systems, one being implemented in some U.S. cities, is *synoptic analysis*, which considers the weather patterns that are most likely to cause heat-related deaths.[74] The indices to assess heat stress incorporate multiple factors, such as forecasted temperature and relative humidity, duration of forecasted heat, nighttime temperature, background temperature, and time of the year

SUMMARY POINTS

- Average ambient temperatures are increasing and heat waves are becoming more frequent, more intense, and longer in duration.

- Heat-related disorders and heat-related complications of underlying medical conditions are becoming more frequent, but they have been underreported.
- High-risk populations include older people, pregnant people, and young children; people with underlying medical conditions; people of low socio-economic status; and those without access to air conditioning.
- People living in urban areas are often at greater risk for heat-related morbidity and mortality because of the urban heat island effect.
- Preventive measures include developing and implementing heat response plans with heat-warning systems, making cool environments available, and educating the public about the risks of heat-related disorders and measures to lower these risks.

DISCLAIMER

The opinions expressed in this chapter are those of the authors and do not represent those of the California Environmental Protection Agency or the Office of Environmental Health Hazard Assessment.

References

1. Yee G. FBI analysis details final hours of family who died on remote Northern California trail. Los Angeles Times, February 18, 2022.
2. Kovats RS, Hajat S. Heat stress and public health: A critical review. Annual Review of Public Health 2008; 29: 9.1–9.15.
3. Pörtner H-O, Roberts DC, Tignor M, et al. (eds.). Climate Change 2022: Impacts, Adaptation, and Vulnerability. Contribution of Working Group II to the Sixth Assessment Report of the Intergovernmental Panel on Climate Change. Cambridge: Cambridge University Press, 2022, doi: 10.1017/9781009325844.
4. World Health Organization. Heatwaves. Available at: https://www.who.int/health-topics/heatwaves#tab=tab_1. Accessed March 15, 2022.
5. Burkart KG, Brauer M, Aravkin AY, et al. Estimating the cause-specific relative risks of non-optimal temperature on daily mortality: A two-part modelling approach applied to the global burden of disease study. Lancet 2021; 398: 685–697.
6. Centers for Disease Control and Prevention. Heat & health tracker. Available at: https://ephtracking.cdc.gov/Applications/heatTracker/. Accessed May 15, 2022.
7. Sarofim MC, Saha S, Hawkins MD, et al. Temperature-related death and illness. In: The Impacts of Climate Change on Human Health in the United States: A Scientific Assessment. Washington, DC: U.S. Global Change Research Program, 2016. Available at: https://health2016.globalchange.gov. Accessed June 7, 2022.
8. Guirguis K, Gershunov A, Tardy A, Basu R. The impact of recent heat waves on human health in California. Journal of Applied Meteorology and Climatology 2014; 53: 3–19.

9. Kent ST, McClure LA, Zaitchik BF, et al. Heat waves and health outcomes in Alabama (USA): The importance of heat wave definition. Environmental Health Perspectives 2014; 122: 151–158.

10. Smith TT, Zaitchik BF, Gohlke JM. Heat waves in the United States: Definitions, patterns and trends. Climate Change 2013; 118: 811–825.

11. Basu R, Samet JM. Relation between elevated ambient temperature and mortality: A review of the epidemiologic evidence. Epidemiologic Reviews 2002; 24: 190–202.

12. National Weather Service, NWS HeatRisk prototype. Available at: https://www.wrh.noaa.gov/wrh/heatrisk/. Accessed August 25, 2022.

13. Anderson BG, Bell ML. Weather-related mortality: How heat, cold, and heat waves affect mortality in the United States. Epidemiology 2009; 20: 205–213.

14. Anderson GB, Bell ML. Heat waves in the United States: Mortality risk during heat waves and effect modification by heat wave characteristics in 43 U.S. communities. Environmental Health Perspectives 2011; 119: 210–218.

15. Semenza JC, Rubin CH, Falter KH, et al. Heat-related deaths during the July 1995 heat wave in Chicago. New England Journal of Medicine 1996; 335: 84–90.

16. Klinenberg E. Heat Wave: A Social Autopsy of Disaster in Chicago. Chicago: The University of Chicago Press, 2003.

17. Naughton MP, Henderson A, Mirabelli MC, et al. Heat-related mortality during a 1999 heat wave in Chicago. American Journal of Preventive Medicine 2002; 22: 221–227.

18. Hoshiko S, English P, Smith D, Trent R. A simple method for estimating excess mortality due to heat waves, as applied to the 2006 California heat wave. International Journal of Public Health 2010; 55: 133–137.

19. Ostro BD, Roth LA, Green RS, Basu R. Estimating the mortality effect of the July 2006 California heat wave. Environmental Research 2009; 109: 614–619.

20. Basu R, Feng WY, Ostro BD. Characterizing temperature and mortality in nine California counties. Epidemiology 2008; 19: 138–145.

21. Gershunov A, Cayan DR, Iacobellis SF. The great 2006 heat wave over California and Nevada: Signal of an increasing trend. Journal of Climate 2009; 22: 6181–6203.

22. Kovats RS, Hajat S, Wilkinson P. Contrasting patterns of mortality and hospital admissions during hot weather and heat waves in Greater London, UK. Occupational and Environmental Medicine 2004; 61: 893–898.

23. Rooney C, McMichael AJ, Kovats RS, et al. Excess mortality in England and Wales, and in Greater London, during the 1995 heatwave. Journal of Epidemiology & Community Health 1998; 52: 482–486. doi: 10.1136/jech.52.8.482.

24. D'Ippoliti D, Michelozzi P, Marino C, et al. The impact of heat waves on mortality in 9 European cities: Results from the EuroHEAT project. Environmental Health 2010; 9: 37.

25. Fouillet A, Rey G, Laurent F, et al. Excess mortality related to the August 2003 heat wave in France. International Archives of Occupational and Environmental Health 2006; 80: 16–24.

26. Michelozzi P, de Donato F, Bisanti L, et al. The impact of the summer 2003 heat waves on mortality in four Italian cities. Eurosurveillance 2005; 10: 161–165.

27. Theoharatos G, Pantavou K, Mavrakis A, et al. Heat waves observed in 2007 in Athens, Greece: Synoptic conditions, bioclimatological assessment, air quality levels and health effects. Environmental Research 2010; 110: 152–161.

28. Osborn A. Moscow smog and nationwide heat wave claim thousands of lives. British Medical Journal 2010; 341: c4360.

29. Bugler W. One hundred thousand deaths in a year: Europe tops mortality league for extreme heat. PreventionWeb February 3, 2021. Available at: https://www.prevention web.net/news/one-hundred-thousand-deaths-year-europe-tops-mortality-league-extreme-heat. Accessed March 15, 2022.

30. Watts N, Amann M, Arnell N, et al. The 2020 report of the Lancet countdown on health and climate change: Responding to converging crises. Lancet 2021; 397:129–170.

31. Red Cross Red Crescent Climate Centre. European summer heatwaves the most lethal disaster of 2019, says international research group. Reliefweb May 5, 2020. Available at: https://reliefweb.int/report/world/european-summer-heatwaves-most-lethal-disaster-2019-says-international-research-group. Accessed March 15, 2022.

32. AON. Global Catastrophe Recap, September 2021. AON 2021. Available at: http://thoughtleadership.aon.com/Documents/20210012-analytics-if-september-global-recap.pdf. Accessed June 7, 2022.

33. Anderson GB, Dominici F, Wang Y, et al. Heat-related emergency hospitalizations for respiratory diseases in the Medicare population. American Journal of Respiratory and Critical Care Medicine 2013; 187: 1098–1103.

34. Basu R, Pearson D, Malig B, et al. The effect of high ambient temperature on emergency room visits. Epidemiology 2012; 23: 813–820.

35. Ha S, Liu D, Zhu Y, et al. Ambient temperature and early delivery of singleton pregnancies. Environmental Health Perspectives 2013; 125: 453–459.

36. Pudpong N, Hajat S. High temperature effects on out-patient visits and hospital admissions in Chiang Mai, Thailand. Science of the Total Environment 2011; 409: 5260–5267.

37. Green RS, Basu R, Malig B, et al. The effect of temperature on hospital admissions in nine California counties. International Journal of Public Health 2010; 55: 113–121.

38. Michelozzi P, Accetta G, De Sario M, et al. High temperature and hospitalizations for cardiovascular and respiratory causes in 12 European cities. American Journal of Respiratory and Critical Care Medicine 2009; 179: 383–389.

39. Schwartz J, Samet JM, Patz JA. Hospital admissions for heart disease: The effects of temperature and humidity. Epidemiology 2004; 15: 755–761.

40. Wang L, Tong S, Toloo G, Yu W. Submicrometer particles and their effects on the association between air temperature and mortality in Brisbane, Australia. Environmental Research 2014; 128: 70–77.

41. Zanobetti A, O'Neill MS, Gronlund CJ, Schwartz JD. Susceptibility to mortality in weather extremes: Effect modification by personal and small-area characteristics. Epidemiology 2013; 24: 809–819.

42. Ren C, O'Neill MS, Park SK, et al. Ambient temperature, air pollution, and heart rate variability in an aging population. American Journal of Epidemiology 2011; 173: 1013–1021.

43. Hansen AL, Bi P, Ryan P, et al. The effect of heat waves on hospital admissions for renal disease in a temperate city of Australia. International Journal of Epidemiology 2008; 37: 1359–1365.

44. Bekkar B, Pacheco S, Basu R, DeNicola N. Association of air pollution and heat exposure with preterm birth, low birth weight, and stillbirth in the US: A systematic review. JAMA Network Open 2020; 3: e208243–e208243.

45. Nguyen AM, Malig BJ, Basu R. The association between ozone and fine particles and mental health-related emergency department visits in California, 2005-2013. PLoS One 2021; 16: e0249675.

46. Lõhmus M. Possible biological mechanisms linking mental health and heat: A contemplative review. International Journal of Environmental Research and Public Health 2018; 15: 1515.

47. Charkoudian N. 2010. Mechanisms and modifiers of reflex induced cutaneous vasodilation and vasoconstriction in humans. Journal of Applied Physiology 1985; 109: 1221-1228.

48. Dreiling CE, Carman FS 3rd, Brown DE. Maternal endocrine and fetal metabolic responses to heat stress. Journal of Dairy Scienc 1991;74: 312-327.

49. Rasmussen DJ, Hu J, Mahmud A, Kleeman MJ. The ozone-climate penalty: Past, present, and future. Environmental Science & Technology 2013; 47: 14258-14266.

50. Kim KH, Jahan SA, Kabir E. A review on human health perspective of air pollution with respect to allergies and asthma. Environment International 2013; 59: 41-52.

51. Li S, Williams G, Jalaludin B, Baker P. Panel studies of air pollution on children's lung function and respiratory symptoms: A literature review. Journal of Asthma 2012; 49: 895-910.

52. Gilliland FD, Berhane K, Rappaport EB, et al. The effects of ambient air pollution on school absenteeism due to respiratory illnesses. Epidemiology 2001; 12: 43-54.

53. Choi M, Curriero FC, Johantgen M, et al. Association between ozone and emergency department visits: An ecological study. International Journal of Environmental Health Research 2011; 21: 201-221.

54. Strickland MJ, Darrow LA, Klein M, et al. Short-term associations between ambient air pollutants and pediatric asthma emergency department visits. American Journal of Respiratory and Critical Care Medicine 2010; 182: 307-316.

55. Moore K, Neugebauer R, Lurmann F, et al. Ambient ozone concentrations cause increased hospitalizations for asthma in children: An 18-year study in Southern California. Environmental Health Perspectives 2008; 116: 1063-1070.

56. Bell ML, Dominici F, Samet JM. A meta-analysis of time-series studies of ozone and mortality with comparison to the national morbidity, mortality, and air pollution study. Epidemiology 2005; 16: 436-445.

57. Basu R. High ambient temperature and mortality: A review of epidemiologic studies from 2001 to 2008. Environmental Health 2009; 8. doi: 10.1186/476-069X-8-40.

58. Luber G, McGeehin M. Climate change and extreme heat events. American Journal of Preventive Medicine 2008; 35: 429-435.

59. Knowlton K. The 2006 California heat wave: Impacts on hospitalizations and emergency department visits. Environmental Health Perspectives 2009; 117: 61-67.

60. Basu R, Malig B. High ambient temperature and mortality in California: Exploring the roles of age, disease, and mortality displacement. Environmental Research 2011; 111: 1286-1292.

61. Hajat SA, Ben G, Gouveia N, Wilkinson P. Mortality displacement of heat-related deaths: A comparison of Delhi, Sao Paulo, and London. Epidemiology 2005; 16: 613-620.

62. Intergovernmental Panel on Climate Change. Climate change 2007: Synthesis report. IPCC 2007.

63. Boden BP, Beachler JA, Williams A, Mueller FO. Fatalities in high school and college football players. American Journal of Sports Medicine 2013; 41: 1108-1116.

64. Basu R, Malig B, Ostro B. High ambient temperature and the risk of preterm delivery. American Journal of Epidemiology 2010; 172: 1108–1117.
65. US Environmental Protection Agency. Heat island effect. Updated July 10, 2023. Available at: https://www.epa.gov/heatislands. Accessed August 25, 2022.
66. Hoffman JS, Shandas V, Pendleton N. The effects of historical housing policies on resident exposure to intra-urban heat: A study of 108 US urban areas. Climate 2020; 8:12. doi:10.3390/cli8010012.
67. Peng RD, Bobb JF, Tebaldi C, et al. Toward a quantitative estimate of future heat wave mortality under global climate change. Environmental Health Perspectives 2011; 119: 701–706.
68. Karwat A, Franzke CLE. Future projections of heat mortality risk for major European cities. Weather, Climate, and Society 2021; 13: 913–931.
69. Yang J, Zhou M, Ren Z, et al. Projecting heat-related excess mortality under climate change scenarios in China. Nature Communications 2021; 12: 1039. doi: 10.1038/s41467-021-21305-1
70. Tong MX, Wondmagegn BY, Williams S, et al. Hospital healthcare costs attributable to heat and future estimations in the context of climate change in Perth, Western Australia. Advances in Climate Change Research 2021; 12: 638–648.
71. Cayan DR, Das T, Pierce DW, et al. Future dryness in the southwest US and the hydrology of the early 21st century drought. Proceedings of the National Academy of Sciences of the United States of America 2010; 107: 21271–21276.
72. Beuhler M. Potential impacts of global warming on water resources in Southern California. Water Science and Technology 2003; 47: 165–168.
73. Cal Fire. Stats and events. Available at: https://www.fire.ca.gov/incidents/2020/. Accessed May 15, 2022.
74. Kalkstein LS, Barthel CD, Greene JS, Nichols MC. A new spatial synoptic classification: Application to air mass analysis. International Journal of Climatology 1996; 16: 983–1004.

Further Reading

World Health Organization. Factsheet: Heat and health. 2018. Available at: https://www.who.int/news-room/fact-sheets/detail/climate-change-heat-and-health. Accessed April 13, 2022.

This fact sheet provides a concise summary of valuable information.

United States Environmental Protection Agency. Excessive heat events guidebook. 2006 (appendix updated in 2016). Available at: https://www.epa.gov/heatislands/excessive-heat-events-guidebook. Accessed July 15, 2022.

This guidebook highlights best practices that have been employed to save lives during excessive heat events and provides a menu for responding to extreme heat events.

California Environmental Protection Agency. Protecting Californians from extreme heat: A state action plan to build community resilience. April 2022. Available at: https://resources.ca.gov/-/media/CNRA-Website/Files/Initiatives/Climate-Resilience/2022-Final-Extreme-Heat-Action-Plan.pdf. Accessed July 6, 2022.

This publication includes recommendations on protecting people from extreme heat that can be used by state and local planners, local governments, emergency response workers, and public health and healthcare professionals and institutions.

Health Canada. Communicating the health risks of extreme heat events: Toolkit for public health and emergency management officials. 2011. Available at: https://ghhin.org/wp-cont ent/uploads/heat-chaleur-eng.pdf. Accessed July 6, 2022.

This toolkit provides guidance for public health and emergency management officials who are developing or updating communication strategies on the adverse health effects of heat.

Centers for Disease Control and Prevention. Extreme heat and your health. Available at: http://www.cdc.gov/extremeheat. Accessed May 31, 2022.

This website provides easily accessible resources for members of the public, local health departments, and other organizations to assist in outreach to those people who are most vulnerable to extreme heat events.

5

Respiratory Disorders

Ioana O. Agache, Vanitha Sampath, Juan Aguilera, and Kari C. Nadeau

A 7-year-old girl with preexisting asthma develops worsening wheezing, cough, and shortness of breath during an extended heat wave.

A 58-year-old man with preexisting chronic obstructive pulmonary disease (COPD) due to cigarette smoking develops increased shortness of breath, cough, and sputum production during an air pollution alert related to a nearby wildfire.

A 34-year-old woman with preexisting allergic rhinitis ("hay fever") develops increased nasal congestion, runny nose, and difficulty breathing due to increased exposure to ragweed pollen.

Climate change is a contributing cause to these cases of asthma, COPD, and allergic rhinitis. And it is increasing the frequency and severity of new cases and exacerbations of these and other respiratory disorders. While climate change is not the most frequent cause of cases of chronic respiratory disorders, many cases are climate-sensitive.[1]

Higher temperatures increase airborne concentrations of ozone if there are sufficiently high levels of nitrogen oxides, volatile organic compounds (VOCs), and other ozone precursors.[2] Higher temperature also contributes to increased probability of wildfires—and, in desert regions, sand and dust storms—which produce particulate matter and other airborne pollutants that contribute to new-onset and exacerbation of respiratory disorders. And higher temperature leads to increased pollen production and longer pollen seasons, increasing the incidence and severity of allergic rhinitis.

Climate change can indirectly affect respiratory health. For example, it can influence the distribution of microorganisms, such as spores of fungi that cause infectious respiratory disorders. And airborne particulate matter, ozone, and nitrogen dioxide—concentrations of which are increased by climate change—can damage the respiratory epithelial barrier, increasing permeability and

Ioana O. Agache, Vanitha Sampath, Juan Aguilera, and Kari C. Nadeau, *Respiratory Disorders* In: *Climate Change and Public Health*. Edited by: Barry S. Levy and Jonathan A. Patz, Oxford University Press. © Oxford University Press 2024. DOI: 10.1093/oso/9780197683293.003.0005 penetration of allergens, toxins, and infectious agents and modulating innate and adaptive immune response with excessive inflammation and tissue damage.

Respiratory diseases are frequently caused by occupational exposures. The frequency of some of these occupational respiratory disorders is likely to increase as a result of climate change (Box 5-1).

This chapter focuses on the main causes of climate-related respiratory disorders: ozone, particulate matter, heat, pollen, and mold—and the prevention of these disorders.

Box 5-1 Occupational Respiratory Disorders

Crystal M. North and David C. Christiani

A 57-year-old man was diagnosed with asthma in childhood. It was usually well controlled with a rescue inhaler, which he needed only in the spring. He had worked for a local utility company since he was 22 years old, spending much time outdoors. Over the years, his asthma became more difficult to control and began earlier in the spring. He missed many days of work in the spring because of breathing difficulty and was hospitalized twice for asthma exacerbations in the past few years.

Globally, more than 262 million people have asthma, which causes an estimated 455,000 deaths annually.[1] Occupational asthma is the most common occupational respiratory disorder in high-income countries. Approximately 16% of adult-onset asthma cases are caused by occupational exposure.[2] More than 400 agents have been linked to occupational asthma, with additional causative agents reported each year. It appears that almost any protein that becomes airborne and is inhaled could be a potential cause of occupational asthma. The incidence of self-reported asthma has increased globally in the past 50 years, from 2-4% in the 1960s to 15-20% in many countries by 2000.[3] Both the incidence and severity of asthma are projected to continue to increase due, in part, to climate change.[4]

The effects of climate change on asthma and allergic respiratory disease are predicted from clinical studies of symptoms associated with exposure to respiratory irritants and allergens and in vitro testing of the effects of allergens. Studies based on the number of asthma-related emergency department visits underestimate the actual burden of disease because many patients with asthma exacerbations do not seek emergency medical care.[5] Without intervention, occupational respiratory disorders will increase dramatically, interfering with work productivity and impairing overall health.

Climate change can contribute to occupational respiratory disorders in several ways. As the Earth's temperature rises, seasonal variations intensify and extreme weather events become more severe and more frequent; both extreme heat and extreme cold are associated with more respiratory symptoms and emergency department visits in patients with asthma.[6] Heat waves are associated with all-cause mortality in patients with chronic respiratory disease.[7] Thunderstorms and floods are associated with asthma exacerbations due to increased pollen dispersion. Smoke from wildfires is associated with asthma exacerbations and hospitalizations, even at distant locations; for example, smoke from more than 500 wildfires burning in Russia during a 2010 heat wave spread 1,860 miles, about the distance from San Francisco to Chicago.[8] Higher temperatures lead to more production of ozone, which increases asthma morbidity and sensitizes the respiratory tract to inhaled allergens.[7]

Climate change increases aeroallergen exposure, thereby increasing the risk of exacerbations for patients with asthma. Increasing temperatures, carbon dioxide levels, and ground-level ozone concentrations lead to both earlier and longer flowering seasons (which, in turn, result in increased pollen production) and substantially increase the potency of pollen in causing allergy symptoms. Affected allergenic fauna, which vary among regions, include oak and birch trees, ragweed, and grasses.[5] Pollen levels are associated with wheezing, asthma exacerbations, and emergency department visits.[7] In addition, climate change may stimulate allergenic fauna to exacerbate chronic respiratory disease through by generation of reactive oxygen species and pollen-associated lipid mediators.[9,10] Changing weather patterns also affect pollen dispersion.

Climate change can also increase levels of some air pollutants, thereby increasing asthma incidence, severity, and exacerbations, especially in urban areas that have relatively high concentrations of air pollutants.[4] The proportion of asthma cases related to climate change is not known; however, much research is being performed on this subject. Respiratory irritants include nitrogen dioxide, sulfur dioxide, ozone, and particulate matter. These and other air pollutants decrease pulmonary function by damaging airway mucosa and impairing mucociliary clearance. These adverse effects increase exposure of immune cells to inhaled allergens and irritants and facilitate subsequent sensitization.[6]

Air pollutants also increase hyperreactivity to respiratory irritants. For example, ozone induces airway inflammation and bronchial hyperreactivity, making asthma more difficult to control and increasing asthma-related hospitalizations.[6] About half of U.S. residents live in regions with ambient air concentrations of ozone above the Environmental Protection Agency standard.[4]

Climate change also affects indoor air quality. Airborne pollen, carbon dioxide, ozone, and other air contaminants can penetrate indoors, decreasing indoor air quality and exacerbating respiratory disorders. Increased carbon dioxide concentrations are associated with decreased cognitive performance. Extreme weather damages building infrastructure, resulting in exposure to aerosolized chemicals. Some of these effects can be reduced with energy-efficient building designs; however, decreased building ventilation is associated with reduced indoor air quality and increased incidence of exacerbations of allergic respiratory disorders and respiratory infections.[11]

Aside from climate change mitigation, methods to prevent occupational respiratory disorders focus on avoiding hazardous exposures. Occupational exposure to poor air quality can be reduced by use of air filtration systems, by requiring workers to wear masks during periods of exposure, and by reducing the duration of exposure to poor air quality or high aeroallergen levels. Improved ventilation of indoor areas may improve air quality and decrease respiratory symptoms.

Box References

1. Vos T, Lim SS, Abbafati C, et al. Global burden of 369 diseases and injuries in 204 countries and territories, 1990–2019: A systematic analysis for the Global Burden of Disease Study 2019. Lancet 2020; 396: 1204–1222.

2. Torén K, Blanc PD. Asthma caused by occupational exposures is common: A systematic analysis of estimates of the population-attributable fraction. BMC Pulmonary Medicine 2009; 9: 7. https://doi.org/10.1186/1471-2466-9-7.

3. Sears MR. Trends in the prevalence of asthma. Chest 2014; 145: 219–225.

4. Wright RJ. Influences of climate change on childhood asthma and allergy risk. Lancet Child and Adolescent Health 2020; 4: 859–860.

5. Darrow LA, Hess J, Rogers CA, et al. Ambient pollen concentrations and emergency department visits for asthma and wheeze. Journal of Allergy and Clinical Immunology 2012; 130: 630–638.

6. Guarnieri M, Balmes JR. Outdoor air pollution and asthma. Lancet 2014; 383: 1581–1592.

7. D'Amato G, Chong-Neto HJ, Monge Ortega OP, et al. The effects of climate change on respiratory allergy and asthma induced by pollen and mold allergens. Allergy 2020; 75: 2219–2228. doi: 10.1111/all.14476.

8. NASA Earth Observatory. Fires and Smoke in Russia, August 5, 2010. Available at: http://earthobservatory.nasa.gov/IOTD/view.php?id=45046. Accessed February 2, 2023.

9. Pachecho SE, Guidos-Fogelbach G, Annesi-Maesano I, et al. Climate change and global issues in allergy and immunology. Journal of Allergy and Clinical Immunology 2021; 148: 1366–1377.

10. Gan RW, Liu J, Ford B, et al. The association between wildfire smoke exposure and asthma-specific medical care utilization in Oregon during the 2013 wildfire season. Journal of Exposure Science and Environmental Epidemiology 2020; 30: 618–628.

11. Mansouri A, Wei W, Alessandrini JM, et al. Impact of climate change on indoor air quality: A review. International Journal of Environmental Research and Public Health 2022; 19: 15616. doi: 10.3390/ijerph192315616.

OZONE

Ozone, a key component of smog, is generated in ambient air from precursor gases, mainly nitrogen oxides and VOCs in the presence of sunlight and heat (Figure 5-1).[3] There is a strong association between high ambient temperatures and ground-level ozone concentrations (Figure 5-2). Ozone levels typically peak during hot and dry days with stagnant air. Ground-level ozone concentrations in ambient air are increasing globally, mainly from increasing levels in densely populated regions of Asia and Africa[4] and will likely continue to increase as global temperatures continue to increase.[5–7]

Figure 5-1 Traffic on I-95 North in Miami during the evening rush hour. Exhaust from gasoline-powered vehicles includes ozone precursors, which, in the presence of heat, combine to form ozone. (Photograph by Daniel J. Macy/Shutterstock.com.)

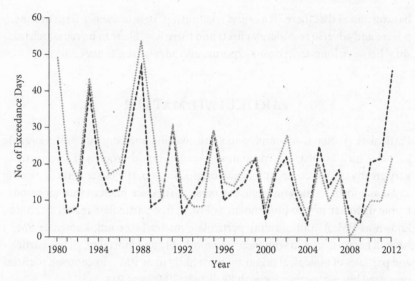

Figure 5-2 Association between days of high ambient temperatures and ground-level ozone concentrations, Chicago, 1980-2012. The black dashed line indicates the number of days, each year, in which the temperature exceeded 90°F. The other line indicates the number of days per year in which the ambient concentration of ground-level ozone was greater than 75 parts per billion. (*Source*: Patz J, Frumkin H, Holloway T, et al. Climate change: Challenges and opportunities for global health. JAMA 2014; 312: 1565–1580.)

While ozone in the stratosphere (the "ozone layer") protects people from the harmful effects of ultraviolet light, ground-level ozone directly causes much morbidity and mortality. An estimated 365,000 people globally died in 2019 from exposure to ground-level ozone, an increase of 16% from 2010. Ozone causes respiratory tract irritation (with symptoms of shortness of breath, wheezing, and cough), acute decreases in lung function, systemic inflammation, and oxidative stress. It also causes increased epithelial barrier permeability, airway hyperresponsiveness (with exacerbations of asthma), and susceptibility to lower respiratory tract infections. India and China have the most ozone-related deaths.[8]

In children and adolescents, ozone exposure for 2 years or more has been associated with decreases in airflow—a 1% decrease in forced expiratory volume in the first second (FEV_1) per 10 ppb of ozone. Ozone exposure for up to 2 weeks in children has also been associated with decreases in FEV_1. These findings suggest that even low levels of ozone, well below the current Environmental Protection Agency (EPA) standard, adversely affect children's lung function. However, the impact in adults of short- and long-term exposure to ozone is unclear.[9] The EPA has concluded that there "is a causal relationship" between short-term ozone exposure and adverse respiratory effects and there is a "likely to be causal relationship" between long-term ozone exposure and adverse respiratory effects.[10]

PARTICULATE MATTER

Particulate matter is a complex mixture that includes organic and inorganic particles such as dust, pollen, soot, smoke, and liquid droplets. These particles vary greatly in size, composition, and origin. Some particles are emitted directly and some form in the atmosphere. Coarse particulate matter (with an aerodynamic diameter of <10 μm, known as PM_{10}), fine particulate matter (<2.5 μm, known as $PM_{2.5}$), and ultrafine particulate matter (<0.1 μm, known as $PM_{0.1}$) have all been associated with increased morbidity and mortality.[11] Dust particles and particles of biological origin are most likely to be PM_{10}; in contrast, particles produced by combustion are most likely to be $PM_{2.5}$ and $PM_{0.1}$.

Particulate matter concentrations in ambient air increase after wildfires, droughts, and sand and dust storms—all of which have increased in frequency and severity due to climate change. Exposure to particulate matter is associated with acute and chronic respiratory symptoms, decreased lung function, worsening of asthma, and development of chronic bronchitis. Coarse particles can irritate eyes, nose, throat, and airways. Fine and ultrafine particles, which can penetrate into lung tissue and enter the blood, are associated with more severe respiratory disorders than coarse particles. Exposure to fine particulate matter is also associated with cardiovascular disease (including ischemic heart disease and stroke), lung cancer, type 2 diabetes, preterm birth, and premature mortality.[12-14]

Climate change is impacting particulate matter, including the sources of particulate matter, in complex ways involving temperature, rainfall, wind speed, and other factors. The number of wildfires and the aggregate area burned by wildfires have increased, partly due to drier soils and vegetation caused by global warming.[15,16] Wildfires also release carbon dioxide into the atmosphere, thereby exacerbating global warming.

Wildfires can emit high concentrations of fine particulate matter as well as acrolein, other aldehydes, and other potent lung irritants; carbon monoxide; and carcinogens, such as formaldehyde and benzene (Figure 5-3).[17] Wildfires have been associated with respiratory disorders. High concentrations of fine particulate matter, VOCs, and gaseous pollutants from wildfires are extremely harmful to children.[18] Wildfire smoke exacerbates asthma[19,20]; fine particulate matter from wildfires can increase childhood asthma symptoms about 10-fold, compared to $PM_{2.5}$ from other ambient sources, such as vehicular emissions.[11]

Figure 5-3 Smoke from a wildfire. (Copyright Associated Press.)

During wildfire seasons, increased particulate matter is associated with emergency department admissions for asthma.[21] A 10 μg/m³ increase in wildfire-derived particulate matter can significantly increase the risk of diagnosed asthma by 9%.[22] After long-term exposure to smog from wildfires, air flow can be reduced.[23] In addition to asthma, in children, wildfire smoke can cause nose, eye, and throat irritation as well as lower respiratory tract effects, including bronchitis, cough, and wheezing.[18]

Exposure to high fire-danger days globally increased 61% from 2001–2004 to 2018–2021.[24] In the western United States, where wildfire activity is expected to increase with climate change, fire-related $PM_{2.5}$ deaths could increase significantly.[25,26] There is also evidence that particulate matter released from wildfires may increase rates of infant mortality[27] and restrict fetal and infant growth.[28,29]

Global warming, increased frequency of droughts, poor management of land and water resources, and inefficient agricultural practices have increased the frequency and severity of sand and dust storms. These storms can contain quartz, other minerals and salts, organic matter, pollutants, and allergens. They contain high levels of coarse and fine particulate matter, including pollen, fungi and fungal spores, bacteria, and viruses. In very strong dust storms, particulate matter can exceed 6,000 μg/m³, spread hundreds of miles from the source, and remain suspended in the air for long periods.[30] In drought-prone areas, increases in dryness will likely increase ambient dust; premature mortality from fine dust particles could increase by 24% to 130%.[31,32] Pathogens in sand and dust storms can increase some infectious diseases, such as tuberculosis and coccidioidomycosis ("valley fever"), which is caused by inhalation of fungal spores in soil.[30,33–35] Sand and dust storms can also increase emergency department visits for asthma and COPD.[36–39]

HEAT

Higher temperatures and increased frequency of heat waves increase morbidity and mortality from respiratory disorders. A study in the United States found an approximately 6% increased risk of respiratory mortality during heat waves.[40] Another study in the United States found that each 5.6°C (10.0°F) temperature increase was associated with a 4.3% increase in same-day emergency hospitalizations for respiratory disorders.[41] A study in China found that about 13% of deaths during a heat wave were due to respiratory disorders, mainly from exacerbations of COPD.[42] A study in Brazil, during hot seasons from 2000 to 2015, found that approximately 7% of COPD hospitalizations could be attributed to heat exposure.[43] There is some evidence that high temperatures act synergistically with both ozone and particulate matter in causing adverse health effects.[44] (See Chapters 3 and 4.)

POLLEN

Pollen production increases with increased temperature and, to a lesser extent, with increased atmospheric concentrations of carbon dioxide.[45] Global warming has expanded the duration of the plant flowering season and has increased human exposure to pollen. Increased carbon dioxide leads to increased pollen production. In the long term, continued global warming is likely to cause changes in patterns of plant habitat and species density—with gradual movement northward in the Northern Hemisphere and southward in the Southern Hemisphere—and lead to increased occurrence of pollen allergy.

Tree, grass, and weed pollen are associated with seasonal asthma exacerbations. In the United States, tree pollen is annually associated with 25,000 or more emergency department visits for pollen-related asthma, while grass pollen is associated with less than 10,000 visits annually.[46] However, in some parts of Europe, weed pollen is more frequently associated with asthma, with ragweed generating about half of pollen production.[47] Pollen allergy is also associated with allergic rhinitis.[48,49] By 2050, airborne concentrations of ragweed pollen in Europe may increase fourfold and sensitization to ragweed may increase more than twofold (Figure 5-4).[47,50,51]

Exposure to pollen is an established risk factor for the incidence and severity of viral infections of the respiratory tract. But some evidence suggests that exposure to pollen might also inhibit or weaken viral infections. Climate change will likely increase the importance of pollen exposure as a cause of respiratory problems and affect the probability of infectious disease outbreaks. At present, however, it is not possible to conclude whether pollen exposure protects against or increases susceptibility to viral infections. Further research is needed to determine if pollen can act as a carrier of viruses and if pollen that carries viral particles can transmit a viral infection.[52]

Figure 5-4   Ragweed. (Olha Solodenko/Shutterstock.com.)

It has been postulated that air pollutants and allergens have a synergistic effect on respiratory health.[53] Air pollutants, such as diesel exhaust particles, adhere to the surface of pollen grains, modifying the morphology and allergenicity of the pollen grains and promoting allergen sensitization. Particulate matter and gaseous pollutants, such as ozone, increase allergen absorption in the lungs by increasing epithelial permeability and airway oxidative stress.[54,55] However, epidemiological studies have not found a synergistic effect between air pollutants and allergens.[56]

During the pollen season, thunderstorms can increase the allergenicity of pollen and, in turn, increase the frequency of asthma exacerbations in the short term, which can overwhelm emergency departments. For example, in 2016, within 30 hours of a thunderstorm in Melbourne, emergency department admissions for respiratory disorders increased more than sixfold—and those for asthma increased almost 10-fold.[57] One possible explanation of this increase in asthma attacks during thunderstorms may be that thunderstorms may rupture pollen particles into smaller fragments that are capable of reaching deeper into the lungs.[58]

Tree planting is one strategy to adapt to climate change and reduce urban heat islands (Chapter 17). However, different types of trees planted in cities can have vastly different impacts on pollen levels. For example, policies that have promoted planting of male trees to avoid fruit waste have led to large increases in pollen. Selection of less allergenic or more diverse trees may reduce exposure to allergic pollen.[59,60] Trees also contribute to ozone formation by producing VOCs. For example, oak trees produce isoprene, which can combine with nitrogen oxides to form ozone. Replacing oak trees with maple trees may reduce ozone concentrations.[61]

MOLD

Mold is a fungal growth that forms and spreads on various kinds of damp or decaying organic matter. As heavy precipitation increases, mold increases in indoor environments, causing new-onset or exacerbation of asthma and allergic rhinitis (Figure 5-5). Damp housing has been recognized as a cause of poor respiratory health and is associated with wheezing and cough in both adults and children; in the United States, exposure to dampness and mold has been estimated to account for about one-fifth of all asthma cases.[44] Exposure to mold is also associated with rhinitis, allergic rhinitis, and rhinoconjunctivitis.[62] As climate change increases the frequency and intensity of severe storms and floods in many regions,[63] dampness and moisture in indoor environments is likely to increase, leading to growth of and exposure to mold.[64,65]

Figure 5-5 Mold growth on the wall of a hotel in New Orleans after Hurricane Katrina in 2005. (Courtesy of Louisiana State University AgCenter.)

VULNERABLE POPULATIONS

People who are vulnerable to climate-induced respiratory disorders include infants and children, older people, and people with respiratory allergies or chronic respiratory diseases. Children are vulnerable because they spend more time outdoors, are still developing, and have lower immunity; they also breathe more air per body weight than adults. Older people have lower immunity, lower mobility, and co-morbid health conditions, all of which place them at greater risk of these disorders.

Increased vulnerability often results from a combination of multiple exposures. For example, changing pollen patterns, damp buildings with increased mold exposure, air pollution, and heat stress all adversely affect people with chronic respiratory disorders.[66] Asthma exacerbations increase with exposure to heat, $PM_{2.5}$, and ozone.[67]

PREVENTIVE MEASURES

By Governments

There are a variety of ways in which local, state, and national governments can help prevent climate-induced respiratory disorders. In addition to working to mitigate greenhouse gas (GHG) emissions, they can:

- Regulate emissions of air pollutants other than GHGs
- Improve standards for indoor air quality and access to indoor air filters
- Plant trees to reduce ambient temperature and decrease air pollution (Chapter 17)
- Plant trees that are not allergenic in order to reduce allergic respiratory disorders
- Provide incentives for people to commute using public transportation or active transportation (walking or bicycling) (Chapter 13)
- Warn the public about the hazards of mold and measures to prevent indoor dampness and mold growth
- Monitor levels of ambient ozone, particulate matter, and pollen and advise the public, especially people with chronic respiratory disorders, to limit outdoor activities when these levels are high
- Manage forests and take other measures to reduce the likelihood of wildfires
- Improve access to healthcare and medications for climate-induced respiratory disorders
- Implement warning systems for imminent heat waves and provide cool indoor environments for people at increased risk of heat-induced respiratory disorders.

Preparing for and mitigating the impact of climate change on pollen and their effects on respiratory health would greatly benefit from collaboration among health departments, healthcare providers, and universities and research institutions, ideally coordinated by the federal government. In New York City, there is a strong seasonal pattern in emergency department visits for asthma. Since some seasonal peaks reflect the influence of environmental factors, there are potential opportunities for intervention. The fall peak in asthma emergency department visits likely reflects rhinovirus infections. The spring peak in late April and early May likely reflects the influence of spring tree pollens. To reduce the number of emergency department visits in the spring, the New York City Department of Health has been issuing alerts to healthcare providers for several years, including in these alerts a figure showing the association between mid-spring tree pollen and both over-the-counter sales of allergy medication and emergency department visits for asthma (Figure 5-6).

By Individuals

People with chronic respiratory disorders can reduce the likelihood of exacerbations by adhering to the recommendations of their physicians, including

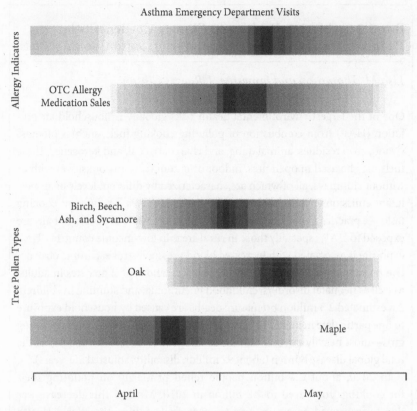

Figure 5-6 Average timing of spring peaks in tree pollen and both over-the-counter sales for allergy medications and emergency department visits, New York City, 2002-2009. (*Source*: New York City Department of Health Advisory to Healthcare Providers, 2014.)

prescribed medications, and by avoiding locations where they are likely to become exposed to ozone, particulate matter, pollen, mold, or excessive heat. They can also avoid smoking cigarettes and sharing households with people who smoke, and avoid places where biomass is used for cooking. If necessary, they can use high-quality masks to reduce exposure to harmful air pollutants, such as from wildfires.[68]

In places where biomass, such as wood, charcoal, or dung, is used as the primary fuel for cooking, availability of clean cookstoves can (a) improve air quality in and around people's homes and reduce GHG emissions and (b) reduce related health effects, including lower respiratory tract infections and chronic lung disease. (See Box 5-2 and Figure 5-7.)

Box 5-2　Achieving Health and Climate Co-Benefits by Reducing Household Air Pollution from Solid Cooking Fuels

---

Lisa M. Thompson and Jamesine V. Rogers Gibson

One of the largest environmental health risks globally is household air pollution (HAP) from combustion of polluting cooking fuels, such as biomass (wood, crop residues, animal dung, and charcoal), coal, and kerosene.[1] These fuels may be used in open fires, indoor or outdoors, or in cookstoves with or without chimneys, all of which are characterized by different levels of air pollutant emissions. Households may use multiple stoves for different cooking tasks—a practice known as *stove stacking*. More than 30% of people globally are exposed to HAP, especially those in rural areas in low-income countries. HAP exposure is associated with increased blood pressure, stroke, chronic obstructive pulmonary disease, lung cancer, cataracts, and type 2 diabetes in adults, as well as neonatal disorders, childhood pneumonia, and stunting in children. An estimated 2.3 million premature deaths are caused by household exposure to fine particulate matter ($PM_{2.5}$), with women and children—especially young girls—most heavily exposed during cooking. HAP causes about 3.6% of the total global disease burden (about 92 million disability-adjusted life-years).[2]

In 2020, about 2.4 billion people relied primarily on polluting fuels for cooking, compared to 2.8 billion in 2010. Most of this decrease was due to improved access in Asia to clean fuels, which the World Health Organization has defined as including liquefied petroleum gas, natural gas, biogas, electricity, and solar power. However, in sub-Saharan Africa, the number of people exposed to HAP has been increasing. Even when stoves using clean fuels are available, stove stacking and the use of polluting fuels often occur due to continued access and affordability challenges for clean fuels. It is unlikely that Sustainable Development Goal 7.1.2—ensuring access to affordable, reliable, sustainable, and modern energy for all—will be met by 2030.[3]

Household Fuel Use and Climate Change

Biomass fuels are a major source of greenhouse gases (GHGs). Because forests act as carbon sinks, the unsustainable harvesting of fuelwood at a rate that exceeds regrowth represents, in a sense, a "non-renewable energy source" that increases net carbon dioxide emissions and can contribute to the degradation of forests. Globally, the fraction of biomass harvested non-renewably varies widely. Other biomass fuels, such as crop residues and animal dung, are, by definition, renewably harvested; in contrast, coal, a fossil fuel, always leads to a net increase in carbon dioxide emissions. Emissions from the burning of biomass and coal contain methane and other pollutants that contribute to climate change.

Fuels that are burned in inefficient stoves produce aerosols, such as black carbon, that can warm the atmosphere, and other aerosols, such as sulfates, that can cool the atmosphere (Figure 2-2 in Chapter 2). The net global climate impact from cooking with biomass is uncertain at present, but a transition away from biomass to liquefied petroleum gas (LPG) and/or LPG and electricity may lead to net cooling in the future.[4] By some estimates, black carbon is the second most important climate-forcing pollutant after carbon dioxide. At least one-fourth of human-generated black carbon emissions result from household combustion of solid fuels, including wood.[5] Use of wood as fuel annually releases a gigaton of carbon dioxide equivalents (including not only carbon dioxide, but also black carbon and other air contaminants), about the same amount as emissions from aviation.[6] Transitioning to clean cooking fuels and technologies can not only reduce global warming, but also provide health co-benefits.

Addressing Household Air Pollution from Polluting Fuels

There are two general approaches to addressing HAP.[7] The first is to "make the available clean" by designing and providing advanced ("improved") cookstoves to efficiently burn available biomass fuels. However, many of these "improved" cookstoves may not reduce HAP to health-protective levels. Health may be improved by the widespread availability and use of advanced cookstoves that meet stringent international standards for increased efficiency, reliability, and lower emissions, along with effective social marketing and improved business models to encourage the use of these cookstoves and repairment when they fail.

The second general approach is to "make the clean available" by providing more households with cleaner fuels, primarily gas and electricity, but also biogas, ethanol, and biomass pellets—as well as solar cookers. Ministries of energy can incentivize stove and fuel companies and work with utilities to make the clean available on a large scale. They can also provide financing mechanisms, subsidies, and distribution networks to make cleaner fuels accessible and affordable to low-income households. In low-income countries, liquefied petroleum gas and electricity (partly supplied by fossil fuels) frequently serve as transitional fuels as households move away from polluting cooking fuels. Cleaner fuel transitions are underway, including in China, Ecuador, India, and other countries. Studies have shown that long-term solutions, such as cooking with electricity generated by renewable sources, need to be prioritized so that they are reliable, affordable, and widely available in order to reduce global warming and protect population health.[7]

Acknowledgment

The content of this textbox is based on a textbox by Kirk R. Smith, M.P.H., Ph.D., which appeared in the first edition of this book. Dr. Smith pioneered the field of household air pollution and tirelessly advocated for marginalized people who lacked access to clean cooking fuels. This textbox is dedicated to his memory.

Box References

1. GBD 2019 Risk Factors Collaborators. Global burden of 87 risk factors in 204 countries and territories, 1990–2019: A systematic analysis for the Global Burden of Disease Study 2019. Lancet Global Health 2020; 396: 1223–1249. doi: 10.1016/S0140-6736(20)30752-2.

2. Bennitt FB, Wozniak SS, Causey K, et al. Estimating disease burden attributable to household air pollution: New methods within the Global Burden of Disease Study. Lancet Global Health 2021; 9: S18. doi: 10.1016/S2214-109X(21)00126-1.

3. Tracking SDG 7: The Energy Progress Report. Progress Towards Sustainable Energy. Available at: https://trackingsdg7.esmap.org/. Accessed January 23, 2023.

4. Floess E, Grieshop A, Puzzolo E, et al. Scaling up gas and electric cooking in low- and middle-income countries: Climate threat or mitigation strategy with co-benefits?" Environmental Research Letters 2023; 18: 034010. https://doi.org/10.1088/1748-9326/acb501.

5. Bond TC, Doherty SJ, Fahey DW, et al. Bounding the role of black carbon in the climate system: A scientific assessment. JGR Atmospheres 2013; 118: 5380–5552. https://doi: 10.1002/jgrd.50171.

6. Bailis R, Drigo R, Ghilardi A, Masera, O. The carbon footprint of traditional woodfuels. Nature Climate Change 2015: 5: 266–272. doi: 10.1038/nclimate2491.

7. Goldemberg J, Martinez-Gomez J, Sagar A, Smith KR. Household air pollution, health, and climate change: Cleaning the air. Environmental Research Letters 2018; 13: 030201. doi: 10.1088/1748-9326/aaa49d.

SUMMARY POINTS

- Climate change increases ground-level ozone concentrations in ambient air, heightening the risk of new-onset and exacerbation of asthma and COPD.
- Climate change increases the frequency and severity of wildfires, thereby raising ambient levels of particulate matter and other air pollutants and

Figure 5-7 Woman cooking over an open wood fire in India. (Photograph by Ajay Pillarisetti.)

increasing respiratory tract irritation and new-onset and exacerbation of asthma and COPD.

- Climate change increases the duration of the pollen season and the allergenicity of pollen, thereby worsening symptoms of allergic rhinitis and asthma.
- Climate change increases the frequency and severity of floods, increasing the likelihood of indoor moisture and growth of and exposure to mold, as well as heightened incidence of fungal respiratory disorders.
- Prevention of climate-induced respiratory disorders can be achieved by mitigation of GHGs and by preventive measures taken by governments and individuals to reduce exposure to ozone, particulate matter, pollen, mold, and excessive heat.

References

1. Cissé GR, McLeman H, Adams P, et al. Health, Wellbeing, and the Changing Structure of Communities. In H-O Pörtner, DC Roberts, M Tignor, et al. (eds.). Climate Change 2022: Impacts, Adaptation and Vulnerability. Contribution of Working Group II to the Sixth Assessment Report of the Intergovernmental Panel on Climate Change. Cambridge: Cambridge University Press, 2022, pp. 1041–1170. doi: 10.1017/9781009325844.009.

2. Jacob D, Winner D. Effect of climate change on air quality. Atmospheric Environment 2009; 43: 51–63.
3. US Environmental Protection Agency. Ground-level Ozone Pollution. Available at: https://www.epa.gov/ground-level-ozone-pollution. Accessed October 19, 2022.
4. DeLang MN, Becker JS, Chang K-L, et al. Mapping yearly fine resolution global surface ozone through the Bayesian Maximum Entropy data fusion of observations and model output for 1990–2017. Environmental Science & Technology 2021; 55: 4389–4398.
5. Silva R, West J, Lamarque JF, et al. Future global mortality from changes in air pollution attributable to climate change. Nature Climate Change 2017; 7: 647–651. doi: 10.1038/nclimate3354.
6. Stowell JD, Kim Y-M, Gao Y, et al. The impact of climate change and emissions control on future ozone levels: Implications for human health. Environment International 2017; 108: 41–50. https://doi.org/10.1016/j.envint.2017.08.001.
7. Wilson A, Reich B, Nolte C, et al. Climate change impacts on projections of excess mortality at 2030 using spatially varying ozone–temperature risk surfaces. Journal of Exposure Science and Environmental Epidemiology 2017; 27: 118–124. https://doi.org/10.1038/jes.2016.14.
8. NOAA Chemical Sciences Laboratory. A Systematic Analysis for the Global Burden of Disease Study 2019. October 15, 2020. Available at: https://csl.noaa.gov/news/2020/293_1015.html. Accessed October 18, 2022.
9. Holm SM, Balmes JR. Systematic review of ozone effects on human lung function, 2013 through 2020. Chest 2022; 161: 190–201.
10. US Environmental Protection Agency. Integrated Science Assessment (ISA) for Ozone and Related Photochemical Oxidants. Final Report, April 2020. Available at: https://cfpub.epa.gov/ncea/isa/recordisplay.cfm?deid=348522. Accessed October 18, 2022.
11. Aguilera R, Corringham T, Gershunov A, et al. Fine particles in wildfire smoke and pediatric respiratory health in California. Pediatrics 2021; 147: e2020027128. doi: 10.1542/peds.2020-027128.
12. Sang S, Chu C, Zhang T, et al. The global burden of disease attributable to ambient fine particulate matter in 204 countries and territories, 1990–2019: A systematic analysis of the Global Burden of Disease Study 2019. Ecotoxicology and Environmental Safety 2022; 238: 113588. https://doi.org/10.1016/j.ecoenv.2022.113588.
13. Brunekeef B, Holgate ST. Air pollution and health. Lancet 2022; 19: 360: 1233–1242.https://doi.org/10.1016/S0140-6736(02)11274-8.
14. Gallagher CL, Holloway T. Integrating air quality and public health benefits in U.S. decarbonization strategies. Frontiers in Public Health 2020; 8: 563358. doi.org/10.3389/fpubh.2020.563358
15. Zhuang Y, Fu R, Santer B, et al. Quantifying contributions of natural variability and anthropogenic forcings on increased fire weather risk over the western United States. Proceedings of the National Academy of Sciences 2021; 118: e2111875118. https://doi.org10.1073/pnas.2111875118.
16. Goss M, Swain D, Abatzoglou J, et al. Climate change is increasing the likelihood of extreme autumn wildfire conditions across California. Environmental Research Letters 2020; 15: 094016. doi.org/10.1088/1748-9326/ab83a7.
17. Bernstein AS, Myers SS. Climate change and children's health. Current Opinion in Pediatrics 2011; 23: 221–226.

18. Shankar HM, Rice MB. Update on climate change: Its impact on respiratory health at work, home, and at play. Clinics in Chest Medicine 2020; 41: 753–761.
19. Wang IJ, Tung TH, Tang CS, Zhao ZH. Allergens, air pollutants, and childhood allergic diseases. International Journal of Hygiene and Environmental Health 2016; 219: 66–71.
20. Reid CE, Maestas MM. Wildfire smoke exposure under climate change: Impact on respiratory health of affected communities. Current Opinions in Pulmonary Medicine 2019; 25: 179–187.
21. Haikerwal A, Akram M, Sim MR, et al. Fine particulate matter (PM2.5) exposure during a prolonged wildfire period and emergency department visits for asthma. Respirology 2016; 21: 88–94.
22. Gan RW, Liu J, Ford B, et al. The association between wildfire smoke exposure and asthma-specific medical care utilization in Oregon during the 2013 wildfire season. Journal of Exposure Science and Environmental Epidemiology 2020; 30: 618–628.
23. Ontawong A, Saokaew S, Jamroendararasame B, Duangjai A. Impact of long-term exposure wildfire smog on respiratory health outcomes. Expert Review of Respiratory Medicine 2020; 14: 527–531.
24. Romanello M, McGushin A, Napoli CD, et al. The 2021 report of the Lancet Countdown on health and climate change: Code red for a healthy future. Lancet 2021; 398: 1619–1662.
25. Neumann JE, Amend M, Anenberg S, et al. Estimating PM2.5-related premature mortality and morbidity associated with future wildfire emissions in the western US. Environmental Research Letters 2021; 16: 035019. doi: 10.1088/1748-9326/abe82b.
26. Ford B, Val Martin M, Zelasky SE, et al. Future fire impacts on smoke concentrations, visibility, and health in the contiguous United States. GeoHealth 2018; 2: 229–247.
27. Jayachandran S. Air quality and early-life mortality: Evidence from Indonesia's wildfires. Journal of Human Resources 2009; 44: 916–954.
28. Dejmek J, Selevan SG, Benes I, et al. Fetal growth and maternal exposure to particulate matter during pregnancy. Environmental Health Perspectives 1999; 107: 475–480.
29. Ritz B, Wilhelm M, Hoggatt KJ, Ghosh JKC. Ambient air pollution and preterm birth in the environment and pregnancy outcomes study at the University of California, Los Angeles. American Journal of Epidemiology 2007; 166: 1045–1052.
30. Aghababaeian H, Ostadtaghizadeh A, Ardalan A, et al. Global health impacts of dust storms: A systematic review. Environmental Health Insights 2021; 15: 11786302211018390. doi: 10.1177/11786302211018390.
31. Achakulwisut P, Anenberg SC, Neumann JE, et al. Effects of increasing aridity on ambient dust and public health in the U.S. Southwest under climate change. GeoHealth 2019; 3: 127–144. doi: 10.1029/2019GH000187.
32. Achakulwisut P, Mickley LJ, Anenberg SC. Drought-sensitivity of fine dust in the US Southwest: Implications for air quality and public health under future climate change. Environmental Research Letters 2018; 13: 054025. doi.org/10.1088/1748-9326/aabf20.
33. Hashizume M, Kim Y, Ng CFS, et al. Health effects of Asian dust: A systematic review and meta-analysis. Environmental Health Perspectives 2020; 128: 66001. doi: 10.1289/EHP5312.
34. Vergadi E, Rouva G, Angeli M, Galanakis E. Infectious diseases associated with desert dust outbreaks: A systematic review. International Journal of Environmental Research and Public Health 2022; 19: 6907. doi: 10.3390/ijerph19116907.

35. Gorris ME, Treseder KK, Zender CS, Randerson JT. Expansion of coccidioidomy-cosis endemic regions in the United States in response to climate change. GeoHealth 2019; 3: 308–327. doi.org/10.1029/2019GH000209.

36. Yitshak-Sade M, Novack V, Katra I, et al. Non-anthropogenic dust exposure and asthma medication purchase in children. European Respiratory Journal 2015; 45: 652–660.

37. Lorentzou C, Kouvarakis G, Kozyrakis GV, et al. Extreme desert dust storms and COPD morbidity on the island of Crete. International Journal of Chronic Obstructive Pulmonary Disease 2019; 14: 1763–1768.

38. Geravandi S, Sicard P, Khaniabadi YO, et al. A comparative study of hos-pital admissions for respiratory diseases during normal and dusty days in Iran. Environmental Science and Pollution Research International 2017; 24: 18152–18159.

39. Thalib L, Al-Taiar A. Dust storms and the risk of asthma admissions to hospitals in Kuwait. Science of the Total Environment 2012; 433: 347–351.

40. Anderson BG, Bell ML. Weather-related mortality: How heat, cold, and heat waves affect mortality in the United States. Epidemiology 2009; 20: 205–213.

41. Anderson GB, Dominici F, Wang Y, et al. Heat-related emergency hospitalizations for respiratory diseases in the Medicare population. American Journal of Respiratory and Critical Care Medicine 2013; 187: 1098–1103.

42. Ma Y, Zhou L, Chen K. Burden of cause-specific mortality attributable to heat and cold: A multicity time-series study in Jiangsu Province, China. Environment International 2020; 144: 105994. https://doi.org/10.1016/j.envint.2020.105994.

43. Zhao Q, Li S, Coelho M, et al. Ambient heat and hospitalisation for COPD in Brazil: A nationwide case-crossover study. Thorax 2019; 74: 1031–1036.

44. Ayres JG, Forsberg B, Annesi-Maesano I, et al. Climate change and respiratory disease: European Respiratory Society position statement. European Respiratory Journal 2009; 34: 295–302.

45. Anderegg WRL, Abatzoglou JT, Anderegg LDL, et al. Anthropogenic climate change is worsening North American pollen seasons. Proceedings of the National Academy of Sciences USA 2021; 118: e2013284118. https://doi.org/10.1073/pnas.2013284118.

46. Neumann JE, Anenberg SC, Weinberger KR, et al. Estimates of present and future asthma emergency department visits associated with exposure to oak, birch, and grass pollen in the United States. GeoHealth 2019; 3: 11–27.

47. Lake IR, Jones NR, Agnew M, et al. Climate change and future pollen allergy in Europe. Environmental Health Perspectives 2017; 125: 385–391.

48. Pang K, Li G, Li M, et al. Prevalence and risk factors for allergic rhinitis in China: A sys-tematic review and meta-analysis. Evidence-Based Complementary and Alternative Medicine 2022; 7165627. doi: 10.1155/2022/7165627.

49. Hamizan A, Rimmer J, Alvarado R, et al. Positive allergen reaction in allergic and nonallergic rhinitis: A systematic review. International Forum of Allergy & Rhinology 2017; 7: 868–877.

50. Urbanowicz C, Hutyra LR, Stinson KA. The effects of urbanization and land use on ragweed distribution. Ecosphere 2018; 9: e02512. https://doi.org/10.1002/ecs2.2512.

51. Hamaoui-Laguel L, Vautard R, Liu L, et al. Effects of climate change and seed dis-persal on airborne ragweed pollen loads in Europe. Nature Climate Change 2015; 5: 766–771.

52. Martikainen M-V, Tossavainen T, Hannukka N, Roponen M. Pollen, respiratory viruses, and climate change: Synergistic effects on human health. Environmental Research 2023; 219: 115149. doi: 10.1016/jenvres.2022.115149.

53. Anenberg SC, Haines S, Wang E, et al. Synergistic health effects of air pollution, temperature, and pollen exposure: A systematic review of the epidemiological evidence. Environmental Health 2020; 19: 130. https://link.springer.com/article/10.1186/s12 940-020-00681-z.

54. Cai Y. Ambient pollen and air quality on children's lung function: Is there a synergy? Thorax 2021; 76: 858–859.

55. Olaniyan TA, Dalvie MA, Roosli M, Jeebhay MF. Air pollution, pollens and childhood asthma - Is there a link? Current Allergy and Clinical Immunology 2016; 29: 252–261.

56. Lam HCY, Jarvis D, Fuertes E. Interactive effects of allergens and air pollution on respiratory health: A systematic review. Science of the Total Environment 2021; 757: 143924. doi: 10.1016/j.scitotenv.2020.143924.

57. Venkatesan P. Epidemic thunderstorm asthma. Lancet Respiratory Medicine 2022; 10: 325–326.

58. D'Amato G, Annesi-Maesano I, Urrutia-Pereira M, et al. Thunderstorm allergy and asthma: State of the art. Multidiscipinary Respiratory Medicine 2021; 16: 806. doi: 10.4081/mrm.2021.806.

59. Sousa-Silva R, Smargiassi A, Kneeshaw D. et al. Strong variations in urban allergenicity riskscapes due to poor knowledge of tree pollen allergenic potential. Scientific Reports 2021; 11: 10196. https://doi.org/10.1038/s41598-021-89353-7.

60. van Dorn A. Urban planning and respiratory health. Lancet Respiratory Medicine 2017; 5: 781–782. doi.org/10.1016/S2213-2600(17)30340-5.

61. Drewniak BA, Snyder PK, Steiner AL, et al. Simulated changes in biogenic VOC emissions and ozone formation from habitat expansion of Acer Rubrum(red maple). Environmental Research Letters 2014; 9: 014006. doi: 10.1088/1748-9326/9/1/014006.

62. Jaakkola MS, Quansah R, Hugg TT, et al. Association of indoor dampness and molds with rhinitis risk: A systematic review and meta-analysis. Journal of Allergy and Clinical Immunology 2013; 132: 1099–1110.

63. Caretta MA, Mukherji A, Arfanuzzaman M, et al. Water. In H-O Pörtner, DC Roberts, M Tignor, et al. (eds.). Climate Change 2022: Impacts, Adaptation and Vulnerability. Contribution of Working Group II to the Sixth Assessment Report of the Intergovernmental Panel on Climate Change. Cambridge: Cambridge University Press, 2022, pp. 551–712. doi: 10.1017/9781009325844.006.

64. Barbeau DB, Grimsley LF, White LE, et al. Mold exposure and health effects following Hurricanes Katrina and Rita. Annual Review of Public Health 2010; 31: 165–178. doi. org/10.1146/annurev.publhealth.012809.103643

65. Rando RJ, Kwon C-W, Lefante JJ. Exposures to thoracic particulate matter, endotoxin, and glucan during post-hurricane Katrina restoration work, New Orleans 2005–2012. Journal of Occupational and Environmental Hygiene 2014; 11: 9–18.

66. Demain JG. Climate change and the impact on respiratory and allergic disease: 2018. Current Allergy and Asthma Reports 2018; 18: 22. doi: 10.1007/s11882-018-0777-7.

67. Fishe J, Zheng Y, Lyu T, et al. Environmental effects on acute exacerbations of respiratory diseases: A real-world big data study. Science of the Total Environment 2022; 806: 150352. doi: 10.1016/j.scitotenv.2021.150352.

68. Kodros JK, O'Dell K, Samet JM, et al. Quantifying the health benefits of face masks and respirators to mitigate exposure to severe air pollution. GeoHealth 2021; 5: 32021GH000482. https://doi.org/10.1029/2021GH000482

Further Reading

D'Amato G, Chong-Neto HJ, Monge Ortega OP, et al. The effects of climate change on respiratory allergy and asthma induced by pollen and mold allergens. Allergy 2020; 75: 2219–2228.

This paper provides an overview of climate-induced respiratory allergy and asthma.

Reid CE, Maestas MM. Wildfire smoke exposure under climate change: Impact on respiratory health of affected communities. Current Opinion in Pulmonary Medicine 2019; 25: 179–187.

This review summarizes the literature on the respiratory effects associated with exposure to wildlife smoke.

Joshi M, Goraya H, Joshi A, Barter T. Climate change and respiratory diseases: A 2020 perspective. Current Opinion in Pulmonary Medicine 2020; 26: 119–127.

This article presents an overview of the adverse effects of climate change on respiratory health.

Pacheco SE, Guidos-Fogelbach G, Annesi-Maesano I, et al. Climate change and global issues in allergy and immunology. Journal of Allergy and Clinical Immunology 2021; 148: 1366–1377.

This article provides a good review of the impact of climate change on allergies and immune responses.

6

Vectorborne Diseases

Christopher M. Barker and William K. Reisen

Vectorborne diseases are caused by pathogens that are transmitted by invertebrates, primarily arthropods. For vectorborne pathogen transmission to occur, susceptible vertebrate hosts, competent vectors, and virulent pathogens must intersect within a suitable environment. Several vectorborne diseases have emerged because climatic conditions have become suitable for vector introduction and subsequent pathogen transmission.

Human population growth and resulting climate change are altering the Earth in ways that extend environmental suitability for transmission in both time and space. As a result, people in northern latitudes, including much of Canada and Europe, are now at risk from mosquitoborne and tickborne diseases that originated in the tropics. In addition, the global expansion of invasive mosquitoes within the genus *Aedes* has increased the risk for introduction of vectorborne viruses, such as dengue, Zika, and chikungunya, into temperate latitudes and higher elevations.

The tripling of the global human population during the past 75 years along with concurrently increasing demands for resources to house, feed, and provide fuel for transportation, cooking, and heating have irreversibly changed the ecology and climate of the Earth,[1] facilitating the emergence and transmission of vectorborne pathogens.[2] In addition, many people have migrated from rural to urban areas, leading to a greater density of people who are linked more closely than ever by transportation and commercial networks. Many of these urban centers are situated in low-lying areas that have become more vulnerable to flooding from storm surges and sea-level rise. In combination, anthropogenic changes have altered the atmosphere and climate of the Earth,[3] leading to general warming (especially at northern latitudes),[4] rising sea levels, increasing storm intensity, and changing rainfall patterns.[1] These long-term trends vary considerably at the daily scale (weather), the seasonal scale (El Niño-Southern Oscillation, or ENSO), and the decadal scale—all of which originate largely from changes in ocean temperature.[5] Within urban centers, impervious and reflective surfaces have altered precipitation run-off and temperature patterns, creating *urban heat islands*.[6] Collectively, these changes have created new challenges for public health. In this chapter, we provide an update on this topic and include

Christopher M. Barker and William K. Reisen, *Vectorborne Diseases* In: *Climate Change and Public Health*.
Edited by: Barry S. Levy and Jonathan A. Patz, Oxford University Press. © Oxford University Press 2024.
DOI: 10.1093/oso/9780197683293.003.0006

recent examples of latitudinal and seasonal changes in the risk of selected vectorborne diseases.

NATURE OF THE PROBLEM

Climate change is resulting in long-term changes in the averages or frequencies of shorter-term weather. These changes are observable at scales of decades to centuries. Shorter-term variability in weather events and the complex nature of vectorborne disease epidemiology make it difficult to isolate singular examples where climate change solely and directly has led to increases in the vectorborne disease burden. However, transmission cannot occur unless environmental temperatures are suitable for blood-feeding by vectors and the propagation of pathogens. As examples, warming temperatures have influenced the invasion and persistence of West Nile virus in North America and Europe, the reemergence and northward expansion of eastern equine encephalomyelitis virus, the emergence and expansion of tick populations and tickborne diseases, and the invasion of temperate latitudes by tropical mosquitoes. Climate change may also be causing some pathogens, such as western equine encephalomyelitis virus and tickborne encephalitis virus, to recede in some areas or become more focal in their distributions.

Biology of Vectors and Vectorborne Pathogens

Among the infectious diseases, those that are vectorborne are especially vulnerable to climate change because the vectors are *poikilotherms* ("cold-blooded" organisms)—that is, their body temperature approaches that of ambient conditions. Minimally, transmission requires the intersection of four factors: susceptible vertebrate hosts, competent vectors, virulent pathogens, and an environment suitable for transmission.[7] Vector and host abundance as well as seasonality are dependent on ecosystem dynamics and therefore climate.

Bloodfeeding by vectors brings vectors and hosts together in time and space, thereby enabling the acquisition and transmission of pathogens. This behavior has evolved independently within arthropods on multiple occasions and has been exploited as a mechanism for transmission among vertebrate hosts by a wide variety of viral, bacterial, and metazoan (multicellular) pathogens. Most arthropod vectors, especially various flies within the order Diptera, utilize blood as a dietary resource for egg maturation. However, other vectors, such as ticks and mites, also require blood meals for metabolic needs that enable molting between stages. The nature of host-vector contact ranges from sustained contact in which the vector remains on the host throughout its entire life cycle, as in the case of lice and certain species of mites, to intermittent contact for purposes of bloodfeeding, as in the case of mosquitoes and several other biting flies. Climate variation and change have the greatest impact on those vectors with intermittent host contact that spend most of their lives away from the host and exposed to the effects of environmental variability.

Typically, pathogen transmission cycles are highly evolved and largely dependent upon host-seeking behavior and resulting host selection by the primary vectors. Because arthropod vectors are poikilotherms, the pathogens they acquire are subjected to ambient conditions during the period of vector infection, although vectors are capable of modifying these temperatures by selecting suitable microhabitats for resting and other daily activities.[8] Most vectorborne pathogens undergo an extrinsic incubation period within the vector (and away from the vertebrate host) in which the pathogen replicates and, for some pathogens, transitions to an infectious stage.

Because many vectorborne pathogens cause *zoonoses* (diseases with their natural lifecycle in non-human animals), the distribution and abundance of both host and vector are closely linked to vegetation and ecosystem dynamics, which are sensitive to climate change and variation. Humans frequently become infected incidentally after pathogen amplification within the natural vector-host zoonotic transmission cycle. However, from a global health perspective, many important pathogens, such as malaria parasites and the dengue fever virus, cause *anthroponoses* (diseases in which pathogens can persist in vector-human life cycles without requiring nonhuman vertebrate hosts). This results in efficient transmission indoors from person to person by endophagic (indoor-feeding) mosquitoes that thrive in and around human dwellings.

Thermodynamics of Pathogen Transmission

In poikilothermic vectors, the rate of most physiological processes, such as the digestion of blood meals or maturation of life stages, proceeds as a function of ambient temperature—that is, when the environment is warmer, reactions proceed more quickly, and, when it is cooler, they proceed more slowly.[9] In mosquitoes, for example, temperature dictates the rate of blood digestion and therefore the rate of egg maturation, oviposition (egg-laying), and refeeding. This gonotrophic cycle (feeding and egg-laying) proceeds rapidly under warm summer temperatures, leading to rapid population growth and more frequent host contact by more mosquitoes and therefore more opportunities for pathogen acquisition and transmission.

The period of pathogen development within the vector, known as the *extrinsic incubation period*, is also temperature-dependent. The duration of the extrinsic incubation period decreases as an exponential function of temperature.[9-11] Therefore, when it is warm, pathogens are transmitted earlier in the reproductive life of the vector, making it more likely that infected vectors will survive to transmit the pathogen. The number of bites or blood meals (gonotrophic cycles) prior to transmission has been used as a temperature-sensitive metric in evaluating the impact of climate change.[12-15] In addition, because the gonotrophic cycle and the rate of immature development is faster, *multivoltine* vector populations (those with multiple generations annually) increase rapidly when the temperature is warm, thereby increasing the number of vectors and host-vector contacts. Conversely, female survival tends to decrease as a function of temperature; however, this decrease is compensated by the rapid extrinsic incubation of pathogens and shorter gonotrophic cycles.[9]

Collectively, these interactions result in more efficient pathogen transmission during warm periods, as expressed by the *vectorial capacity equation*, which represents the number of future infectious bites that would be expected to arise from vectors feeding on a single infectious host per day (Figure 6-1).[16]

Most of the parameters in this equation are temperature-sensitive. Abundance of the vector in relation to the host (m) is related to reproductive success (the number of new vectors produced each generation). Generation times are based on the rate of larval development and the duration of the gonotrophic cycle, which is shorter when temperatures are warm, resulting in more vectors produced per day. At temperate latitudes, the number of generations produced per year by multivoltine species is directly related to the intensity and duration

$$c = ma^2\, p^n V / - \log_e p$$

c = New cases (or infected hosts) per day per infectious case

ma = Mosquito biting rate in vectors per host per day

a = HI/GC, where HI = host selection index and **GC** = duration of the gonotrophic cycle

p = Probability of daily survival

n = Duration of the extrinsic incubation period

V = Vector competence

Figure 6-1 Vectorial capacity equation estimating the number of new cases per case per day based on entomological parameters of the Ross-Macdonald malaria model. Factors affected by temperature are shown in boldface type. (More information on the Ross-Macdonald malaria model can be found in: Smith DL, Battle KE, Hay SI, et al. Ross, Macdonald, and a theory for the dynamics and control of mosquito-transmitted pathogens. PLoS Pathogens 2012; 8: e1002588.)

of temperature anomalies, which extend the vector reproduction season and therefore the transmission season. In mosquitoes, the rate of population growth and geographic expansion also is related to precipitation patterns and how they affect available habitats for larval development. Too little precipitation restricts oviposition success; too much can produce floods that reduce edge habitat and flush away developing larvae. Adult female survival (p) tends to decrease with increasing temperature. The expectation of infectious life of the vector also tends to decrease at the highest temperatures ($p^n/\text{-log}_e\, p$), where n is the duration of the extrinsic incubation period.

Overwintering

The distribution of vectorborne pathogens depends on the presence of suitable vector populations. At temperate or arctic latitudes or at higher elevations, cold winter temperatures suppress transmission and may lead to seasonal extinction. Most commonly, pathogens either persist as long-term infections within vector populations or are reintroduced annually from warmer latitudes by migratory vertebrate hosts. The ability of vectors and therefore the pathogens they transmit to invade and persist in new areas depends largely on the ability of the vector to survive unfavorable weather periods that are either too cold or too dry.

In some species, winter survival is enabled by physiological preparation in response to seasonal triggers such as photoperiods that induce and eventually terminate diapause (hibernation) or by arresting activity without antecedent physiological changes (quiescence).[17] Generally, species that enter a true diapause are better able to tolerate colder winter conditions than those that enter quiescence. In some Aedes species, such as Aedes albopictus, adult females respond to shortening day length and lay desiccation-resistant eggs that enter diapause and can tolerate freezing conditions,[18,19] allowing the species to persist at higher latitudes than closely related Aedes aegypti, which lacks the ability to diapause and is limited by winter temperatures.[20,21] This has left a marked clinal difference in the distribution of these invasive species in North America, with Aedes aegypti confined to southern latitudes and Aedes albopictus able to persist into northern latitudes.[22] Historically, Aedes aegypti, transported by sailing ships, would recolonize seaports in the northeastern United States, thereby enabling periodic outbreaks of yellow fever and dengue,[23,24] even though the vector could not persist and would become locally extinct during the harsh winters at northern latitudes.

Similarly, the fourth-instar larvae of some Culex mosquito species (such as Culex pipiens) respond to shortening photoperiod and cooling temperatures and emerge as cold-hardy adult females that postpone bloodfeeding and egg development, imbibe sugar for metabolic needs, produce increased fat reserves, and seek refugia (areas in which a population of organisms can survive through a period of unfavorable conditions).[25] Conversely, other *Culex* species, such as the southern sibling species within this complex, *Culex quinquefasciatus*, lack this diapause ability and produce females that are not cold-tolerant and usually do not survive winter at northern latitudes.[26] These two species readily interbreed where their distributions overlap, producing viable hybrids.[27] Interestingly, warming winters in the Central Valley of California have enabled the northern intrusion of *Culex quinquefasciatus* genes into *Culex pipiens* populations, thereby eliminating their ability to diapause.[28]

Species that exhibit quiescence may survive during unfavorable conditions by seeking refugia and then resuming activity during warm periods. *Culiseta melanura*, for example, overwinters as quiescent larvae that can survive under ice by cuticular respiration.[29] The rate of larval development and the timing of spring emergence is strongly influenced by vernal temperatures. Similarly, ticks that exhibit multiyear and multihost life cycles survive winter off-host within forest litter under snow cover, allowing northern range extensions in a warming climate.[30]

EXAMPLES OF HOW CLIMATE CHANGE MAY ALTER TRANSMISSION PATTERNS

Because it is difficult to generalize about how climate change and variation impact multiple vectorborne transmission systems at different scales of time and space, the following representative scenarios have been selected to illustrate some of the different effects of climate and environmental change. In the figures showing transmission cycles, arrows show the direction of interactions created by blood-feeding behavior, but the frequency of these interactions, and therefore the rate of transmission, is driven by temperature and host availability.

Temperature and West Nile Virus in North America

In 1937, West Nile virus (*Flaviviridae*, *Flavivirus*, or WNV) was discovered in tropical Uganda during a survey to determine the etiology of febrile illness. WNV seemed restricted to tropical Africa until the 1950s, after which outbreaks were documented in the Nile Delta, the Mediterranean region, South Africa, Europe, and then the Western Hemisphere. In periurban settings in these locations, WNV is amplified in a zoonotic cycle that involves a variety of passeriform (perching) birds and several *Culex* mosquito species, and then spills over to tangentially infect humans as well as a wide variety of other mammals and birds (Figure 6-2). Although these are *dead-end hosts* for the virus (that is, they do not contribute to the transmission cycle of the virus), infected humans and horses occasionally develop WNV neuroinvasive disease, which can be fatal, especially in older individuals.

The invasion of North America by WNV has been facilitated by climate change, which has produced some of the warmest summers on record during the past 20 years. Hot, dry summers often increase the abundance of urban *Culex* mosquitoes, which frequently develop in drainage systems that receive a steady supply of residual water from landscape irrigation, and the efficiency of transmission is enhanced through various effects of warmer temperatures. The receptivity of New York City for the initial WNV invasion in 1999 probably was enhanced by the warmest July, based on mean daily temperature, ever recorded there.[31] This heat wave produced a major public health emergency,[32] which initially overshadowed the onset of what was to become the largest mosquitoborne encephalitis outbreak in the Western Hemisphere and the largest WNV outbreak globally.[33]

Within only 5 years, WNV extended its distribution westward to the Pacific Coast, northward into Canada, and southward to Argentina. Based on a comparison of maps of temperature anomalies[34] and case incidence,[35] WNV outbreaks during this westward expansion were associated with abnormally warm summers. During these initial outbreaks, the mean June-to-September temperatures averaged about 1.0°–3.0°C (about 1.8°–5.4°F) above the 1971–2000 average, and the incidence of neuroinvasive disease was greater than 10 per 100,000 population.[10] The importance of summer temperatures in enzootic (bird-mosquito-bird) transmission also was demonstrated spatially in different parts of California, where transmission was confined mostly to the warmer Central Valley and southern California.[31] In contrast, outbreaks at southern latitudes in the United States, such as in Louisiana or parts of southeastern California, where mean summer temperatures remain higher than 25°C (77°F), seemed to lack a clear temperature signal, perhaps because temperatures were consistently warm enough to support efficient transmission.

Very hot temperatures, greater than 30°C (86°F), limit mosquito survival and total reproductive capacity,[36] but it is unclear whether such limits present critical constraints on the ecological niche in areas where mosquitoes can seek refuge in microhabitats with more moderate temperatures. This is especially true in areas where rainfall or anthropogenic irrigation provides adequate humidity to moderate the effects of high temperatures on mosquito survival. Mosquitoes with access to moderate microclimates are capable of behavioral thermoregulation that can allow them to avoid extreme heat or other environmental stressors.

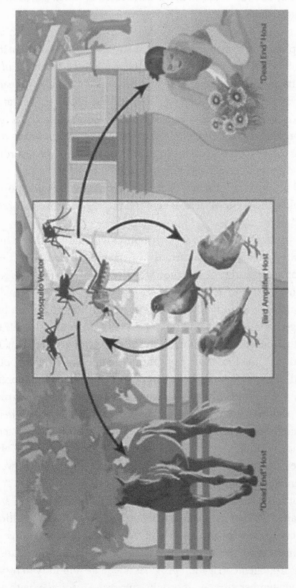

Figure 6-2 West Nile virus transmission cycle. Shown is the basic amplification cycle among peridomestic birds and mosquitoes with tangential transmission to human and equine hosts. Not shown are important corvid amplification hosts or other modes of possible transmission, including avian predation or scavenging and vertical transmission by mosquitoes. (*Source:* Centers for Disease Control and Prevention. West Nile Virus: Transmission. Available at: http://www.cdc.gov/westnile/transmission. Accessed September 22, 2022.)

In Canada, at the northern boundary of WNV transmission, the two largest outbreaks of WNV disease, during 2003 and 2007, were centered in Saskatchewan and were linked closely with exceptionally warm summers, when the daily mean temperatures in Saskatoon were up to 10°C (18°F) above the 40-year average.[37] Considering 14.3°C (57.7°F) as the threshold for WNV replication,[10] the transmission season during those summers was extended from 2 to 5 months. Similarly, the 1.0°–2.0°C (1.8°–3.6°F) temperature increase in Europe has been associated with outbreaks of WNV in Mediterranean countries as well as Romania, Serbia, and eastern Russia.[38] Although a variety of factors have contributed to the rapid expansion of WNV in the western hemisphere[39] and Europe,[40,41] summer temperatures seem to have been a key factor, especially at northern latitudes. Here, warm spring temperatures have led to an early onset of transmission and have shaped the pattern of transmission during summer.[41]

Eastern Equine Encephalomyelitis Virus

Eastern equine encephalomyelitis virus (*Togaviridae: Alphavirus*, or EEEV) is a mosquitoborne zoonosis that causes focal disease outbreaks in the United States east of the Mississippi River. Like WNV, natural transmission cycles of EEEV involve mosquitoes and birds (Figure 6-2). The mosquito *Culiseta melanura* is the primary enzootic vector among avian hosts within forested habitats of the northeastern and midwestern United States,[42-44] with a variety of floodwater *Aedes* species and *Coquillettidia perturbans* serving as bridge vectors—acquiring the virus from birds and transmitting it to equines and humans, who may experience disease following infection.[45]

Within the past 20 years, there has been a resurgence of EEEV amplification and tangential transmission to equines and humans in the northeastern United States, concurrent with a northward expansion of transmission.[46] Sequencing studies on viral isolates have indicated that EEEV strains persist locally for multiple years and then are replaced by new strains, most likely introduced by passerine (perching) birds migrating north during spring via the eastern flyway.[46,47] Warm vernal temperatures that drive the early emergence of overwintering *Culiseta melanura* seem critical in ensuring the abundance of vectors during the arrival of migrant birds and the nesting season.[42] In Florida, EEEV transmission can be detected year-round, with the greatest numbers of equine cases reported during mild winters.[48] The quiescent overwintering behavior of *Culiseta melanura* results in variable transmission patterns, with a cessation of transmission during cold northern winters but apparent continuous transmission during mild winters in Florida.

Precipitation and Rift Valley Fever Virus

Rift Valley fever virus (*Bunyaviridae, Phlebovirus,* or RVFV) is distributed widely in eastern and southern Africa, where outbreaks typically follow periods of increased rainfall and flooding (Figure 6-3). RVFV is transmitted by *Aedes* and *Culex* mosquitoes among domestic and wild ungulates, in which it causes epizootics (epidemics of disease in animals) characterized by widespread abortion in cattle, sheep, and some wildlife.[49,50] The virus purportedly persists between outbreaks within infected, desiccation-resistant eggs of *Aedes* mosquitoes (especially *Aedes mcintoshi*). These mosquitoes typically oviposit (lay eggs) at the margins of receding pools of water, known locally as *dambos*, where the eggs remain viable in the soil until the next rain inundates them and stimulates hatching—an event that may occur intermittently every few years.[51,52]

A series of years with low rainfall may result in mosquitoes ovipositing at progressively lower strata along these pool edges. Such dry years are followed episodically by periods of very high rainfall, linked to warm Indian Ocean temperatures, and these extreme rainfall events lead to the hatching of multiple generations of eggs that were oviposited but not flooded during previous years, leading to the emergence of many *Aedes* mosquitoes infected with several strains of RVFV.[53] Rains also produce new grass associated with these dambos, concentrating wildlife (especially water buffalo) and domestic animals (cattle, sheep, and goats) for grazing. Offspring of many species are produced during this period, bringing vertically infected mosquitoes together with immunologically naïve vertebrate hosts, thereby enhancing the efficiency of RVFV amplification. *Culex* mosquitoes, which emerge after *Aedes* mosquitoes have seeded RVFV into these host populations, purportedly serve as bridge vectors that transmit virus from animals to humans.[54]

The distribution and intensity of RVFV transmission seems closely tied to the intensity of the rainy season, especially in East Africa.[55] There is an association among (a) strong El Niño-Southern Oscillation (ENSO) events; (b) increased sea surface temperatures, leading to the development of rain clouds over the Indian Ocean; (c) increased intensity of rain in East Africa (especially in the Rift Valley); and (d) the flooding of grasslands.[56] These interrelationships have been observed through satellite imagery, which has enabled the development of predictive models that can provide warning of heavy rainfall and flooding several weeks in advance of potential outbreaks.[57] Such early warning can lead to vaccination campaigns to protect livestock and limit transmission. The impact of climate change on the frequency and intensity of high-rainfall events probably increases the risk of RVFV outbreaks and the impact of RVFV on wildlife, veterinary, and human health in Africa.

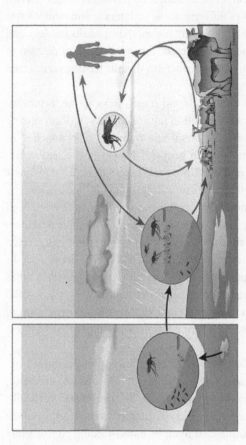

Figure 6-3 Rift Valley fever virus transmission cycle, emphasizing the importance of rainfall in virus emergence. Not shown are possible maintenance transmission cycles among reservoir hosts. The far left portion of the figure demonstrates the enzootic cycle, in which local enzootic transmission of RVF occurs at low levels in nature during periods of average rainfall. The virus is maintained through transovarial transmission from the female *Aedes* mosquito to her eggs and through occasional amplification cycles in susceptible livestock. The remainder of the figure shows the epizootic-epidemic cycle. Abnormally high rainfall and flooding stimulate hatching of the infected *Aedes* mosquito eggs, resulting in a massive emergence of *Aedes*, including RVF virus–infected *Aedes* adults. The infected *Aedes* then feed on vulnerable livestock, triggering virus amplification and an epizootic. Epizootics cause abortion storms, with more than 90% mortality in newborns and 10 to 30% mortality in adults. Secondary vectors include other mosquito genera, such as *Culex*, which can pass the virus to humans and animals, producing disease. Human exposure to viremic livestock blood and tissue can occur during slaughtering or birthing activities. (*Source:* Centers for Disease Control and Prevention. Rift Valley Fever [RVF]. Available at: https://www.cdc.gov/vhf/rvf/index.html. Accessed September 14, 2023.)

Climate and the Complexity of Tickborne Encephalitis Virus Transmission

The vectors of tickborne encephalitis virus (*Flaviviridae, Flavivirus,* or TBEV) in the *Ixodes ricinus* complex have long life cycles which frequently require more than 1 year for completion at temperate latitudes. Different life stages (larvae, nymphs, and adults) infest hosts of different sizes, with larvae and nymphs frequently found on small mammals (especially rodents) and adults on deer and large mammals.[58] Larvae may become infected transovarially (by infection of eggs in ovaries), but most tick infections are acquired while feeding on infected mammals.[59]

Because of the long intervals between blood meals, ticks are an important reservoir for TBEV, and most transmission in nature appears to occur during co-feeding by infected and uninfected larval and nymphal ticks attached in close proximity on the same mammalian host.[60,61] However, the seasonality of the off-host immature stages and the timing and height at which they quest on vegetation are affected by temperature, rainfall, and humidity, thereby impacting transmission efficiency.[62] In combination with a wide variety of anthropogenic factors ranging from landscape changes to increased outdoor activities during warmer summers, microclimate variation has created a spatial mosaic of TBEV infection risk throughout Europe.[63] The geographical distribution and health impact of TBEV generally has increased in areas with varying climate change signals, indicating the complexity of transmission of TBEV to humans.[58] However, the elevational and northward expansion of both the tick vector and TBEV would not be possible without permissive warm temperatures.[63,64]

A very similar scenario in North America has allowed the northward expansion of the Lyme disease bacterium, *Borrelia burgdorferi,* and its tick vector, *Ixodes scapularis.*[65] Because this bacterium cannot be transmitted vertically, transmission depends on the temporal overlap of infected overwintering larvae blood-feeding on rodents and uninfected nymphs blood-feeding concurrently on bacteremic rodents.[65] Changes in tick phenology theoretically could reduce the overlap of these life stages and therefore transmission.

Increasing Temperature and the Disappearance of Western Equine Encephalomyelitis Virus

In contrast to the expanding distribution of WNV, western equine encephalomyelitis virus (*Togaviridae, Alphavirus,* or WEEV) seems to have disappeared from formerly endemic areas in North America. As an example, WEEV has not been detected in mosquitoes in California since 2007, while, concurrently, WNV has become increasingly prevalent in the same mosquito vectors and avian hosts, indicating that the components of the enzootic cycle have remained intact (Figure 6-4). Laboratory infection studies have not shown decadal changes in either avian host or vector competence,[66] although such studies can evaluate only variation associated with past viral strains. A 2007 isolate from California seems to be attenuated in mice,[67] perhaps explaining the decrease in equine neurological disease. Because there is no cross-immunity between WEEV and WNV, negative trends in WEEV activity may be due to increasing temperatures and prolonged droughts in California in recent decades that could have reduced the potential for reestablishment of local transmission cycles.[68]

As measured by the dose of the WEEV required to infect 50% of engorged females, the susceptibility of the primary vector, *Culex tarsalis*, varies seasonally, with females least susceptible to infection during the hot mid-summer transmission season,[69] especially in the hotter southeastern deserts of California.[70] In the laboratory, infected female *Culex tarsalis* mosquitoes from Kern County (in the southern Central Valley of California) essentially stopped transmitting WEEV after 10 days when incubated at warm temperatures, although they remained infected for life.[71] Mosquitoes of another strain of *Culex tarsalis* cleared their infection at elevated temperatures.[72] Based on these laboratory studies, the

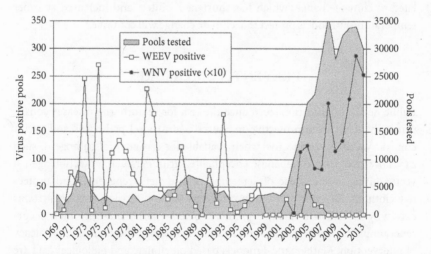

Figure 6-4 Temporal decline in the prevalence of western equine encephalomyelitis virus (WEEV) in groups (pools) of mosquitoes collected and tested for virus in California, 1969-2013. West Nile virus (WNV)-positive pools (shown as ×10) have predominated during the past decade, although WEEV utilizes the same mosquito vector (*Culex tarsalis*) and avian hosts. (*Source*: Data from annual reports of the California Mosquitoborne Encephalitis Surveillance Program.)

previous decline in WEEV activity in the United States has been attributed to climate warming, but attempts to elucidate the mechanisms have been unsuccessful.[73,74] In addition to temperature, the loss of wetlands, drought, the change in agricultural practices and crop rotation, and intensive mosquito control to interrupt WNV transmission may have contributed to lower vector abundance and thereby impeded WEEV transmission. Further studies on WEEV genetics have not resolved this enigma.[75]

FUTURE PROJECTIONS

Continued human population growth and associated environmental change will further impact the atmosphere, climate, and ecosystem dynamics. Collectively, these changes may extend the current distributions of vectors and the pathogens they transmit in both time (longer transmission seasons) and space (wider variations in latitude and elevation). Landscape change and increased connectivity by rapid transportation will continue to facilitate expansion of the distributions of important vector species, creating new environments for pathogens and increasing the risk of outbreaks.[76] The movement of tropical pathogens into temperate environments historically has resulted in outbreaks of important pathogens, such as WNV.[77] The severity of these outbreaks has been related to climate change, which has shortened winters and increased summer temperatures.[76] Models project that climate change will continue.[78]

PREVENTIVE MEASURES

Public health interventions to reduce the risk for vectorborne diseases consist primarily of personal protective measures against viral pathogens and vector control. Vaccines are effective when available for certain viral pathogens, such as yellow fever virus or Japanese encephalitis virus. Personal protection against vector biting includes use of screening or bed nets, repellents, and protective clothing. Effective prevention programs require decision-support systems driven by surveillance data that predict problems, verify and track their occurrence, implement control measures in a timely manner, and evaluate the efficacy of intervention. Mathematical models based on climate and biological data are useful to anticipate the possible future distribution of vectors and the pathogens that they transmit.[79] These models must account not only for vector and host movements,[80] but also for anthropogenic factors and social interactions.[81,82]

Understanding dispersal mechanisms may provide methods of tracking and mitigating invasions. For example, the first invasion of Los Angeles by the Asian tiger mosquito (*Aedes albopictus*) was facilitated by the importation from China and sale of "Lucky Bamboo" plants (*Dracaena* species), which initially were shipped in standing water, which allowed oviposition of eggs during packing and development of immature forms during transit. Packing the plants in an aqueous gel prevented further releases of these mosquitoes.[83] Stringent humidity requirements[84] seem to have limited the dispersal of *Aedes albopictus* into the drier areas of California, whereas more drought-tolerant *Aedes aegypti* mosquitoes have recently invaded and become established in the southeastern California deserts, the Los Angeles basin, and the Central Valley.

Although both statistical and mathematical models are useful for predicting problems, surveillance in real time is necessary to inform the models, track vectors and pathogens in time and space, measure risk, and inform decision-support systems that direct intervention.[85,86] Because outbreaks occur intermittently in time and space, surveillance must be performed indefinitely, and preparedness for implementing intervention measures must be maintained so that they can be implemented rapidly to prevent outbreaks.

Managing endemic pathogens, especially those that cause zoonoses, requires ongoing intervention with source reduction, insecticide application, and vaccination. Surveillance systems that rely solely on human case detection are sufficient when humans are the only vertebrate hosts, such as for malaria and dengue fever, but these measures are insufficient for zoonoses such as WNV and EEEV. Vaccination programs in humans may prevent human cases but not those in animals where enzootic activity persists. Where vaccines are not available, intervention for vectorborne diseases, such as dengue fever and malaria, must rely on mosquito control and case management. Interrupting host-vector contact relies on use of residual insecticides, ultra-low-volume adulticides, and/or barriers such as insecticide-treated bed nets; most of these programs have not focused on larval control, although this has been important in successful eradication campaigns.[87]

Successful intervention against zoonoses requires area-wide control of the immature stages to limit vector population size or repeated aerial applications of adulticides to kill infected females during the viremia period in the intermediate hosts.[88] These programs typically are expensive, difficult to sustain, and compromised by vector resistance to insecticides and by pathogen resistance to medications. In addition to possible long-lasting health impacts, outbreaks of vectorborne diseases are socially devastating, difficult to manage, and costly to healthcare systems and the people affected. Effective vector control measures are generally far less expensive than healthcare costs.[89]

There are two main debates concerning approaches to control vectorborne disease: (a) one between eradication and control and (b) the other between preventive and reactive control. Eradication of the vector[90] or the pathogen[91] after its introduction into a new area is costly, but possible. Eradication can save future generations from disease and eliminate the endless cost of control—although success can be difficult to maintain if there are repeated introductions, such as malaria into Sri Lanka.[92] However, as seen globally for malaria and dengue, eradication from large endemic areas is strategically and logistically difficult, increasingly costly to complete, and likely to revert to ongoing management[93] until socioeconomic conditions improve.[94]

When eradication is deemed impossible, management programs often are faced with the choice of performing preventive or reactive intervention. Because of the time required to scale up widespread vaccination or vector control programs, public health officials must make decisions based on predictive models, antecedent climate information, and limited surveillance data. If intervention is successful and there are no cases, then the need for the expensive intervention may be questioned. If, on the other hand, intervention is reactive and delayed until human cases are detected, then it is often impossible to prevent many of the cases,[95] and the efficacy of the response may be questioned.

With more people clustered in large cities, a changing environment facilitating transmission, and rapid travel linking populations,[96] vectorborne pathogens will remain a wildlife, veterinary, and public health problem of global proportions for years to come. Although molecular biology has produced new rapid and sensitive tools to track pathogens and modern computers have improved models, informatics, data management, and mapping, intervention technology has not kept pace, and prevention and control still rely on bed nets and spraying of insecticides in homes—technologies that have been used for decades.[97] Although vaccines hold great promise, there have been serious problems with vaccine development and with the evaluation trials that are necessary for licensing. For example, malaria and dengue vaccines have been under development for more than 50 years but still have not been widely deployed.

We are beginning to understand the impacts that global population growth and associated anthropogenic changes are having on ecosystems and the transmission of vectorborne pathogens, but these challenges are difficult to overcome, in part because there is no coherent international agreement on how to manage them. Although some countries have slowed population growth, the trajectory in global human population increase is a concern, and many people throughout the world live in areas of increasing population density, often with little or no sanitation. Unless current global patterns change,[94] the most vulnerable populations with the greatest per-capita growth rates in low-income countries will continue to bear the greatest burden of vectorborne diseases.

SUMMARY POINTS

- Vectorborne diseases result from infection with pathogens that have intricate transmission cycles in which pathogens are transmitted between humans or other vertebrate hosts by arthropod vectors. These cycles are strongly affected by climate and other environmental conditions.
- The effects of climate on vectorborne diseases are complex. Warming temperatures, changing precipitation patterns, and increasing frequency of extreme weather events have resulted in the emergence or reemergence of many vectors and vectorborne diseases in recent decades.
- Warming temperatures enable earlier and longer seasons of annual vector activity and have enabled the expansion of important tick and mosquito vectors into temperate latitudes.
- Effects of climate change have coincided with other major anthropogenic changes, such as urbanization and globalized travel and trade, which increase contact between humans and vectors and enable long-range vector and pathogen movement into new areas.
- Prevention of vectorborne diseases relies on early detection of emerging threats through surveillance and timely data sharing. Climate-aware models that can reliably predict outbreak risk hold promise for guiding future intervention.

References

1. Masson-Delmotte V, Zhai P, Pirani A, et al. (eds.). Climate Change 2021: The Physical Science Basis. Contribution of Working Group I to the Sixth Assessment Report of the Intergovernmental Panel on Climate Change. Cambridge: Cambridge University Press, 2021. Available at: https://www.ipcc.ch/report/ar6/wg1/. Accessed September 13, 2022.
2. Kilpatrick AM, Randolph SE. Drivers, dynamics, and control of emerging vectorborne zoonotic diseases. Lancet 2012; 380: 1946–1955.
3. O'Neill BC, Dalton M, Fuchs R, et al. Global demographic trends and future carbon emissions. Proceedings of the National Academy of Sciences 2010; 107: 17521–17526.
4. Hansen J, Sato M, Ruedy R, et al. Global temperature change. Proceedings of the National Academy of Sciences 2006; 103: 14288–14293.
5. Franzke CLE, Barbosa S, Blender R, et al. The structure of climate variability across scales. Reviews of Geophysics 2020; 58. https://doi.org/10.1029/2019RG000657.
6. Oke TR. The energetic basis of the urban heat island. Quarterly Journal of the Royal Meteorological Society 1982; 108: 1–24.
7. Reisen WK. Landscape epidemiology of vector-borne diseases. Annual Review of Entomology 2009; 55: 461–483.
8. Meyer RP, Hardy JL, Reisen WK. Diel changes in adult mosquito microhabitat temperatures and their relationship to the extrinsic incubation of arboviruses in

mosquitoes in Kern County, California. Journal of Medical Entomology 1990; 27: 607–614.

9. Walton WE, Reisen WK. Influence of climate change on mosquito development and blood-feeding patterns. In SK Singh (ed.). Viral Infections and Global Change. Hoboken, NJ: John Wiley & Sons, 2013, pp. 35–56.

10. Reisen WK, Fang Y, Martinez VM. Effects of temperature on the transmission of West Nile virus by *Culex tarsalis* (Diptera: Culicidae). Journal of Medical Entomology 2006; 43: 309–317.

11. Detinova TS. Age-grading methods in Diptera of medical importance. World Health Organization Monograph Series 1962; 47: 1–216.

12. Hartley DM, Barker CM, Le Menach A, et al. Effects of temperature on emergence and seasonality of West Nile virus in California. American Journal of Tropical Medicine and Hygiene 2012; 86: 884–894.

13. Brown HE, Comrie AC, Drechsler DM, et al. Human health. In G Garfin, A Jardine, R Merideth, et al. (eds.). Assessment of Climate Change in the Southwest United States. A report prepared for the National Climate Assessment. Washington, DC: Island Press, 2013. pp. 312–339.

14. Barker CM, Reisen WK. Mosquito-borne diseases. In: L Mazur, C Milanes, K Randles. (eds.). Indicators of Climate Change in California. Sacramento: California Environmental Protection Agency, Office of Environmental Health Hazards Assessment, 2012, pp. 30–41.

15. Hoover KC, Barker CM. West Nile virus, climate change, and circumpolar vulnerability. WIREs Climate Change 2016; 7: 283–300.

16. Garrett-Jones C. Prognosis for interruption of malaria transmission through assessment of the mosquito's vectorial capacity. Nature 1964; 204: 1173–1175.

17. Denlinger DL, Armbruster PA. Mosquito diapause. Annual Review of Entomology 2014; 59: 73–93.

18. Hanson SM, Craig GB Jr. Cold acclimation, diapause, and geographic origin affect cold hardiness in eggs of *Aedes albopictus* (Diptera: Culicidae). Journal of Medical Entomology 1994; 31: 192–201.

19. Hanson SM, Craig GB Jr. *Aedes albopictus* (Diptera: Culicidae) eggs: Field survivorship during northern Indiana winters. Journal of Medical Entomology 1995; 32: 599–604.

20. Kraemer MUG, Sinka ME, Duda KA, et al. The global distribution of the arbovirus vectors Aedes aegypti and Ae. albopictus. elife 2015; 4: e08347. https://doi.org/10.7554/eLife.08347.

21. Hawley WA. The biology of *Aedes albopictus*. Journal of the American Mosquito Control Association Supplement 1988; 1: 1–39.

22. Armbruster PA. Photoperiodic diapause and the establishment of *Aedes albopictus* (Diptera: Culicidae) in North America. Journal of Medical Entomology 2016; 53: 1013–1023.

23. Bryan CS, Moss SW, Kahn RJ. Yellow fever in the Americas. Infectious Disease Clinics of North America 2004; 18: 275–292.

24. Lounibos LP. Invasions by insect vectors of human disease. Annual Review of Entomology 2002; 47: 233–266.

25. Spielman A, Wong J. Environmental control of ovarian diapause in *Culex pipiens*. Annals of the Entomology Society of America 1973; 66: 905–907.

26. Eldridge BF. The effect of temperature and photoperiod on blood-feeding and ovarian development in mosquitoes of the *Culex pipiens* complex. American Journal of Tropical Medicine and Hygiene 1968; 17: 133–140.

27. Kothera L, Zimmerman EM, Richards CM, Savage HM. Microsatellite characterization of subspecies and their hybrids in *Culex pipiens* complex (Diptera: Culicidae) mosquitoes along a north-south transect in the central United States. Journal of Medical Entomology 2009; 46: 236–248.

28. Nelms BM, Kothera L, Thiemann T, et al. Phenotypic variation among *Culex pipiens* complex (Diptera: Culicidae) populations from the Sacramento Valley, California: Horizontal and vertical transmission of West Nile virus, diapause potential, autogeny, and host selection. American Journal of Tropical Medicine and Hygiene 2013; 89: 1168–1178.

29. Andreadis TG, Shepard JJ, Thomas MC. Field observations on the overwintering ecology of Culiseta melanura in the northeastern USA. Journal of the American Mosquito Control Association 2012; 28: 286–291.

30. Ogden NH, Ben Beard C, Ginsberg HS, Tsao JI. Possible effects of climate change on ixodid ticks and the pathogens they transmit: Predictions and observations. Journal of Medical Entomology 2021; 58: 1536–1545.

31. Brault AC, Reisen WK. Environmental perturbations that influence arboviral host range. In: SK Singh. (ed.). Viral Infections and Global Change. Hoboken, NJ: John Wiley & Sons, 2013, pp. 57–75.

32. Revkin AC. Heat wave toll climbs to 27 dead in New York City. New York Times, July 10, 1999.

33. Kramer LD, Styer LM, Ebel GD. A global perspective on the epidemiology of West Nile virus. Annual Review of Entomology 2008; 53: 61–81.

34. NOAA Physical Sciences Laboratory. NOAA/NCEI U.S. Climate division data: Plotting and analysis. Available at: https://psl.noaa.gov/data/usclimdivs/. Accessed September 13, 2022.

35. California Department of Public Health. Statistics & maps. Available at: https://www.cdc.gov/westnile/statsmaps/index.html. Accessed September 13, 2022.

36. Reisen WK. Effect of temperature on *Culex tarsalis* (Diptera: Culicidae) from the Coachella and San Joaquin Valleys of California. Journal of Medical Entomology 1995; 32: 636–645.

37. European Centre for Disease Prevention and Control. Surveillance and disease data for West Nile virus infections. Available at: https://www.ecdc.europa.eu/en/west-nile-fever/surveillance-and-disease-data. Accessed September 13, 2022.

38. Reisen W. Ecology of West Nile virus in North America. Viruses 2013; 5: 2079–2105.

39. Paz S, Malkinson D, Green MS, et al. Permissive summer temperatures of the 2010 European West Nile fever upsurge. PLoS One 2013; 8:e56398. https://doi.org/10.1371/journal.pone.0056398.

40. Marcantonio M, Rizzoli A, Metz M, et al. Identifying the environmental conditions favouring West Nile virus outbreaks in Europe. PLoS One 2015; 10:e0121158. https://doi.org/10.1371/journal.pone.0121158.

41. Marini G, Manica M, Delucchi L, et al. Spring temperature shapes West Nile virus transmission in Europe. Acta Tropica 2021; 215:105796.

42. Molaei G, Thomas MC, Muller T, et al. Dynamics of vector-host interactions in avian communities in four eastern equine encephalitis virus foci in the northeastern U.S.

PLoS Neglected Tropical Diseases 2016; 10: e0004347. https://doi.org/10.1371/jour nal.pntd.0004347.

43. Stobierski MG, Signs K, Dinh E, et al. Eastern equine encephalomyelitis in Michigan: Historical review of equine, human, and wildlife involvement, epidemi- ology, vector associations, and factors contributing to endemicity. Journal of Medical Entomology 2022; 59: 27–40.

44. Corrin T, Ackford R, Mascarenhas M, et al. Eastern equine encephalitis virus: A scoping review of the global evidence. Vector Borne Zoonotic Diseases 2021; 21: 305–320.

45. Armstrong PM, Andreadis TG. Ecology and epidemiology of eastern equine enceph- alitis virus in the northeastern United States: An historical perspective. Journal of Medical Entomology 2022 12; 59: 1–13.

46. Armstrong PM, Andreadis TG, Anderson JF, et al. Tracking eastern equine enceph- alitis virus perpetuation in the northeastern United States by phylogenetic analysis. American Journal of Tropical Medicine and Hygiene 2008; 79: 291–296.

47. Crans WJ, Cassamise DF, McNelly JR. Eastern equine encephalomyelitis in relation to the avian community of a coastal cedar swamp. Journal of Medical Entomology 1994; 31: 711–728.

48. Heberlein-Larson LA, Tan Y, Stark LM, et al. Complex epidemiological dynamics of eastern equine encephalitis virus in Florida. American Journal of Tropical Medicine and Hygiene 2019; 100: 1266–1274.

49. Bird BH, Ksiazek TG, Nichol ST, MacLachlan NJ. Rift Valley fever virus. Journal of the American Veterinary Medical Association 2009; 234: 883–893.

50. Meegan JM, Bailey CL. Rift Valley fever. In: TP Monath (ed.). The Arboviruses: Epidemiology and Ecology. Boca Raton, FL: CRC Press, 1989, pp. 51–76.

51. Linthicum KJ, Davies FG, Kairo A, Bailey CL. Rift Valley fever virus (family Bunyaviridae, genus Phlebovirus). Isolations from Diptera collected during an inter- epizootic period in Kenya. Journal of Hygiene 1985; 95: 197–209.

52. Logan TM, Linthicum KJ, Thande PC, et al. Egg hatching of *Aedes* mosquitoes during successive floodings in a Rift Valley fever endemic area in Kenya. Journal of the American Mosquito Control Association 1991; 7: 109–112.

53. Bird BH, Githinji JWK, Macharia JM, et al. Multiple virus lineages sharing recent common ancestry were associated with a large Rift Valley Fever outbreak among live- stock in Kenya during 2006–2007. Journal of Virology 2008; 82: 11152–11166.

54. Turell MJ, Presley SM, Gad AM, et al. Vector competence of Egyptian mosquitoes for Rift Valley fever virus. American Journal of Tropical Medicine and Hygiene 1996; 54: 136–139.

55. Linthicum KJ, Bailey CL, Tucker CJ, et al. Application of polar-orbiting, meteorolog- ical satellite data to detect flooding of Rift Valley Fever virus vector mosquito habitats in Kenya. Medical and Veterinary Entomology 1990; 4: 433–438.

56. Linthicum KJ, Anyamba A, Tucker CJ, et al. Climate and satellite indicators to fore- cast Rift Valley fever epidemics in Kenya. Science 1999; 285: 397–400.

57. Anyamba A, Chretien J-P, Small J, Linthicum KJ. Prediction of a Rift Valley fever out- break. Proceedings of the National Academy of Sciences 2009; 106: 955–959.

58. Estrada-Peña A, Hubálek Z, Rudolf I. Tick-transmitted viruses and climate change. In: SK Singh (ed.). Viral Infections and Global Change. Hoboken, NJ: John Wiley & Sons, 2013, pp. 573–602.

59. Nuttall PA, Jones LD, Labuda M, Kaufman WR. Adaptations of arboviruses to ticks. Journal of Medical Entomology 1994; 31: 1–9.

60. Labuda M, Jones LD, Williams T, et al. Efficient transmission of tick-borne encephalitis virus between cofeeding ticks. Journal of Medical Entomology 1993; 30: 295–299.

61. Randolph SE. Transmission of tick-borne pathogens between co-feeding ticks: Milan Labuda's enduring paradigm. Ticks and Tick-borne Diseases 2011; 2: 179–182.

62. Léger E, Vourc'h G, Vial L, et al. Changing distributions of ticks: Causes and consequences. Experimental and Applied Acarology 2013; 59: 219–244.

63. Jaenson TGT, Hjertqvist M, Bergström T, Lundkvist A. Why is tick-borne encephalitis increasing? A review of the key factors causing the increasing incidence of human TBE in Sweden. Parasites & Vectors 2012; 5:184. https://doi.org/10.1186/1756-3305-5-184.

64. Medlock JM, Hansford KM, Bormane A, et al. Driving forces for changes in geographical distribution of Ixodes ricinus ticks in Europe. Parasites & Vectors 2013; 6:1. https://doi.org/10.1186/1756-3305-6-1.

65. Ogden NH, Maarouf A, Barker IK, et al. Climate change and the potential for range expansion of the Lyme disease vector Ixodes scapularis in Canada. International Journal of Parasitology 2006; 36: 63–70.

66. Reisen WK, Fang Y, Brault AC. Limited interdecadal variation in mosquito (Diptera: Culicidae) and avian host competence for Western equine encephalomyelitis virus (Togaviridae: Alphavirus). American Journal of Tropical Medicine and Hygiene 2008; 78: 681–686.

67. Logue CH, Bosio CF, Welte T, et al. Virulence variation among isolates of western equine encephalitis virus in an outbred mouse model. Journal of General Virology 2009; 90: 1848–1858.

68. Cordero EC, Kessomkiat W, Abatzoglou J, Mauget SA. The identification of distinct patterns in California temperature trends. Climate Change 2011; 108: 357–382.

69. Hardy JL, Meyer RP, Presser SB, Milby MM. Temporal variations in the susceptibility of a semi-isolated population of Culex tarsalis to peroral infection with western equine encephalomyelitis and St. Louis encephalitis viruses. American Journal of Tropical Medicine and Hygiene 1990; 42: 500–511.

70. Reisen WK, Hardy JL, Presser SB, Chiles RE. Seasonal variation in the vector competence of Culex tarsalis (Diptera: Culicidae) from the Coachella Valley of California for western equine encephalomyelitis and St. Louis encephalitis virus. Journal of Medical Entomology 1996; 33: 433–437.

71. Reisen WK, Meyer RP, Presser SB, Hardy JL. Effect of temperature on the transmission of western equine encephalomyelitis and St. Louis encephalitis viruses by Culex tarsalis (Diptera: Culicidae). Journal of Medical Entomology 1993; 30: 151–160.

72. Kramer LD, Hardy JL, Presser SB. Effect of temperature of extrinsic incubation on the vector competence of Culex tarsalis for western equine encephalomyelitis virus. American Journal of Tropical Medicine and Hygiene 1983; 32: 1130–1139.

73. Reisen WK, Hardy JL, Presser SB. Effects of water quality on the vector competence of Culex tarsalis (Diptera: Culicidae) for western equine encephalomyelitis (Togaviridae) and St. Louis encephalitis (Flaviviridae) viruses. Journal of Medical Entomology 1997; 34: 631–643.

74. Hardy JL, Reeves WC. Experimental studies on infection in vectors. In WC Reeves (ed.). Epidemiology and Control of Mosquito-Borne Arboviruses in California,

1943–1987. Sacramento: California Mosquito Vector Control Association, 1990, pp. 145–250.

75. Bergren NA, Haller S, Rossi SL, et al. "Submergence" of Western equine encephalitis virus: Evidence of positive selection argues against genetic drift and fitness reductions. PLoS Pathogens 2020; 16: e1008102. https://doi.org/10.1371/journal.ppat.1008102.

76. Beard CB, Eisen RJ, Barker CM, et al. Vectorborne diseases. In AJ Crimmins, J Balbus, JL Gamble, et al. (eds.). The Impacts of Climate Change on Human Health in the United States: A Scientific Assessment. Washington, DC: U.S. Global Change Research Program, 2016. pp. 129–156. Available at: https://health2016.globalchange.gov/. Accessed September 20, 2022.

77. Weaver SC, Reisen WK. Present and future arboviral threats. Antiviral Research 2010; 85: 328–345.

78. Field CB, Barros VR, Dokken DJ, et al. (eds.). Climate Change 2014: Impacts, Adaptation, and Vulnerability. Working Group II Contribution to the Fifth Assessment Report of the International Panel on Climate Change. 2014. Available at: https://www.ipcc.ch/report/ar5/wg2/. Accessed September 20, 2022.

79. Fischer D, Thomas SM, Suk JE, et al. Climate change effects on Chikungunya transmission in Europe: Geospatial analysis of vector's climatic suitability and virus' temperature requirements. International Journal of Health Geographics 2013; 12: 51. https://doi.org/10.1186/1476-072X-12-51.

80. Rappole JH, Compton BW, Leimgruber P, et al. Modeling movement of West Nile virus in the Western hemisphere. Vector-Borne and Zoonotic Diseases 2006; 6: 128–139.

81. Stoddard ST, Forshey BM, Morrison AC, et al. House-to-house human movement drives dengue virus transmission. Proceedings of the National Academy of Sciences 2013; 110: 994–999.

82. Stoddard ST, Morrison AC, Vazquez-Prokopec GM, et al. The role of human movement in the transmission of vector-borne pathogens. PLoS Neglected Tropical Diseases 2009; 3: e481. https://doi.org/10.1371/journal.pntd.0000481.

83. Metzger ME, Hardstone Yoshimizu M, Padgett KA, et al. Detection and establishment of *Aedes aegypti* and *Aedes albopictus* (Diptera: Culicidae) mosquitoes in California, 2011–2015. Journal of Medical Entomology 2017; 54: 533–543.

84. Juliano SA, O'Meara GF, Morrill JR, Cutwa MM. Desiccation and thermal tolerance of eggs and the coexistence of competing mosquitoes. Oecologia 2002; 130: 458–469.

85. VectorSurv. Vectorborne disease surveillance system. Available at: https://vectorsurv.org/. Accessed July 31, 2022.

86. Barker CM. Models and surveillance systems to detect and predict West Nile virus outbreaks. Journal of Medical Entomology 2019; 56: 1508–1515. http://dx.doi.org/10.1093/jme/tjz150

87. Gubler DJ, Trent DW. Emergence of epidemic dengue/dengue haemorrhagic fever as a public health problem in the Americas. Infectious Agents and Disease 1994; 2: 383–393.

88. California Department of Public Health, Mosquito and Vector Control Association of California, University of California. California mosquito-borne virus surveillance and response plan. June 2022. Available at: https://westnile.ca.gov/pdfs/CAMosquitoSurveillanceResponsePlan.pdf. Accessed September 20, 2022.

89. Barber LM, Schleier JJ 3rd, Peterson RKD. Economic cost analysis of West Nile virus outbreak, Sacramento County, California, USA, 2005. Emerging Infectious Diseases 2010; 16: 480–486.

90. Packard RM, Gadelha P. A land filled with mosquitoes: Fred L. Soper, the Rockefeller Foundation, and the Anopheles gambiae invasion of Brazil. Parassitologia 1994; 36: 197–213.

91. Gray HF, Fontaine RE. A history of malaria in California. Proceedings and Papers of the Annual Conference of the California Mosquito Control Association 1957; 25: 18–39.

92. Galappaththy GNL, Fernando SD, Abeyasinghe RR. Imported malaria: A possible threat to the elimination of malaria from Sri Lanka? Tropical Medicine & International Health 2013; 18: 761–768.

93. Guerra CA, Gikandi PW, Tatem AJ, et al. The limits and intensity of *Plasmodium falciparum* transmission: Implications for malaria control and elimination worldwide. PLoS Medicine 2008; 5: e38. doi: 10.1371/journal.pmed.0050038.

94. Bonds MH, Dobson AP, Keenan DC. Disease ecology, biodiversity, and the latitudinal gradient in income. PLoS Biology 2012; 10: e1001456. https://doi.org/10.1371/journal.pbio.1001456.

95. Carney RM, Husted S, Jean C, et al. Efficacy of aerial spraying of mosquito adulticide in reducing incidence of West Nile Virus, California, 2005. Emerging Infectious Diseases 2008; 14: 747–754.

96. Tatem AJ. The worldwide airline network and the dispersal of exotic species: 2007–2010. Ecography 2009; 32: 94–102.

97. Reisen WK. Medical entomology: Back to the future? Infection, Genetics and Evolution 2013; 28: 573–582.

Further Reading

Brault AC, Reisen WK. Environmental perturbations that influence arboviral host range: Insights into emergence mechanisms. In: SK Singh (ed.). Viral Infections and Global Change. Hoboken, NJ: John Wiley and Sons, 2014, pp. 57–75.

Case studies describe how environmental change due to the impact of human population increase has altered the landscape and climate, enabling the transmission of arboviral pathogens.

Beard CB, Eisen RJ, Barker CM, et al. Vectorborne diseases. In: AJ Crimmins, JL Balbus, CB Gamble, et al. The Impacts of Climate Change on Human Health in the United States: A Scientific Assessment. Washington, DC: U.S. Global Change Research Program, 2016, pp. 129–156. doi:10.7930/J0765C7V.

This chapter describes observed and anticipated changes in vectorborne diseases in the United States that are attributable to climate change. The information includes traceable accounts from the literature to support key findings, featuring Lyme disease and West Nile virus as case studies.

Paz S. Climate change impacts on West Nile virus transmission in a global context. Philosophical Transactions of the Royal Society B: Biological Sciences 2015; 370. doi:10.1098/rstb.2013.0561.

This paper describes how recent changes in temperature and rainfall have enabled global expansion and persistence of West Nile virus.

Anyamba A, Linthicum KJ, Small JL, et al. Climate teleconnections and recent patterns of human and animal disease outbreaks. PLoS Neglected Tropical Diseases 2012; 6: e1465. doi: 10.1371/journal.pntd.0001465.

This paper details how the El Niño-Southern Oscillation drives climatic changes that lead to extreme weather, including rainfall or drought in East Africa, which enable outbreaks of chikungunya and Rift Valley fever.

Reisen WK. Landscape epidemiology of vector-borne diseases. Annual Review of Entomology 2010; 55: 461–483.

This paper describes how the temporal dynamics of host, vector, and pathogen populations interact spatially within a permissive environment to enable vectorborne disease transmission.

7

Waterborne Diseases

Jennifer R. Bratburd and Sandra L. McLellan

In 1993, Milwaukee experienced its heaviest rainfall in 50 years. As a result, the protozoan *Cryptosporidium* entered the water treatment plant from the Lake Michigan source water and, because of a series of failures at the plant, was not removed in the treatment process. These events caused an outbreak of cryptosporidiosis affecting more than 400,000 people and costing an estimated $90 million in healthcare expenses and productivity losses. An estimated 69 deaths occurred during the outbreak, mainly among immunocompromised individuals, such as people with AIDS.[1-3]

In January 2010, a 7.0-magnitude earthquake struck Haiti, killing more than 300,000 people, and leaving over a million homeless. In October 2010, United Nations peacekeeping troops who came to Haiti from Nepal set up a camp lacking proper sewage disposal and, in the process, brought cholera to Haiti. Over the next 10 years, a cholera outbreak with more than 820,000 cases and more than 10,000 deaths occurred. By 2015, Haiti had the highest rate of cholera in the world.[4] In 2016, in the midst of the outbreak, Hurricane Matthew hit Haiti, stressing water infrastructure, worsening sanitation, and creating a huge increase in cholera cases.[4-6]

These two outbreaks of waterborne disease, although different in many respects, illustrate how climate change can cause and contribute to waterborne disease. Extreme weather events, like hurricanes, combined with inadequate infrastructure can lead to waterborne disease with substantial health consequences.

This chapter covers climate impacts on waterborne disease, including impacts from increased extreme weather events, increased rainfall and flooding (Figure 7-1), decreased rainfall and drought, sea-level rise, and increased ambient temperature—all of which can increase pathogens in drinking water and recreational water.[7] It describes various waterborne pathogens, some of which occur in the environment (such as *Vibrio cholerae*) and some of which are mainly associated with humans or other animals (such as norovirus and *Salmonella* species).

Jennifer R. Bratburd and Sandra L. McLellan, *Waterborne Diseases* In: *Climate Change and Public Health*. Edited by: Barry S. Levy and Jonathan A. Patz, Oxford University Press. © Oxford University Press 2024. DOI: 10.1093/oso/9780197683293.003.0007

Figure 7-1 Women carrying their belongings salvaged from their flooded home after monsoon rains in Sindh Province, Pakistan, September 2022. During the flood, more than 1,300 people were killed and millions of people lost their homes in flooding caused by unusually heavy monsoon rains that were attributed to climate change. (AP Photo/Fareed Khan.)

MAGNITUDE AND IMPACT

Diarrhea, a primary symptom of many waterborne diseases, is the eighth leading cause of death globally, causing about 1.5 million deaths annually.[8] Two of the leading risk factors for diarrhea are unsafe water and unsafe sanitation. Approximately 2.0 billion people lack safely managed drinking water, and 3.6 billion people lack safely managed sanitation services.[9,10]

Many cases of waterborne disease are not recognized or reported. For example, a study in Ontario estimated that for every case of infectious gastrointestinal illness reported, the median number of community cases was 285.[11] The U.S. Centers for Disease Control and Prevention (CDC) has estimated that about 7.2 million cases of waterborne disease occur in the United States annually, associated with $3.3 billion in direct healthcare costs.[12] Others estimate that there are between 12 and 19 million incident cases of waterborne disease due to contaminated drinking water in the United States annually.[13–15] In addition, it has been estimated that there could be in the United States annually as many as 90 million cases of recreational waterborne disease—gastrointestinal,

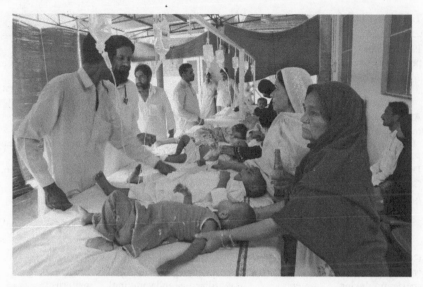

Figure 7-2 Children being treated for waterborne disease during severe flooding in Pakistan in 2010. (AP Images/Khalid Tanveer.)

respiratory, ear, eye, and skin-related illnesses with a wide range of severity related to swimming/wading, paddling, motorboating, and fishing in surface waters.[16]

Those at highest risk of waterborne disease include children (Figure 7-2), older people, pregnant women, and people who are immunocompromised. Communities lacking consistent access to safely managed drinking water are at increased risk.[17,18]

CAUSATIVE AGENTS OF DISEASE

A wide variety of microorganisms cause waterborne disease, including enteric viruses, enteric bacteria, protozoan parasites, and toxin-producing algae (Table 7-1). To identify common patterns of transmission and risk, microbial pathogens can be ecologically categorized as host-associated or environmental. *Host-associated pathogens* are normally carried by humans and other animals and incidentally shed into the environment; often, some part of the pathogen's life cycle and transmission must occur within a host. In contrast, *environmental pathogens* spend most of their life cycles in the environment and typically end up in hosts incidentally.[19]

Table 7-1 Climate-sensitive agents of water-related illness

Pathogen or toxin producer	Exposure pathway	Selected health outcomes and symptoms	Major climate correlation or driver (strongest driver[s] listed first)
Algae: Toxigenic marine species of *Alexandrium, Pseudo-nitzschia, Dinophysis, Gambierdiscus; Karenia brevis*	Shellfish Fish Recreational waters (aerosolized toxins)	Gastrointestinal and neurologic illnesses caused by shellfish poisoning (paralytic, amnesic, diarrhetic, neurotoxic) or fish poisoning (ciguatera). Asthma exacerbation, eye irritation caused by contact with aerosolized toxins (*K. brevis*).	Temperature (increased water temperature), ocean surface currents, ocean acidification, hurricanes (*Gambierdiscus* species and *K. brevis*)
Cyanobacteria (multiple freshwater species producing toxins including microcystin)	Drinking water Recreational waters	Liver and kidney damage, gastroenteritis (diarrhea and vomiting), neurological disorders, and respiratory arrest.	Temperature, precipitation patterns
Enteric bacteria and protozoan parasites: *Salmonella enterica; Campylobacter* species; Toxigenic *Escherichia coli; Cryptosporidium; Giardia*	Drinking water Recreational waters Shellfish	Enteric pathogens generally cause gastroenteritis. Some cases may be severe and may be associated with long-term and recurring effects.	Temperature (air and water; both increase and decrease), heavy precipitation, and flooding
Enteric viruses: enteroviruses; rotaviruses; noroviruses; hepatitis A and E viruses	Drinking water Recreational waters Shellfish	Most cases result in gastrointestinal illness. Severe outcomes may include paralysis and infection of the heart or other organs.	Heavy precipitation, flooding, and temperature (air and water; both increase and decrease)
Leptospira and *Leptonema* bacteria	Recreational waters	Flu-like illness, meningitis, and kidney and liver failure.	Flooding, temperature (increased water temperature), heavy precipitation
Vibrio bacteria species	Recreational waters Shellfish	Varies by species, but includes gastroenteritis (*V. parahaemolyticus, V. cholerae*); septicemia via ingestion or wounds (*V. vulnificus*); and skin, eye, and ear infections (*V. alginolyticus*).	Temperature (increased water temperature), sea-level rise, precipitation patterns (as they affect coastal salinity)

Source: U.S. Global Change Research Program. Impacts of climate change on human health in the United States: A scientific assessment. Available at: https://health2016.globalchange.gov/. Accessed March 8, 2023.

Host-Associated Pathogens

Host-associated pathogens often are transmitted through the fecal-oral route, from animal or human waste, where they can contaminate water and food and then be ingested into the next host. These pathogens include enteric bacteria, such as species of *Campylobacter* and *Salmonella* and strains of *Escherichia coli*; species of protozoa, such as *Giardia* and *Cryptosporidium*; and enteric viruses, including enteroviruses, rotaviruses, noroviruses, and hepatitis A and hepatitis E viruses. Host-associated pathogens can be further subdivided into (a) those that are human-associated, which are often transmitted from person to person; and (b) those that are harbored in a wide range of animals, which can cause disease in animal populations or be passed to humans. (Infections transmitted from animals to humans are also called *zoonotic diseases*, or *zoonoses*.)

Management of human waste by sanitation services prevents the transmission of many waterborne diseases. Contamination of waterways by untreated sewage may be the exposure pathway that poses the highest risk to people. Particular risks associated with climate change include inadequate sanitation and disruption of sanitation due to poor maintenance or extreme weather events.

Agricultural waste is a major source of contamination, especially for host-associated pathogens because livestock can often carry and shed many pathogens, including species of *Campylobacter* and *Salmonella* as well as strains of *Escherichia coli* that cause human disease. Cattle can also carry the parasitic protozoan *Cryptosporidium*. Large livestock operations (*concentrated animal feeding operations*, or CAFOs), can produce between 2,800 and 1,600,000 tons of manure per year. Overall, livestock create three to five times more manure than human sanitary waste.[20,21] Runoff, especially in floods, can move contaminants into nearby water bodies, representing a risk to recreational water and sources of drinking water.[18] A study on industrial hog farms in North Carolina found an 11% increase in acute gastrointestinal illness in areas with high hog waste exposure, with a stronger association during the week following heavy rain.[22] Agricultural contamination from animal waste and fertilizer also introduces excess nutrients, which can lead to harmful algal blooms.

While waste from humans and animals is of highest concern, wildlife can also contribute to contamination. Many strains of waterborne pathogens have been isolated in animals, including strains of *Cryptosporidium* and *Giardia*,[23] but it is not clear how often wildlife transmit to humans—and how often humans and livestock transmit these pathogens to wildlife.[24] A study found that 97% of *Campylobacter jejuni* infections in humans are attributable to livestock.[25] Even when humans and livestock are the source of pathogens, these pathogens can circulate in an animal reservoir and continue to contaminate bodies of water. Climate change is expected to affect where wildlife range, and it may affect pathogens carried by wildlife. And both the changing locations in which wild-life range and increasing urbanization are expected to increase opportunities for transmission of pathogens from wildlife to humans.[26,27]

Environmental Pathogens

Pathogens that propagate in the environment and can cause disease (especially in water) include *Pseudomonas aeruginosa*, species of *Legionella* and *Vibrio,* and some species of *Mycobacterium*. Some environmental microorganisms do not need to infect hosts to cause harm; they can produce toxins, such as those from harmful algal blooms (HABs) that ultimately can cause disease in humans and other animals. HABs can be divided into marine (seawater) toxic-producing species and freshwater species. Marine toxin-producing algae include *Karenia brevis* and species of *Alexandrium*, *Pseudo-nitzschia*, *Dinophysis*, and *Gambierdiscus*. In freshwater, toxins are produced by species of *Cyanobacteria* (blue-green algae). Exposure to cyanotoxins or marine algal toxins can occur through ingestion of water, ingestion of fish or shellfish, and breathing small droplets of water.[18]

Additional Contaminants of Water

Chemical contaminants in water, including lead, arsenic, pesticides, excess nutrients (such as nitrogen and phosphorus), and pharmaceuticals, can cause disease. These contaminants may be affected by climate change. Increased rain-fall can cause runoff or sewage overflow, which can increase human exposure to these chemicals. So can breakdowns in drinking water systems.

Exposure to waterborne pathogens and hazardous chemicals may occur via ingestion of contaminated food or drinking water, ingestion of (or contact with) recreational water, and inhalation of harmful algal toxins. Foodborne disease can occur when food has been grown or prepared with water contaminated with microorganisms or chemicals.

IMPACTS OF CLIMATE CHANGE ON WATERBORNE DISEASE

The incidence of waterborne disease is increasing because of increased occur-rence of extreme weather events, increased rainfall and flooding, decreased rain-fall and drought, sea-level rise, and increased ambient temperature. All of these impacts can increase human exposure to pathogens.[28,29]

Extreme Weather Events, Increased Rainfall, and Flooding

Climate change is affecting the water cycle. More intense precipitation events are expected to occur (Figure 7-3). Intense rainfall has already occurred in North America, Europe, and Asia.[30] Increased rainfall is associated with increased incidence of (and increased hospitalization for) gastrointestinal illnesses.[31-35] Extreme rainfall can introduce pathogenic viruses, bacteria, and protozoa into surface water and groundwater, such as by overwhelming sewage treatment infrastructure. For example, widespread contamination with human sewage has been detected after high rainfall in a Lake Michigan estuary, which is a pattern that likely occurs in most major cities (Figure 7-4).[36] Rates of diarrhea have been associated with floods.[37]

After hurricanes Katrina and Rita hit Louisiana and nearby areas in 2005, more than 1,000 drinking water systems were adversely affected. So much damage was done that, a month after Hurricane Rita, 100,000 people were still advised to boil their water because water supply systems were not operating. Some neighborhoods had no safe drinking water for more than a year.[38]

Rainfall can increase transport of pathogens in soil into surface water or groundwater. This transport can be affected by the type of precipitation (rain, snow, or snow melt) and by the degree of dryness preceding precipitation. It has been hypothesized that during dry periods pathogens accumulate in soil, and intense precipitation following dry periods increases the number of pathogens in water and the risk of waterborne disease.[33,37,39]

Rainfall increases the turbidity of drinking water, which is associated with acute gastrointestinal illness. This association may be due to increased delivery of bacteria and particulate matter from soil and sewers into water, transport of pathogens attached to the particulate matter, protection of pathogens from inactivation by particulate matter (such as by blocking ultraviolet light), and blockages in filtration systems.[40,41]

Another way rainfall can increase the burden of pathogens in water is by overwhelming water and sewage infrastructure. Even in countries with extensive sewer infrastructure, much of it was not designed to last for many decades—and little or none of it was designed for climate change.[42] When systems that combine rainwater runoff, sewage, and industrial wastewater are inundated with rain, they can overflow and release the runoff and untreated wastewater together in what are known as *combined sewage outflow events*.[37] Studies have documented that higher bacterial levels that occur during heavy rainfall events often can be attributed to the release of untreated sewage.[43,44]

Private on-site wastewater treatment systems (septic systems) and private wells are often unregulated and vulnerable to heavy rainfall and resultant contamination. Septic systems generally have drainfields (leach fields), in which

(A)

(B)

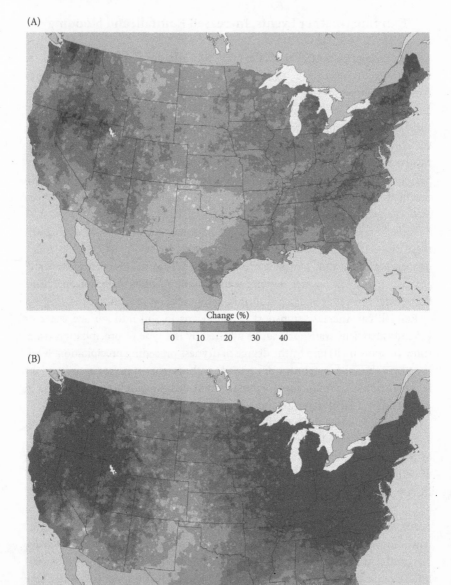

Figure 7-3 Predicted future changes by late 21st century in heavy precipitation events—bouts of heavy rain or snow ranking among the top 1% (99th percentile) of daily events in the United States — (A) under lower emissions and (B) under higher emissions. Percentage changes below zero (decreases) are white, and increases appear increasingly darker. (*Source*: National Oceanic and Atmospheric Administration.)

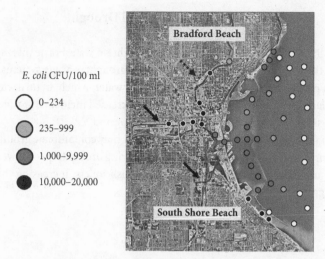

Figure 7-4 *Escherichia coli* levels in nearshore Lake Michigan, off the urbanized coast of Milwaukee. *E. coli* is used as an indicator of fecal pollution from animals and humans. One of the three rivers on the map receives agricultural runoff (dashed arrow), while the other two rivers receive urban stormwater (solid arrows). When there is heavy rainfall, combined sewer overflows also occur. Nearby beaches, such as South Shore Beach and Bradford Beach, can be impacted by contaminated river water and by localized sources of pollution. (CFU: Colony-forming units) (*Source*: Adapted from Google Earth Imagery, April 1, 2005 to August 5, 2022. NOAA TerraMetrics.)

pretreated wastewater filters through soil. When drainfields are waterlogged, pathogens can migrate rapidly through the soil and into groundwater.[45] In areas where climate change causes groundwater levels to rise, such as in coastal areas, septic systems can become less effective, increasing the likelihood of contamination of groundwater with pathogens.[46,47]

Groundwater is the water source for 35% of water used by households globally.[48] Climate change is expected to reduce the quantity and quality of available groundwater.[49,50] Groundwater can be adversely impacted by extreme precipitation, often with long lag periods between rain events and waterborne disease.[31] Soil helps filter rainwater before it enters groundwater, but soil saturated with water allows pathogens to pass quickly, potentially limiting filtering and deactivation of pathogens. In Ohio, extreme precipitation brought about massive contamination of groundwater with microorganisms from wastewater treatment facilities and septic systems, causing an outbreak of waterborne disease that affected 1,450 people.[51] In Walkerton, Ontario, heavy rainfall, which led to contamination of well water with *Campylobacter* and *Escherichia coli* from nearby farms, in turn led to 2,300 cases of gastroenteritis, 7 of which were fatal.[52]

Decreased Rainfall and Drought

There is limited research on the effects of drought on waterborne disease, especially in the context of climate change. During drought, sewage systems cannot function properly because of inadequate flow of water, which, in turn, can result in pipe damage.[53] Several studies have found increased incidence of waterborne disease associated with precipitation after a dry period.[33,37,39]

During droughts, changes in human behavior may contaminate groundwater. For example, a study in Botswana found that, during droughts, people were more likely to use unlined pit latrines rather than flush toilets, increasing the likelihood of groundwater contamiation.[54]

Sea-Level Rise

Sea-level rise has considerable impact. About 11% of people live in low-elevation coastal zones globally. Sea-level rise magnifies the adverse impacts of extreme weather events, such as by increasing the vulnerability of water and sewage infrastructure.[55] Sea-level rise may affect water quality by increasing the salinity of groundwater, which is associated with hypertension and preeclampsia.[56-58] However, removal of large quantities of groundwater from a source and agricultural pollution of groundwater are believed to play larger roles in the salinization of groundwater than sea-level rise from climate change.[56]

Increased Ambient Temperature

For most host-associated pathogens, increased temperature is associated with diarrheal disease; however, some pathogens, such as rotavirus, may be adversely affected by higher temperature.[37] Some environmental pathogens are especially sensitive to temperature. For example, illnesses from *Vibrio* species have tripled since 1996.[59] Seawater with relatively high surface temperature promotes the growth of certain *Vibrio* species and can have higher concentrations of nutrients that *Vibrio* requires. Likewise, higher water temperatures are associated with cyanobacteria blooms.[60] (Increased availability of nutrients from runoff of animal waste and agricultural fertilizers can also contribute to algal blooms.[61-63]) For host-associated pathogens, temperature may affect host range and determine which people are at increased risk.[64,65] Heat waves can increase the number of people using recreational waters, thereby increasing potential exposure to pathogens.

PREVENTION
Sewage and Drinking Water Treatment

Globally, 3.6 billion people lack *safely managed sanitation services*, which are defined by the World Health Organization (WHO) as "provision of facilities and services for safe disposal of human urine and feces," including piped sewer systems, septic tanks, and composting toilets.[9] This situation may be getting worse; for example, a study of eight African countries found decreasing trends in access to basic sanitation services.[66] Adequate sanitation infrastructure must be built, upgraded, and maintained. Sewage systems with combined sewer overflows should be upgraded to avoid water contamination following extreme weather events.

Wastewater treatment plants are vulnerable to climate-induced extreme weather events and accompanying floods, which can lead to waterborne disease.[67] When sources of drinking water become contaminated, water treatment may need to become more intense. For example, in Vancouver, introduction of new filtration systems appeared to prevent predicted increases in incidence of cases of *Giardia* and *Cryptosporidium* from extreme rainfall events.[39,68]

Climate change may increase cyanobacterial blooms, which produce harmful algal toxins that can contaminate sources of drinking water. However, increased water treatment for Cyanobacteria and algal toxins can help reduce the risk of toxic effects. Treatment can remove 60–99.9% of Cyanobacteria.[69,70]

The American Society of Civil Engineers has given U.S. drinking water infrastructure a C– grade and its wastewater infrastructure a D+ grade.[71] Construction and maintenance of water infrastructure—water and wastewater treatment systems, dams, levees, aqueducts, and sewers—can prevent waterborne disease. Water infrastructure capacity has been based on climate conditions from as long as 60–100 years ago. Construction of new infrastructure should consider the expected increases in rainfall intensity, occurrence of extreme weather events, and temperatures. The costs of reconstruction and maintenance for U.S. drinking water infrastructure is estimated to cost trillions of dollars over 25 years.[72]

Drinking Water Systems: A Primary Exposure Route for Waterborne Pathogens

Sustainable access to affordable clean drinking water, which is essential for preventing waterborne disease, requires technological capabilities, financing, and support of government.[73]

In the United States, although the Safe Drinking Water Act regulates public drinking water, private sources of water, such as wells, are not regulated. These private sources of water, which are common in rural and low-income areas, serve approximately 15% of the U.S. population. Many people, especially those residing in underserved and low-income communities, lack resources for testing their drinking water and often do not know about the need for testing.[74]

Access to safe drinking water can be abruptly lost during extreme weather events. Planning for disasters and interruptions in water service can help prevent waterborne disease associated with climate change. In areas without treatment of drinking water or when water treatment has been rendered ineffective (such as sewage overflows during flooding), at-home water treatment is necessary. Boiling water can inactive most, but not all, pathogens.[75] Other methods to treat water include chemicals, filtration, and disinfection with ultraviolet (UV) light.[76] Limited resources and lack of accurate information may make adequate treatment of drinking water difficult.

Human Behaviors

Many behaviors can reduce the transmission of waterborne disease, including handwashing, household treatment of water, and improved water storage practices, which are especially important in low- and middle-income countries.[76] Handwashing with soap can reduce diarrheal disease. But in many countries there is limited access to clean water and soap. In some situations, other alternatives, such as alcohol-based hand sanitizers, may be helpful.[76] Educating the public on the importance of following advisories to boil water or avoid recreational water, when necessary, can help prevent waterborne disease.

Other Societal-Level Interventions

Beyond water infrastructure, many societal-level interventions can improve outcomes of waterborne disease or prevent cases altogether. Access to healthcare, certain vaccinations, adequate nutrition, and other factors can reduce the incidence of waterborne disease. And treatment can decrease the case-fatality rate; for example, treatment of cholera (primarily with oral rehydration therapy) can reduce the case-fatality rate from 50% to 1%.

Agricultural Management Practices

Agricultural management practices can reduce the risk of contaminating surface water and groundwater. Better timing of manure application and fertilizer use and improved handling of animal waste can reduce transport of pathogens and nutrients that lead to harmful algal blooms.[18,77]

Assessments and Monitoring of Recreational Water

The WHO recommends that countries perform systematic assessments and monitoring of recreational water and related public communication.[78] Communication on potentially increased health risks is important, especially after extreme weather events that cause sewage overflows or when conditions favor harmful algal blooms.[79,80]

Surveillance

Public health surveillance for waterborne disease and monitoring of the quality of drinking water and recreational water can help prevent waterborne disease. Public health surveillance can rely on reports from healthcare providers of gastrointestinal illness, laboratory reports of gastrointestinal pathogens, and other sources, including the news media and even social media.

SUMMARY POINTS

- A wide variety of host-associated and environmental pathogens can cause waterborne disease.
- Waterborne disease is being affected by changing climate factors, including extreme weather events, increased rainfall and flooding, decreased rainfall and drought, sea-level rise, and increased ambient temperature.
- Many countries and communities have inadequate infrastructure for water treatment and supply that is vulnerable to the impacts of climate change.
- Prevention of waterborne disease depends on maintaining water infrastructure, ensuring sustainable access to affordable clean drinking water, maintaining sewage and drinking water treatment, supporting

prevention based on human behaviors, improving agricultural management, monitoring recreational water, and performing public health surveillance.

- Because different pathogens are affected differently by climate change, some interventions to prevent waterborne disease need to be adapted to specific pathogens.

References

1. Patz JA, Vavrus SJ, Uejio CK, McLellan SL. Climate change and waterborne disease risk in the Great Lakes region of the U.S. American Journal of Preventive Medicine 2008; 35: 451–458.
2. Mac Kenzie WR, Hoxie NJ, Proctor ME, et al. A massive outbreak in Milwaukee of cryptosporidium infection transmitted through the public water supply. New England Journal of Medicine 1994; 331: 161–167.
3. Corso PS, Kramer MH, Blair KA, et al. Costs of illness in the 1993 waterborne cryptosporidium outbreak, Milwaukee, Wisconsin. Emerging Infectious Diseases 2003; 9: 426–431.
4. Ivers LC. Eliminating cholera transmission in Haiti. New England Journal of Medicine 2017; 376: 101–103.
5. Orata FD, Keim PS, Boucher Y. The 2010 cholera outbreak in Haiti: How science solved a controversy. PLOS Pathogens 2014; 10: e1003967.
6. Hulland E, Subaiya S, Pierre K, et al. Increase in reported cholera cases in Haiti following Hurricane Matthew: An interrupted time series model. American Journal of Tropical Medicine and Hygiene 2019; 100: 368–373.
7. Mora C, McKenzie T, Gaw IM, et al. Over half of known human pathogenic diseases can be aggravated by climate change. Nature Climate Change 2022; 12: 869–875.
8. World Health Organization. The top 10 causes of death. Available at: https://www.who.int/news-room/fact-sheets/detail/the-top-10-causes-of-death.   Accessed January 12, 2023.
9. World Health Organization, United Nations Children's Fund (UNICEF). Progress on Household Drinking Water, Sanitation, and Hygiene 2000–2020: Five Years into the SDGs [Internet]. Geneva: World Health Organization, 2021. Available at: https://apps.who.int/iris/handle/10665/345081. Accessed October 24, 2022.
10. World Health Organization, United Nations Children's Fund (UNICEF), World Bank. State of the World's Drinking Water: An Urgent Call to Action to Accelerate Progress on Ensuring Safe Drinking Water for All. Geneva: World Health Organization, 2022. Available at: https://apps.who.int/iris/handle/10665/363704. Accessed October 24, 2022.
11. Majowicz SE, Edge VL, Fazil A, et al. Estimating the under-reporting rate for infectious gastrointestinal illness in Ontario. Canadian Journal of Public Health 2005; 96: 178.
12. Collier SA, Deng L, Adam EA, et al. Estimate of burden and direct healthcare cost of infectious waterborne disease in the United States. Emerging Infectious Diseases 2021; 27: 140–149.

13. Eisenberg JNS, Hubbard A, Wade TJ, et al. Inferences drawn from a risk assessment compared directly with a randomized trial of a home drinking water intervention. Environmental Health Perspectives 2006; 114: 1199–1204.

14. Messner M, Shaw S, Regli S, et al. An approach for developing a national estimate of waterborne disease due to drinking water and a national estimate model application. Journal of Water and Health 2006; 4: 201–240.

15. Reynolds KA, Mena KD, Gerba CP. Risk of waterborne illness via drinking water in the United States. In: DM Whitacre (ed.). Reviews of Environmental Contamination and Toxicology. New York: Springer, 2008, pp. 117–158.

16. DeFlorio-Barker S, Wing C, Jones RM, Dorevitch S. Estimate of incidence and cost of recreational waterborne illness on United States surface waters. Environmental Health 2018; 17: 3.

17. Balazs CL, Ray I. The drinking water disparities framework: On the origins and persistence of inequities in exposure. American Journal of Public Health 2014; 104: 603–611.

18. Trtanj J, Jantarasami L, Brunkard J, et al. Climate Impacts on Water-Related Illness. Washington, DC: U.S. Global Change Research Program, 2016. Available at: http://dx.doi.org/10.7930/J03F4MH4. Accessed October 24, 2022.

19. Percival SL. Microbiology of Waterborne Diseases: Microbiological Aspects and Risks. London: Elsevier Academic Press, 2004.

20. U.S. Government Accountability Office. Concentrated animal feeding operations: EPA needs more information and a clearly defined strategy to protect air and water quality from pollutants of concern. September 2008. Available at: https://www.gao.gov/assets/gao-08-944.pdf. Accessed December 9, 2022.

21. Hribar C. Understanding concentrated animal feeding operations and their impact on communities. National Association of Local Boards of Health, 2010. Available at: https://www.cdc.gov/nceh/ehs/docs/understanding_cafos_nalboh.pdf. Accessed December 9, 2022.

22. Quist AJL, Holcomb DA, Fliss MD, et al. Exposure to industrial hog operations and gastrointestinal illness in North Carolina, USA. Science and the Total Environment 2022; 830: 154823. doi: 10.1016/j.scitotenv.2022.154823.

23. Appelbee AJ, Thompson RCA, Olson ME. Giardia and Cryptosporidium in mammalian wildlife – current status and future needs. Trends in Parasitology 2005; 21: 370–376.

24. Thompson RCA. Parasite zoonoses and wildlife: One health, spillover and human activity. International Journal of Parasitology 2013; 43: 1079–1088.

25. Wilson DJ, Gabriel E, Leatherbarrow AJH, et al. Tracing the source of campylobacteriosis. PLoS Genetics 2008; 4: e1000203. doi: 10.1371/journal.pgen.1000203.

26. Pauchard A, Milbau A, Albihn A, et al. Non-native and native organisms moving into high elevation and high latitude ecosystems in an era of climate change: New challenges for ecology and conservation. Biological Invasions 2016; 18: 345–353.

27. Zahedi A, Paparini A, Jian F, et al. Public health significance of zoonotic Cryptosporidium species in wildlife: Critical insights into better drinking water management. International Journal for Parasitology: Parasites and Wildlife 2016; 5: 88–109.

28. Pörtner H-O, Roberts DC, Tignor MMB, et al. (eds.). Climate Change 2022: Impacts, Adaptation, and vulnerability. Working Group II Contribution to

the Sixth Assessment Report of the Intergovernmental Panel on Climate Change. Cambridge: Cambridge University Press, 2022.

29. Levy K, Smith SM, Carlton EJ. Climate change impacts on waterborne diseases: Moving toward designing interventions. Current Environmental Health Reports 2018; 5: 272–282.

30. Caretta MA, Mukherji A, Arfanuzzaman M, et al. Water. In: H-O Pörtner, DC Roberts, MMB Tignor, et al. (eds.). Climate Change 2022: Impacts, Adaptation, and Vulnerability. Cambridge: Cambridge University Press, 2022. p. 551–712.

31. Curriero FC, Patz JA, Rose JB, Lele S. The association between extreme precipitation and waterborne disease outbreaks in the United States, 1948–1994. American Journal of Public Health 2001; 91: 1194–1199.

32. Thomas KM, Charron DF, Waltner-Toews D, et al. A role of high impact weather events in waterborne disease outbreaks in Canada, 1975–2001. International Journal of Environmental Health Research 2006; 16: 167–180.

33. Nichols G, Lane C, Asgari N, et al. Rainfall and outbreaks of drinking water related disease and in England and Wales. Journal of Water and Health 2008; 7: 1–8.

34. Gleason JA, Fagliano JA. Effect of drinking water source on associations between gastrointestinal illness and heavy rainfall in New Jersey. PLoS One 2017; 12: e0173794. doi: 10.1371/journal.pone.0173794.

35. Drayna P, McLellan SL, Simpson P, et al. Association between rainfall and pediatric emergency department visits for acute gastrointestinal illness. Environmental Health Perspectives 2010; 118: 1439–1443.

36. Olds HT, Corsi SR, Dila DK, et al. High levels of sewage contamination released from urban areas after storm events: A quantitative survey with sewage specific bacterial indicators. PLOS Med 2018; 15: e1002614. doi: 10.1371/journal.pmed.1002614.

37. Levy K, Woster AP, Goldstein RS, Carlton EJ. Untangling the impacts of climate change on waterborne diseases: A systematic review of relationships between diarrheal diseases and temperature, rainfall, flooding, and drought. Environmental Science & Technology 2016; 50: 4905–4922.

38. Patterson CL, Impellitteri C, Fox KR, et al. Emergency response for public water supplies after Hurricane Katrina. World Environmental and Water Resources Congress, Tampa, FL, 2007.

39. Chhetri BK, Takaro TK, Balshaw R, et al. Associations between extreme precipitation and acute gastro-intestinal illness due to cryptosporidiosis and giardiasis in an urban Canadian drinking water system (1997–2009). Journal of Water and Health 2017; 15: 898–907.

40. Atherholt TB, LeChevallier MW, Norton WD, Rosen JS. Effect of rainfall on giardia and crypto. Journal AWWA 1998; 90: 66–80.

41. De Roos AJ, Gurian PL, Robinson LF, et al. Review of epidemiological studies of drinking-water turbidity in relation to acute gastrointestinal illness. Environmental Health Perspectives 2017; 125:086003. doi: 10.1289/EHP1090.

42. Cutter SL, Solecki W, Bragado N, et al. Urban systems, infrastructure, and vulnerability. In Climate Change Impacts in the United States: The Third National Climate Assessment. U.S. Global Change Research Program, 2014. Available at: https://nca2014.globalchange.gov/downloads. Accessed October 24, 2022.

43. McLellan SL, Hollis EJ, Depas MM, et al. Distribution and fate of Escherichia coli in Lake Michigan following contamination with urban stormwater and combined sewer overflows. J Journal of Great Lakes Research 2007; 33: 566.

44. Sercu B, Van De Werfhorst LC, Murray JLS, Holden PA. Sewage exfiltration as a source of storm drain contamination during dry weather in urban watersheds. Environmental Science & Technology 2011; 45: 7151–7157.

45. Yates MV. On-site wastewater treatment. In JO Nriagu (ed.). Encyclopedia of Environmental Health. New York: Elsevier, 2011, pp. 256–263. Available at: https://linkinghub.elsevier.com/retrieve/pii/B9780444522726000465. Accessed October 28, 2022.

46. Mihaly E. Avoiding septic shock: How climate change can cause septic system failure and whether New England states are prepared. Ocean Coast Law Journal 2018; 23:1.

47. Lusk MG. Public health threats of diminished treatment of onsite sewage. Lancet Planet Health 2022; 6: e707–e708.

48. Döll P, Hoffmann-Dobrev H, Portmann FT, et al. Impact of water withdrawals from groundwater and surface water on continental water storage variations. Journal of Geodynamics 2012; 59–60: 143–156.

49. McDonough LK, Santos IR, Andersen MS, et al. Changes in global groundwater organic carbon driven by climate change and urbanization. Nat Communications 2020; 11:1279. doi: 10.1038/s41467-020-14946-1.

50. Rodell M, Famiglietti JS, Wiese DN, et al. Emerging trends in global freshwater availability. Nature 2018; 557: 651–659.

51. Fong T-T, Mansfield LS, Wilson DL, et al. Massive microbiological groundwater contamination associated with a waterborne outbreak in Lake Erie, South Bass Island, Ohio. Environmental Health Perspectives 2007; 115: 856–864.

52. Hrudey SE, Huck PM, Payment P, et al. Walkerton: Lessons learned in comparison with waterborne outbreaks in the developed world. Journal of Environmental Engineering and Science 2002; 1: 397–407.

53. Chapelle C, McCann H, Jassby D, et al. Managing wastewater in a changing climate. Public Policy Institute of California, 2019. Available at: https://www.ppic.org/wp-content/uploads/managing-wastewater-in-a-changing-climate.pdf. Accessed December 13, 2022.

54. McGill BM, Altchenko Y, Hamilton SK, et al. Complex interactions between climate change, sanitation, and groundwater quality: A case study from Ramotswa, Botswana. Hydrogeology Journal 2019; 27: 997–1015.

55. Allen TR, Crawford T, Montz B, et al. Linking water infrastructure, public health, and sea level rise: Integrated assessment of flood resilience in coastal cities. Public Works Management & Policy 2019; 24: 110–139.

56. Howard G, Calow R, Macdonald A, Bartram J. Climate change and water and sanitation: Likely impacts and emerging trends for action. Annual Review of Environment and Resources 2016; 41: 253–276.

57. Thompson DA, Cwiertny DM, Davis HA, et al. Sodium concentrations in municipal drinking water are associated with an increased risk of preeclampsia. Environmental Advances 2022; 9: 100306.

58. Scheelbeek PFD, Khan AE, Mojumder S, et al. Drinking water sodium and elevated blood pressure of healthy pregnant women in salinity-affected coastal areas. Hypertension 2016; 68: 464–470.

59. Newton A, Kendall M, Vugia DJ, et al. Increasing rates of vibriosis in the United States, 1996–2010: Review of surveillance data from 2 systems. Clinical Infectious Diseases 2012; 54(suppl_5): S391–S395.

60. Griffith AW, Gobler CJ. Harmful algal blooms: A climate change co-stressor in ma-rine and freshwater ecosystems. Harmful Algae 2020; 91: 101590. https://doi.org/10.1016/j.hal.2019.03.008.

61. Anderson DM, Cembella AD, Hallegraeff GM. Progress in understanding harmful algal blooms: Paradigm shifts and new technologies for research, monitoring, and management. Annual Review of Marine Science 2012; 4: 143–176.

62. Moore SK, Trainer VL, Mantua NJ, et al. Impacts of climate variability and future climate change on harmful algal blooms and human health. Environmental Health 2008; 7: S4. doi:10.1186/1476-069X-7-S2-S4.

63. Hallegraeff GM. Ocean climate change, phytoplankton community responses, and harmful algal blooms: A formidable predictive challenge. Journal of Phycology 2010; 46: 220–235.

64. Haake DA, Levett PN. Leptospirosis in humans. Current Topics in Microbiology and Immunology 2015; 387: 65–97.

65. Douchet L, Goarant C, Mangeas M, Menkes C, Hinjoy S, Herbreteau V. Unraveling the invisible leptospirosis in mainland Southeast Asia and its fate under climate change. Science of the Total Environment 2022; 832: 155018.

66. Nhamo G, Nhemachena C, Nhamo S. Is 2030 too soon for Africa to achieve the water and sanitation sustainable development goal? Science of the Total Environment 2019; 669: 129–139.

67. Zouboulis A, Tolkou A. Effect of climate change in wastewater treatment plants: Reviewing the problems and solutions. In S Shrestha, AK Anal, PA Salam, M van der Valk (eds.). Managing Water Resources Under Climate Uncertainty: Examples from Asia, Europe, Latin America, and Australia. Cham: Springer International Publishing, 2015, pp. 197–220.

68. Chhetri BK, Galanis E, Sobie S, et al. Projected local rain events due to climate change and the impacts on waterborne diseases in Vancouver, British Columbia, Canada. Environmental Health 2019; 18: 116. https://doi.org/10.1186/s12940-019-0550-y.

69. Zamyadi A, MacLeod SL, Fan Y, et al. Toxic cyanobacterial breakthrough and accu-mulation in a drinking water plant: A monitoring and treatment challenge. Water Research 2012; 46: 1511–1523.

70. Zamyadi A, Dorner S, Sauvé S, et al. Species-dependence of cyanobacteria removal efficiency by different drinking water treatment processes. Water Research 2013; 47: 2689–2700.

71. American Society of Civil Engineers. 2021 report card for America's infrastructure. 2021. Available at: https://infrastructurereportcard.org/wp-content/uploads/2020/12/National_IRC_2021-report.pdf. Accessed December 13, 2022.

72. Lall U, Johnson T, Colohan P, et al. Water. In DR Reidmiller, CW Avery, DR Easterling, et al. (eds.). The Fourth National Climate Assessment, Volume II: Impacts, Risks, and Adaptation in the United States. Washington, DC: U.S. Global Change Research Program, 2018, pp. 145–173. Available at: https://nca2018.globalchange.gov/chapter/3/. Accessed August 28, 2023.

73. Herrera V. Reconciling global aspirations and local realities: Challenges facing the Sustainable Development Goals for water and sanitation. World Development 2019; 118: 106–117.

74. Heaney CD, Wing S, Wilson SM, et al. Public infrastructure disparities and the microbiological and chemical safety of drinking and surface water supplies in

a community bordering a landfill. Journal of Environmental Health 2013; 75: 24–36.

75. Fagerli K, Trivedi KK, Sodha SV, et al. Comparison of boiling and chlorination on the quality of stored drinking water and childhood diarrhoea in Indonesian households. Epidemiology and Infection 2017; 145: 3294–3302.

76. Chan EYY, Tong KHY, Dubois C, et al. Narrative review of primary preventive interventions against water-borne diseases: Scientific evidence of health-EDRM in contexts with inadequate safe drinking water. International Journal of Environmental Research and Public Health 2021; 18: 12268. doi: 10.3390/ijerph182312268.

77. Alegbeleye OO, Sant'Ana AS. Manure-borne pathogens as an important source of water contamination: An update on the dynamics of pathogen survival/transport as well as practical risk mitigation strategies. International Journal of Hygiene and Environmental Health 2020; 227: 113524. doi: 10.1016/j.ijheh.2020.113524.

78. World Health Organization. WHO Guidelines on Recreational Water Quality: Volume 1: Coastal and Fresh Waters. Geneva: World Health Organization, 2021. Available at: https://apps.who.int/iris/handle/10665/342625. Accessed October 24, 2022.

79. Hardy FJ, Bouchard D, Burghdoff M, et al. Education and notification approaches for harmful algal blooms (HABs), Washington State, USA. Harmful Algae 2016; 60: 70–80.

80. Ekstrom JA, Moore SK, Klinger T. Examining harmful algal blooms through a disaster risk management lens: A case study of the 2015 U.S. West Coast domoic acid event. Harmful Algae 2020; 94:101740. https://doi.org/10.1016/j.hal.2020.101740.

Further Reading

Lall U, Johnson T, Colohan P, et al. Water. In DR Reidmiller, CW Avery, DR Easterling, et al. (eds.). The Fourth National Climate Assessment, Volume II: Impacts, Risks, and Adaptation in the United States. Washington, DC: U.S. Global Change Research Program, 2018, pp. 145–173. Available at: https://nca2018.globalchange.gov/chapter/3/. Accessed August 28, 2023.

   A comprehensive resource.

Levy K, Woster AP, Goldstein RS, Carlton EJ. Untangling the impacts of climate change on waterborne diseases: A systematic review of relationships between diarrheal diseases and temperature, rainfall, flooding, and drought. Environmental Science & Technology 2016; 50: 4905–4922.

   This systematic review was based on 141 studies describing the relationship between diarrheal diseases and four meteorological conditions expected to increase with climate change: ambient temperature, heavy rainfall, drought, and flooding. It found a positive association between ambient temperature and diarrheal diseases, except for viral diarrhea, and an increase in diarrheal disease following heavy rainfall and flooding events.

Levy K, Smith SM, Carlton EJ. Climate change impacts on waterborne diseases: Moving toward designing interventions. Current Environmental Health Reports 2018; 5: 272–282.

   This review summarizes literature on potential impacts of climate change on waterborne diseases. It stated that a growing body of evidence suggests that climate change may alter the incidence of waterborne diseases and diarrheal diseases in particular.

Semenza JC. Cascading risks of waterborne diseases from climate change. Nature Immunology 2020; 21: 479–487.

This paper reviews the association between heavy rainfall, flooding, and hot weather with waterborne diseases and how early warning systems could intercept these cascading risks.

8

Food Insecurity and Malnutrition

Jessica Fanzo, Kate R. Schneider, and Stanley Wood

Nutrition, health, and climate are closely interconnected. Climate-related shocks, which are becoming more frequent and extreme, are increasingly disrupting agri-food systems (the processes that food products go through from agriculture production to consumption), especially in low- and middle-income countries (LMICs).[1]

Climate change impacts nutrition and health by:

- Disrupting the quantity, availability, and quality of food that is produced
- Causing volatility in food production and disrupting market logistics (transportation and distribution networks), which reduce the affordability of and access to food
- Decreasing food safety and increasing food loss
- Increasing the risk of malnutrition and vulnerability to infectious disease among both animals and humans.[2,3] (Figure 8-1).

"*Food security* exists when all people, at all times, have physical, social, and economic access to sufficient, safe and nutritious food that meets their dietary needs and food preferences for an active and healthy life."[1,4] Food security consists of four components: availability, access, utilization, and stability.[5] Each of the impacts listed above compromise food security, which is necessary for adequate nutrition and good health. In addition, the broad impacts of climate change on migration, livelihoods, and the status of women, adversely affect diet and nutrition indirectly.[6] As this chapter describes, measures can be implemented in agri-food systems to improve food security, nutrition, and health—measures that are most urgent for LMICs.

This chapter reviews how climate change impacts food production, food security, and ultimately nutrition and health. Specifically, it addresses how climate change is altering outcomes in the following areas:

1. <u>Food production</u>: How climate change impacts food-producing systems
2. <u>Food security</u>: How changes in food availability and stress on resources reduce food access and food utilization

Jessica Fanzo, Kate R. Schneider, and Stanley Wood, *Food Insecurity and Malnutrition* In: *Climate Change and Public Health*. Edited by: Barry S. Levy and Jonathan A. Patz, Oxford University Press. © Oxford University Press 2024. DOI: 10.1093/oso/9780197683293.003.0008

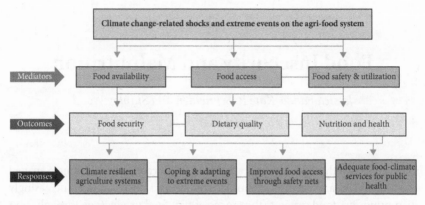

Figure 8-1 The mediator, outcomes, and responses to climate-related extreme events on agri-food systems (Diagram by Jessica Fanzo, Kate Schneider, and Stanley Wood.)

3. <u>Nutrition and health</u>: How climate change is expected to alter diets, thereby adversely impacting nutritional status, susceptibility to infectious diseases, and mortality.

It then discusses interventions and policies in agri-food systems that can help maintain and improve food supply chains, food security, and diet quality.

CLIMATE CHANGE IMPACTS ON AGRI-FOOD PRODUCTION SYSTEMS

Anthropogenic greenhouse gas (GHG) emissions are changing climate patterns at an accelerating pace and increasing the frequency and intensity of weather extremes throughout the world.[7,8] Climate change is posing serious threats to agri-food systems that adversely affect labor productivity as well as productivity of land and animals. These threats are exacerbated by acute disruptions from extreme weather events that are more frequent and more severe. There is a consensus that climate change is:

- Stressing all major food production sectors—crops, livestock, and fisheries
- Stifling recent progress addressing poverty, malnutrition, and disease
- Altering patterns and timing of biological processes, such as flowering and insect emergence, thereby affecting crop yields and food quality
- Limiting agriculture, livestock, and fishery productivity, jeopardizing both food security and livelihoods.

- Presenting far greater risks to vulnerable groups in the agri-food system—such as small-scale producers, women, people with low incomes, and Indigenous Peoples.[9]

Changing Temperature and Rainfall

Changes in temperature and precipitation patterns cause shifting patterns of agroecological suitability for rainfed crops and for pasture species. Rising mean temperature is inevitable, but the pace and scale of rising temperature varies by latitude and climatological context. Globally, temperatures are rising much faster in the Arctic than in any other place—a phenomenon known as the *Arctic amplification*. However, within the world's major agricultural production zones, temperatures are rising most rapidly in mid-latitude areas during warm seasons. Precipitation patterns are also shifting, but less predictably, resulting in either more or less total rainfall. New precipitation norms are characterized by fewer, more-intense rainfall events and more dry periods. As temperature rises, the increased rate of evapotranspiration dries out soils more quickly, shortening growing seasons and constraining production options. The Intergovernmental Panel on Climate Change (IPCC) projects that, without adaptation between now and the end of this century, the yield for four staple crops will generally decrease—particularly in the lower latitudes of the world—in several scenarios (Figure 8-2).

In already warmer, drier production areas, such as the semi-arid tropics of West Africa and India, even modest temperature increases can diminish productivity of crops and pasture below viable economic thresholds. Although these evolving climate-induced impacts on rainfed crop and pasture systems predominate in global agriculture, there are additional adverse impacts that are already occurring to locally and regionally important agroecosystems.[10]

Agricultural areas dependent on mountain systems for water are threatened. For example, the Hindu-Kush Himalaya (HKH) mountain system is both the source and regulator of glacier/snowmelt-derived water to 1.9 billion people and an extensive array of agricultural systems. In the HKH system, as in other mountain areas, temperature is rising faster than the global average—0.20°C (0.36°F) per decade compared to 0.13°C (about 0.23°F) per decade. Rising temperature is accelerating irreversible glacial retreat and progressively depleting baseflows, which feed aquifers and rivers used extensively for irrigation of food crops. Wheat and rice production predominates in this region, and the aquifers and rivers fed by the HKH system irrigate these crops across 10 river basins in eight countries, which are already experiencing high levels of poverty and undernutrition.

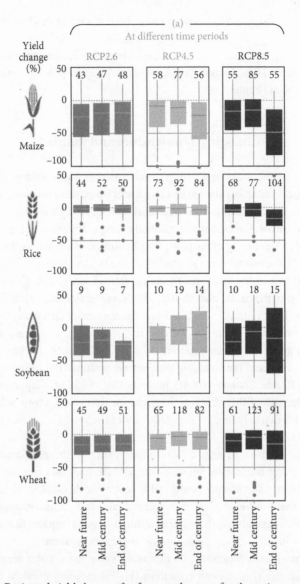

Figure 8-2 Projected yield changes for four staple crops for three time periods (relative to the 2001–2010 baseline period without adaptation) for three climate-change scenarios. Each box represents the interquartile range (IQR) and the middle line in each box represents the median. The number above each box represents the number of model simulations run for each scenario. The upper and lower end of whiskers are median 1.5 × IQR ± median. Open circles are values outside the 1.5 × IQR. The time periods are: near future = baseline to 2039; mid-century = 2040-2069; and end-century = 2070-2100. The scenarios are IPCC Representative Concentration Pathways 2.6RCP, 4.5 RCP, and 8.5 RCP. (*Source*: Bezner Kerr R, Hasegawa T, Lasco R, et al. Food, Fibre, and Other Ecosystem Products. In H-O Pörtner, DC Roberts, M Tignor, et al. [eds.]. Climate Change 2022: Impacts, Adaptation and Vulnerability. Contribution of Working Group II to the Sixth Assessment Report of the Intergovernmental Panel on Climate Change. Cambridge: Cambridge University Press, 2022, pp. 713–906.)

Similar threats to sustainability are evident in other major mountain-reliant agricultural production regions, such as the Po Basin (reliant on the Alps for water) and the Colorado Basin (reliant on the Rocky Mountains).[11,12]

Other insidious effects of increasing global temperature are mediated through sea-level rise. While the existential threats to the livelihood and food security of Small Island Developing States have raised concern, the challenges to coastal food production systems more globally are already significant and worsening.[13] Coastal areas of South Asia, Southeast Asia, and West Africa used extensively for food production—often in rice- and aquaculture-based systems—are at increasing risk from both encroachment by coastal waters as sea level rises and increased salinity of groundwater and surface water. Salinization of soil and groundwater further limits the physical extent of land suitable for rainfed production and the potential for irrigation to a limited number of salt-tolerant crops.[14]

Beyond direct productivity impacts, shifting agroecological conditions also alter the prevalence and severity of pests, weeds, and diseases that can threaten crop and livestock health and productivity. Warmer temperatures extend the geographic range of many pests, increase overwintering survival, accelerate generation cycles, and increase the likelihood of the spread of invasive species (Box 8-1). Taken together, these dynamics impose new challenges to maintaining both the productivity and the quality of farm output and derived food products.[10]

Extreme Weather Events

Extreme weather events further exacerbate the climate threats already described as droughts, heat waves, floods, and tropical storms (and associated coastal storm surges) create more acute disruption of production and markets, although typically in more localized areas (Figure 8-3). Many low-income farm households rely heavily on farm outputs to meet their nutrition needs, while most depend on crop and livestock sales for the majority of their annual income. Extreme weather events can cause acute crises in diet adequacy and nutrition through rapid shortfalls in harvest quantity and quality, significant reduction in farm household income, and the loss or liquidation of productive assets, such as animals and farm implements. These immediate losses have carry-over effects, which can further diminish production and livelihood potential in subsequent seasons. In addition, there is mounting empirical evidence that greater risk leads to short-term decision-making among farmers and livestock keepers exposed to repeated shocks, which, in turn, reduces their willingness to adopt proven and profitable climate-adapted production technologies and practices.

Box 8-1 Threats from Plant Pathogens

Caitilyn Allen

Plant diseases result from complex interactions among diverse factors, including pathogenic and protective microbes, the plant hosts, insect vectors that transmit pathogens, farming systems, and the environment. Although climate change may sometimes reduce the impact of specific plant diseases, overall climate change will increase plant diseases and related losses in crop yields.[1]

Manifestations of climate change—specifically higher temperatures, floods, and droughts—threaten food security *directly* by reducing crop productivity because stressed plants provide lower yields. These manifestations of climate change also threaten food security *indirectly* by decreasing crop yields because of plant diseases caused by fungi, bacteria, viruses, nematodes, and oomycetes (water molds). Globally, yield reductions associated with 137 pathogens and pests have been estimated at 22% for wheat, 30% for rice, 23% for maize, 17% for potatoes, and 21% for soybeans. Analyses have suggested that the highest reductions in yield are associated with food-deficit regions with fast-growing populations and often with emerging or reemerging pests and diseases.[2] Yield reductions due to plant pathogens are expected to increase as climate change continues—often dramatically affecting diets and family incomes.[3,4]

Heat-stressed plants are generally less able to defend against pathogens because several immunological systems in plants do not function optimally at higher temperatures. In addition, increased temperatures, drought, and high rainfall can favor specific pathogens, thereby tipping the equilibrium between a crop plant and its pathogens so that what had previously been a minor nuisance causes widespread catastrophic plant disease.

Increased temperature and humidity are the two most important environmental factors for predicting plant disease epidemics, including stem rust, the most destructive disease of wheat. The wheat stem rust fungus, *Puccinia graminis*, produces more spores and completes its life cycle faster as temperatures rise, and its spores can only germinate on wet leaves. Therefore, increases in the rate and intensity of stem rust outbreaks are predicted as wheat-growing regions become warmer and wetter—a grave concern because stem rust destroys about $5 billion of wheat annually.[5]

Extreme weather events, such as hurricanes and heavy downpours, are becoming more frequent. Some plant pathogens are spread by rainsplash or floods. For example, the citrus canker bacterium, which causes ugly scabs that reduce or destroy the value of citrus fruits, is disseminated by the impact of raindrops. As a result, sharp increases in citrus canker disease often follow major storms. Long-range movement of this bacterium to new areas has been associated with hurricanes. Therefore, citrus canker disease is expected to increase as a result of climate change.[6]

Many significant crop pathogens move from plant to plant by hitchhiking on insect vectors. As winters become warmer, these insect vectors are more likely to survive, thereby increasing plant diseases, some of which are appearing in new regions because their insect vectors can now survive the warmer winters. For example, Stewart's wilt, a disease of corn, is caused by the bacterium *Pantoea stewartii*, which overwinters in the gut of its vector, the corn flea beetle. Outbreaks of Stewart's wilt can be accurately predicted by winter temperatures; a warm winter increases both vector survival and wilt disease losses, with infected young seedlings quickly wilting and dying. Since older corn plants are more resistant, there is less Stewart's wilt if plants are infected later in the winter by beetles migrating up from warmer southern regions. Historically, corn growers in the northeastern and upper midwestern regions of the United States did not worry about Stewart's wilt—but now they do because this disease is more prevalent due to recent winters that have not been cold enough to kill corn flea beetles.[7]

One of the best-documented examples of a crop epidemic linked to climate change is the current outbreak of coffee rust disease in Central America. Coffee trees thrive under shade in the highland tropics, where days are warm but nights are cool. However, coffee trees can be rapidly defoliated and killed by coffee rust disease, which is caused by the wind-borne fungus *Hemileia vastatrix*. This fungus was first described in 1867, when it caused an explosive epidemic of coffee rust disease in the dense colonial-era coffee plantations of Sri Lanka and ruined its economy. Subsequently, coffee rust disease spread rapidly in Africa and Asia. It continues to be the most serious disease of coffee plants throughout the world.

For more than 100 years, strict quarantines kept coffee plantations in the Americas free from coffee rust disease. The arrival of the rust fungus in 1970 forced industrial-scale growers to grow lower-quality, sun-tolerant coffee varieties so they could control it with aerial fungicide sprays. Flavorful and expensive Arabica coffee, which grows under shade, has been cultivated for centuries on volcanic mountain slopes in Central America. *H. vastatrix* cannot survive temperatures below 10°C (50°F). Therefore, until recently, highland coffee plantations remained largely free of coffee rust disease because they were in an environmental sweet spot. It was too cool for the coffee rust pathogen, but not too cool for the coffee trees. Because they did not need fungicides to control the disease, growers could farm organically, making their crop even more valuable. Demand for organic shade-grown coffee made many small fair-trade coffee growers economically secure.

This security disappeared after several years of unusually high temperatures allowed *H. vastatrix* to spread and multiply at higher elevations than ever before. In 2013, highland growers lost up to 80% of their crops because of coffee rust disease. Nicaragua, Costa Rica, Honduras, and Guatemala declared states of emergency.[8]

Coffee rust disease can cause yield losses of up to 35%, which can, in turn, greatly impact yields for subsequent years. It threatens the livelihoods of not only coffee growers and exporters, but also those of low-income coffee pickers who depend on their seasonal wages to feed their families and pay for their children's education. Records from the 1500s to the early 1800s document consistently cool temperatures and low incidence of coffee rust disease on many Central American coffee plantations. As temperatures increased during recent decades, the incidence of coffee rust disease also increased. All of these data unequivocally demonstrate that the current epidemic of coffee rust disease is due to climate change.

Box References

1. Food and Agriculture Organization. Five ways climate change is intensifying the threats to plant health. December 5, 2022. Available at: https://www.fao.org/fao-stories/article/en/c/1507753/. Accessed February 10, 2023.

2. Savary S, Willocquet L, Pethybridge SJ, et al. The global burden of pathogens and pests on major food crops. Nature Ecology & Evolution 2019; 3: 430–439.

3. Savary S, Mila A, Willocquet L, et al. Risk factors for crop health under global change and agricultural shifts: A framework of analyses using rice in tropical and subtropical Asia as a model. Phytopathology 2011; 101: 696–709. https://doi.org/10.1094/PHYTO-07-10-0183.

4. Juroszek P, Racca P, Link S, et al. Overview on the review articles published during the past 30 years relating to the potential climate change effects on plant pathogens and crop disease risks. Plant Pathology 2020; 69: 179–193. https://doi.org/10.1111/ppa.13119.

5. Figueroa M, Hammond-Kosack KE, Solomon PS. A review of wheat diseases-a field perspective. Molecular Plant Pathology 2018; 19: 1523–1536. doi: 10.1111/mpp.12618.

6. Neri FM, Cook AR, Gibson GJ, et al. Bayesian analysis for inference of an emerging epidemic: Citrus canker in urban landscapes. PLoS Computational Biology 2014 10: e1003587. doi:10.1371/journal.pcbi.1003587.

7. Esker PD, Harri J, Dixon PM, Nutter FW Jr. Comparison of models for forecasting of Stewart's disease of corn in Iowa. Plant Disease 2006; 90: 1353–1357.

8. Cressey D. Coffee rust regains foothold. Nature 2013; 493: 587.

9. Talhinhas P, Batista D, Diniz I, et al. The coffee leaf rust pathogen Hemileia vastatrix: One and a half centuries around the tropics. Molecular Plant Pathology 2017; 18: 1039–1051. doi: 10.1111/mpp.12512.

Figure 8-3 Woman checks her land in Latifiyah, Iraq, about 20 miles south of Baghdad in 2009. Below-average rainfall and insufficient water in the Euphrates and Tigris rivers left Iraq bone-dry for a second consecutive year. (AP Photo/Hadi Mizban.)

Extreme weather events in tropical and subtropical regions are increasingly associated with periodic El Niño-Southern Oscillation (ENSO) phases, which intensify alternating droughts and monsoons/floods. Climate change has altered both the frequency of El Niño and La Niña phases and made their impacts more severe and less predictable. For example, in the Horn of Africa, departure from prior ENSO patterns led to an unprecedented sequence of three drought years that were attributed to a remarkable persistence of La Niña conditions. This sequence of drought years transformed a succession of poor harvests into famine conditions for 20 million people in the region.[15]

Climate-induced extreme weather events can deliver repeated shocks to the livelihoods and welfare of all households and present major challenges in finding appropriate responses. These extreme weather events are less predictable, more varied, and, in the short- to medium-term, potentially more damaging to nutrition and health outcomes than gradually shifting climate means. The nature and pace of the evolution of climate change offer opportunities for systematic planning and deployment of progressive waves of adaption solutions. More localized and short-term extreme weather events are challenging not only because they are difficult to anticipate, but also because they can manifest in different forms in the same locations over time—such as intense heat waves and severe droughts alternating with catastrophic monsoon rains and flooding. Making farm households, farming systems, and value chains more resilient to withstand and recover from such a range and intensity of shocks requires policies, instruments, and investments for preparedness, risk management, response, recovery, and social safety net protections. These needs, in turn, require levels of institutional responsiveness and capacity that have even been challenging for high-income countries to achieve.

Data and Analytical Needs

The work of the IPCC and other initiatives, such as the Agricultural Model Intercomparison and Improvement Project, have vastly improved capabilities to anticipate the impacts of climate change. Specifically, research has improved the robustness and predictive skill of a range of climatological, biophysical, and economic models to translate projections of critical climate variables into likely productivity, production, and market outcomes for major food staple crops at the regional and national scales. (The next section discusses some findings from these studies.) However, the application of contemporary agricultural impact models remains centered on assessing long-term climate impacts, with less focus on much-needed improvements in risk-based assessments of the potential seasonal and cross-seasonal impact of extreme weather events. These capabilities need to be extended to encompass impacts on (and opportunities for) a more diverse range of farm products, including more nutrient-dense commodities. Other priorities include bringing, in addition to short-term humanitarian assistance, a development perspective for addressing the needs of the agri-food system and determining with greater geographic and subpopulation specificity the likely impacts from extreme weather events.

ADVERSE EFFECTS OF CLIMATE CHANGE

Food security includes the following four components:

- Food availability
- Food access
- Food utilization
- Food stability.

The impact of climate on food security is mediated through markets and prices. Most countries depend on food imports as a key part of domestic food security; other countries depend on food exports for a large share of national income. As the impacts of climate change worsen, the combination of reduced food availability and food access, suboptimal utilization of food, and lack of food stability are expected to place more people at risk of food insecurity.

Food Availability

Increasingly linked global markets have led to a greater impact of global and regional production systems and supply chains on domestic food availability. The benefits of trade include increased availability of some goods and more product diversity; however, the disadvantages of trade include less ability to protect domestic consumers from volatility of global prices and reduced diversity of domestic production, which lower resilience to food shocks.[16,17] When crop yields decrease because of climate change, food-producing countries can use food imports as a buffer to compensate for lost local production, thereby protecting their food security. However, production shortfalls in large food-exporting countries or shortfalls in many locations at the same time increase global food prices and reduce global food availability. Production shortfalls reduce domestic food availability most for import-dependent low-income countries with limited hard currency because they cannot afford to spend more (or borrow) to purchase enough food at higher prices. Unless other actions are taken to protect food security, such as by releasing grain stocks, domestic food availability in these situations will decrease and food insecurity will increase.[18]

Much research has focused on improving models to help assess the magnitude, direction, and temporal and spatial variation of climate impacts on food production, availability, and prices as well as food security outcomes. These approaches bring climate change projections together with (a) agricultural models to assess biophysical impacts on food production and (b) economic models that assess the likely consequences for aggregate food availability, food prices, and trade. Some models go further and attempt to predict how changes in food availability and prices will likely affect diets (nutrient adequacy) and nutrition outcomes (such as stunting, wasting, and being underweight).

Global food security is projected to improve under the most optimistic and most technological scenarios, while the population at risk of insufficient food is projected to be greatest under a noncooperative scenario. However, predictions of future global food security also depend on assumptions regarding population growth, income, and agricultural productivity embedded in projection models.[19]

Food Access

Reduced food availability leads to higher prices, which reduce *economic access to food*—when a person or household has sufficient income to spend on food for a desirable diet at prevailing prices. Low-income households spend a greater proportion of their budgets on food than high-income households. Marginalized and vulnerable groups are most likely to be adversely affected sooner and more severely by higher prices. As prices rise, they must choose to buy either less food or lower-quality food (with insufficient calories or nutrients), thereby compromising their nutritional status. (Alternatively, they can reduce spending on other items in order to increase their food expenditure.[20])

Markets in agriculture-based rural areas of low-income countries are less efficient and very costly to access, making it expensive to transport food to or away from these areas—especially perishable, bulky, or low-value food. If production is disrupted, market prices rise as local demand responds to diminished supply; but high marketing costs limit opportunities to meet food shortfalls by importing food from distant sources. As a result, food access relies primarily on local production and local markets, except where access to regional or world markets exists, such as near major road and rail connections, in coastal cities, and in Small Island Developing States that do not—and often cannot—grow enough food to feed their populations. Because of climate change, remote populations are more likely to be exposed to greater volatility in prices and food access and to have more limited food choices.

Other populations whose food access is more vulnerable due to climate change are those whose incomes are heavily reliant on agriculture—especially rainfed agriculture—and those who do not have enough food to store and/or few options for safe storage.[18] In addition, most smallholder farmers, even those who consume a large share of their own agricultural production, remain net food buyers—that is, they annually purchase more food from markets than they consume from their own production.[20] Therefore, they are also at risk of not being able to afford food when prices rise—even if they also increase their income from the sale of higher-priced agricultural output.[21]

In addition, social structures can impede food access for certain population subgroups. An example of a social barrier to food access is inequitable intrahousehold distribution—social norms that determine which family members are allowed to eat which foods or which are required to wait and eat after others have finished, when there may not be enough remaining food.[22] Another example is women's lack of access to land or income-earning opportunities. While these social dimensions may not be directly affected by climate change, as competition for scarce resources increases those with the least power in their households and society will face the greatest impediments to securing access to these resources.

Food Utilization

Reduced economic access directly impacts diet quality and nutrition. Starchy staple foods, such as grains, roots and tubers, and other carbohydrate-dense foods, are the least expensive source of life-sustaining energy (calories). They are prioritized when food budgets are squeezed due to higher prices. Diets too rich in starchy staples not only contain excessive macronutrients but are also likely to contain inadequate amounts of micronutrients, contributing to malnutrition. Unmitigated higher prices of staple foods put the people with the lowest incomes at greatest disadvantage, least able to afford sufficient calories—let alone the diverse diet necessary for adequate nutrition and long-term health.[23]

Poor utilization of food can take other forms. For example, when food is scarce, people may consume spoiled or contaminated food, such as moldy food that is potentially contaminated with carcinogenic aflatoxins—a situation that occurs more frequently in crops exposed to climate stress during the growing period. Therefore, food safety risks will likely increase as climate stress increases.[24] Another form of poor utilization occurs when the body cannot adequately absorb nutrients from food due to repeated exposure to pathogens.

Climate change is increasing the risk of foodborne disease,[25] such as by (a) changing the conditions that could promote (or hinder) survival of microorganisms that cause foodborne disease and (b) facilitating emergence of new pathogens that cause foodborne disease. Greater use of antibiotics in livestock and poultry production can accelerate development of antimicrobial resistance and greater use of pesticides can increase the presence of chemical residues on foods.[26]

Food Stability

The final dimension of food security is stability, the ability to withstand shocks or to recover from shocks so that the long-term equilibrium or trend can persist over time.[27] As climate becomes less stable, food production shocks are expected to increase in frequency, duration, and magnitude. Coupled with the poorly integrated, high-cost markets typical of many low-income countries, these shocks will cause greater disequilibrium between food supply and demand, yielding more price spikes and long-term volatility of food prices. These shocks may be compounded by others affecting food markets, such as trade disputes and armed conflict (Chapter 10), which may also be associated with climate change and increasing competition for resources.[28,29] These challenges to food safety threaten the resilience of agri-food systems, making it less likely that they will recover over the long term to provide sufficient access to food.

AGRI-FOOD SYSTEM-RELATED CLIMATE IMPACTS ON DIET, NUTRITION, AND HEALTH

The impacts of climate change extend beyond food security to affect malnutrition, both directly and indirectly. Direct pathways link climate change to all forms of malnutrition—to undernourishment, childhood stunting and wasting, and, less clearly, overweight and obesity. Climate change also impacts (a) the immediate determinants of malnutrition, such as access to food, intake of nutrients, and exposure to communicable disease, and (b) longer-term determinants, such as political and economic factors, socioeconomic status, and environmental conditions.[30] Challenges posed to agri-food systems by climatic change adversely impact diet, and, in turn, nutrition. And extreme weather events impact the health of agri-food system workers, especially farmworkers.

Globally, by 2030, there will be a 60% reduction in work hours in the agriculture sector due to heat stress—with 90% of the lost work hours occurring in agriculture in Central Africa and East Africa.[31] In the United States, the average number of days worked in extreme heat by agricultural workers will double by mid-century and, without mitigation, triple by the end of the century.[32] (See Chapter 3.)

Diet and Health

An unhealthy diet is among the leading risk factors for deaths and for lost disability-adjusted life-years (DALYs). Globally, 11 million deaths and 255 million DALYs are annually attributable to various dietary risk factors for cardiovascular disease, diabetes, and some types of cancer and other noncommunicable diseases.[33,34] Malnutrition in all its forms causes more lost DALYs globally than tobacco smoking, high blood pressure, and high fasting plasma glucose.[33] Poor diet and malnutrition combined is the leading contributor to disease and death, regardless of country income status.[34,35]

Diets contributing most to deaths and lost DALYs are those that are high in sodium and low in whole grains, fruits, legumes and other vegetables, nuts, and seeds.[34] Many diets now contain a significant share of packaged, highly processed foods, such as sugar-sweetened beverages, baked goods, dairy products, processed meats, chips and crackers, cake mixes, pies, pastries, and sweets. These highly or ultra-processed foods are detrimental to a range of human health outcomes. Emerging evidence indicates that these foods also have impacts on environmental sustainability.[36] As supply chains and transportation become strained by extreme weather events, processed foods may possibly be eaten more frequently by some populations.

By 2050, global GHG emissions from food production are expected to increase by 80%. This projection does not include increases in GHG emissions from increased population growth and dietary shifts. One modeling study suggests that climate-induced changes in diets and weight could cause more than 500,000 deaths in 2050 due to risk factors related to reduced fruit and vegetable consumption.[37] In addition to changes in diets, there are associations between extreme weather events and worsening nutritional status (as described in the next section), with the severity of impacts on nutritional status mediated by socioeconomic status.[38]

Undernutrition

Climate change can exacerbate undernutrition through

- Household food security (availability of and access to safe, affordable, and healthy diets)
- Unhealthy childcare and feeding practices
- Reduced water quality and availability, leading to waterborne disease[39](Chapter 13)
- Reduced access to health services[40]
- Reduced crop yields and crop nutrients, such as protein and micronutrients (Chapter 14).

Climate change affects what food is available and at what price, thereby impacting calorie consumption, types of food consumed, and quality of diet.[37,41,42] Insufficient food, exposure to waterborne and foodborne pathogens, and poor dietary consumption can have adverse effects on child growth and lead to micronutrient deficiencies.[43]

Responses of women to extreme weather events have implications for how they allocate their time, which, in turn, influences childcare and feeding practices; however, well-managed disaster response can protect child nutritional status.[44] Climate shocks can disrupt access to and delivery of healthcare, adversely affect nutrient utilization, and exacerbate undernutrition, such as by reduced access to routine supplementation of vitamin A for children under age 5 and of iron and folic acid for pregnant women.

Undernutrition causes profound long-term consequences, especially for children under age 5 (Figure 8-4).[45] *Stunting* (chronic undernutrition) affects approximately 22% of children under age 5 globally, limiting their potential to

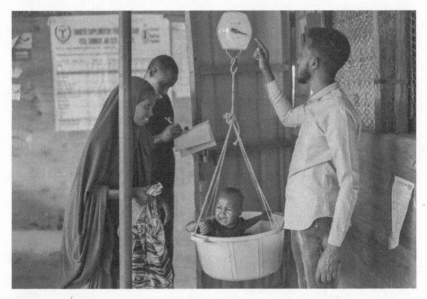

Figure 8-4 A child is weighed at a camp for displaced people on the outskirts of Dollow, Somalia, in 2022. In many Middle Eastern and African nations, climate shocks have worsened food shortages and killed and displaced many people every year. With limited resources, people in these nations are among the world's poorest and most vulnerable to climate change impacts. (AP Photo/Jerome Delay.)

thrive, earn money, and contribute to society.[45] *Wasting* (acute malnutrition) affects about 6.7% of children under age 5 globally.[46,47] Under medium-high climate change scenarios, with limited mitigation, it is estimated that an additional 4.8 million children under age 5 will be undernourished by 2050 (Table 8-1).

Climate-related extreme weather events can lead to seasonal effects with important implications for nutrition and health. Floods, droughts, rainfall, seasonality, and rising temperatures have all been associated with increased childhood stunting because of the impact of seasonality on birth outcomes, thereby setting out a course for child growth in the first several years of life.[48] A study in five countries in West Africa found that lifetime exposures to temperatures above 35°C (95°F) increases the prevalence in young children of stunting by 18% and wasting by 16%. It is estimated that a permanent 2.0°C (3.6°F) rise in average global temperatures above preindustrial levels will increase stunting by 7.4%.[49]

Both extreme drought and extreme precipitation are associated with childhood malnutrition, including stunting.[50,51] Changes in rainfall can exacerbate crop failures and flooding, which can affect nutritional status, especially stunting.[39,48] Flooding has short- and long-term effects on child height attainment due to decreased food consumption and diet quality as well increased incidence of

Table 8-1 Number (in millions) of undernourished children under age 5 years, in 2010 and 2050, using the National Center for Atmospheric Research Climate Model and the A2 Family of Scenarios*

Region	Number of undernourished children under age 5			Additional number of children undernourished because of climate change 2010–2050
	2010, base climate	2050		
		Without climate change	With climate change	
Sub-Saharan Africa	40.9	37.0	39.3	2.3
South Asia	77.1	50.4	51.9	1.5
East Asia and the Pacific	21.9	7.8	8.2	0.4
Latin America and Caribbean	4.3	1.5	1.8	0.3
Middle East and North Africa	4.0	1.7	1.9	0.2
Europe and Former Soviet Union	1.8	1.5	1.6	0.1
Total	150.0	99.9	104.8	4.9

* The A2 family of scenarios is characterized by a world of independently operating, self-reliant nations; continuously increasing population; and regionally oriented economic development.

Source: International Food Policy Research Institute. 2017 Global food policy report. Washington, DC: IFPRI, 2017. Available at: https://ebrary.ifpri.org/digital/collection/p15738coll2/id/131085. Accessed September 9, 2022.

infectious diseases.[52,53] The combination of increased temperature, decreased precipitation, and unstable food production, without intervention, could result in an increased risk of low-birthweight infants.[53] Drought can lead to food shortage and loss of income, slowing height attainment in young children.[54] A study of children's diets in 19 countries found that higher long-term temperatures were associated with lower diet diversity, while higher rainfall was associated with higher diet diversity.[55]

Infectious Diseases

Climate-related infectious diseases can adversely affect nutritional status and increase mortality. Infectious disease outbreaks often follow extreme weather events because microorganisms, disease-carrying vectors, and animal reservoirs of disease may thrive, thereby disrupting social and environmental conditions.[56]

A study found that (a) 58% of known infectious diseases were worsened by climate-related extreme weather events and (b) of the approximately 1,000 pathways in which climate hazards impact pathogenic diseases (including both those caused by transmissible microbes and those caused by nonmicrobial and nontransmissible agents), 5% were foodborne.[57] Floods and storms can adversely affect management of wastewater, which, in turn, can increase foodborne transmissions of microorganisms. In addition, in places with high prevalence of undernutrition and/or immunodeficiency, there are increased risks of outbreaks of measles, cholera, and other infectious diseases.[58]

MINIMIZING ADVERSE IMPACTS ON FOOD SECURITY AND NUTRITION

Measures to minimize climate-related risks to food security and nutrition include a broad range of legislative and regulatory actions, shifts in institutional priorities and responsiveness, and improved access to technologies, products, and services. This section reviews options for reducing these risks in all four dimensions of food security, emphasizing how a comprehensive response strategy can increase the climate and nutrition resilience of the entire agri-food system.[7] Strong linkages exist between food availability and food accessibility. Resilience-focused action targeted to any single dimension of food security will likely have spillover impacts—either positive or negative—elsewhere in the agri-food system, which should be anticipated and accounted for.

Ensuring Agri-Food Systems Are Climate Resilient

For the agri-food sector, progress could be accelerated by aligning sectoral development plans with nationally determined contributions and national adaptation plans of the 2015 Paris Agreement, so that dedicated global climate funds could be used to improve national resilience.[59] Other measures to incentivize progress toward agri-food system resilience include

- Accelerating the development and use of stress-tolerant or resistant seeds for crops of nutritious foods
- Investing in improved technologies and practices for storing food and reducing post-harvest loss[60]
- Improving management and governance of water resources for more equitable and sustainable allocation and use for food production[9,60]
- Transforming public extension programs, which are rapidly evolving, to better acquire and curate knowledge concerning climate-smart agri-food

systems and support its timely and increasingly customized dissemination through digital platforms[60,61]

- Promoting access to affordable financial instruments and insurance products to assist farmers in better managing risk and building confidence in maintaining productive farms[62,63]
- Establishing early-warning systems and climate information services to better enable farmers and others working in the agri-food system to incorporate climate and weather information and analyses into their decision-making[64,65]
- Promoting more open and efficient food and agriculture markets that lower marketing costs, thereby dampening the impacts of locally idiosyncratic climate shocks, moderating food supply and price volatility, and minimizing losses in food quantity and quality[65,66]
- Where climate impacts have exceeded the coping and adaptive capacities of local farming and pastoral communities, providing safety-net programs to support farmers' livelihoods or enabling them to transition to other work.[67,68] (See Chapter 14.)

Ensuring Agri-Food Systems Are Nutrition Resilient

Nutrition resilience can be achieved by improving food productivity and minimizing food loss, reducing GHG emissions from agriculture, and implementing adaptation strategies for those who are nutritionally vulnerable. These measures span agriculture inputs and availability of, access to, and utilization of food. (See Table 8-2.)[69]

Protecting Access to Food with Social Protection

Short-term financial shocks that interrupt access to food can have long-term health consequences, especially for children.[70] Social protection programs (safety nets or social assistance programs) aim to increase incomes and food security and achieve other related goals. Programs can take many forms, including providing cash transfers, vouchers, or benefit cards that can be used to purchase or exchange for food, meals at school, food for preparation at home, and food in exchange for work. For example, in the United States, the Supplemental Nutrition Assistance Program provides households with a monthly cash-equivalent card that can be spent at any approved retailer on all grocery items except alcohol and prepared foods. In Ethiopia, the Productive Safety Nets Programme provides cash or food transfers to low-income households in exchange for short-term work, usually on infrastructure and other public works

Table 8-2 Recommendations for action across agri-food systems

Theme	Action
1. Food supply-chain inputs	Increase access to seed varieties and livestock breeds that are diverse and resilient to variable weather conditions (heat and drought), pests, and diseases
	Use agricultural extension programs to improve access to information and training about these varieties and breeds
	Improve soil quality through the use of cover crops, crop rotation, balanced use of fertilizers, and manure
	Increase irrigation systems to protect crops and livestock from loss due to changes in seasonal precipitation and extreme weather events
2. Agriculture production	Invest in and provide education on integrated land-use policies and mixed crop and livestock systems
	Expand access to services and financing to support farmers, including farmer risk-management tools, insurance, and loans
3. Post-harvest storage and processing	Improve infrastructure, especially in rural areas, including roads, warehouses, and processing plants
	Provide training on safe storage and processing techniques, such as drying
4. Food availability and access	Improve retailer access to water, electricity, and cold storage
	Create networks of food producers to increase market access and help limit food waste
	Improve transportation infrastructure in areas where the effects of climate change will limit people's ability to access markets
5. Food consumption and utilization	Expand access to social protection services, including unconditional cash transfers and supplementary food allowances
	Increase consumption of animal-source foods in low- and middle-income countries while educating the public about the health risks associated with overconsumption of these foods
	Improve access to safe and energy-efficient cookstoves
	Increase access to healthcare for vulnerable populations, especially rural poor people, by increasing healthcare facilities and staff

Source: Fanzo J, Davis C, McLaren R, Choufani J. The effect of climate change across food systems: Implications for nutrition outcomes. Global Food Security 2018; 18: 12–19.

programs.[71] (Pregnant women, older people, and people with disabilities are exempted from the work requirement.) In India, low-income households are issued identification cards that allow them to shop in "fair price shops," where the cost of food items is subsidized by the government.

As climate change continues and its consequences become more severe, social protection is increasingly recognized as an important mechanism to protect food security, nutrition, and health.[67,68,72] Programs must be able to respond quickly to various livelihoods and various emergencies that disrupt livelihoods, such as by quickly enrolling newly eligible beneficiaries to reduce the duration of hardship between the shock and receipt of assistance. They must also be able to respond when a shock simultaneously affects many people, such as by tying weather-insurance disbursements to climatic events rather than to evidence of harvest losses.[73] And they must also be designed to address cumulative impacts that may exhaust a household's capacity to cope.[70]

Climate Services

Public health and nutrition practitioners need access to climate services—information, tools, and training—in order to improve their ability to address the complex health challenges of climate disruption for the populations they serve. Climate services, such as the Famine Early Warning System Network, assist individuals and organizations in improving public policy in response to, or in anticipation of, climate-related impacts. Using these services to respond to immediate programming needs and long-term investment and planning can help address Sustainable Development Goal 2 (end hunger, achieve food security and improved nutrition, and promote sustainable agriculture) and Sustainable Development Goal 3 (ensure healthy lives and promote well-being for all).[74,75]

SUMMARY POINTS

- Climate change is placing stress on food production areas, adversely affecting environments for future agriculture and livestock production and threatening food security and nutrition.
- Changes in temperature and rainfall are increasing the potential for insect, pest, and disease threats to food production, food safety, and water quality as well as the risk of malnutrition.
- Climate change is challenging the ability of production systems in LMICs to raise productivity in the face of growing food demand, thereby raising food prices and jeopardizing nutrition.

- Climate change can exacerbate undernutrition by reducing household food security, environmental health, and access to health services.
- Specific actions that can increase the resilience of agri-food systems and households to climate change include research and development of stress-tolerant and resilient crop varieties and animal breeds, more efficient food markets, agricultural information services, improved social protection measures, and early-warning systems.

References

1. Food and Agriculture Organization of the United Nations (FAO). The State of Food Security and Nutrition in the World 2022: Repurposing Food and Agricultural Policies to Make Healthy Diets More Affordable. Rome: FAO, 2022. Available at: https://www.fao.org/publications/sofi/2022/en/. Accessed November 1, 2022.
2. Fanzo J, Bellows AL, Spiker ML, et al. The importance of food systems and the environment for nutrition. American Journal of Clinical Nutrition 2020; 113: 7–16.
3. Myers SS, Smith MR, Guth S, et al. Climate change and global food systems: Potential impacts on food security and undernutrition. Annual Review of Public Health 2017; 38: 259–277.
4. Food and Agriculture Organization of the United Nations (FAO). Rome Declaration on World Food Security. World Food Summit, November 13–17, 1996. Available at: http://www.fao.org/3/w3613e/w3613e00.htm. Accessed on November 1, 2022.
5. Wheeler T, von Braun J. Climate change impacts on global food security. Science 2013; 341: 508–513.
6. Watts N, Adger WN, Agnolucci P, et al. Health and climate responses to protect public health. Lancet 2015; 386: 1861–1914.
7. Pörtner H-O, Roberts DC, Poloczanska K, et al. (eds.) Summary for policymakers. In H-O Pörtner, DC Roberts, M Tignor, et al. (eds.). Climate Change 2022: Impacts, Adaptation, and Vulnerabilty. Contribution of Working Group II to the Sixth Assessment Report of the Intergovernmental Panel on Climate Change. Cambridge: Cambridge University Press, 2022, pp. 3–33. Available at: https://www. ipcc.ch/report/ar6/wg2/. Accessed November 1, 2022.
8. Masson-Delmotte V, Zhai P, Pirani A, et al. (eds.). Climate Change 2021: The Physical Science Basis. Contribution of Working Group I to the Sixth Assessment Report of the Intergovernmental Panel on Climate Change. Cambridge: Cambridge University Press, 2021. Available at: https://www.ipcc.ch/report/ar6/wg1/. Accessed November 1, 2022.
9. Pörtner H-O, Roberts DC, Tignor M, et al. (eds.). Climate Change 2022: Impacts, Adaptation, and Vulnerability. Contribution of Working Group II to the Sixth Assessment Report of the Intergovernmental Panel on Climate Change. Cambridge: Cambridge University Press, 2022. Available at: https://www.ipcc.ch/rep ort/ar6/wg2/. Accessed November 1, 2022.
10. Skendžić S, Zovko M, Živković IP, et al. The impact of climate change on agricultural insect pests. Insects 2021; 12: 440. doi: 10.3390/insects12050440.
11. Molden DJ, Shrestha AB, Immerzeel WW, et al. The great glacier and snow-dependent rivers of Asia and climate change: Heading for troubled waters. In AK

Biswas, C Tortajada (eds.). Water Security Under Climate Change. Water Resources Development and Management. Singapore: Springer Singapore, 2022, pp. 223–250.

12. Qin Y, Abatzoglou JT, Siebert S, et al. Agricultural risks from changing snowmelt. Nature Climate Change 2020; 10: 459–465.

13. Mycoo M, Wairiu M, Campbell D, et al. Small islands. In H-O Pörtner, DC Roberts, M Tignor, et al. (eds.). Climate Change 2022: Impacts, Adaptation, and Vulnerability. Contribution of Working Group II to the Sixth Assessment Report of the Intergovernmental Panel on Climate Change. Cambridge: Cambridge University Press, 2022, pp. 2043–2121. Available at: https://www.ipcc.ch/report/ar6/wg2/. Accessed November 1, 2022.

14. Tran DD, Park E, Tuoi JTN, et al. Climate change impacts on rice-based livelihood vulnerability in the lower Vietnamese Mekong Delta: Empirical evidence from Can Tho City and Tra Vinh Province. Environmental Technology & Innovation 2022; 28: 102834. https://doi.org/10.1016/j.eti.2022.102834.

15. McPhaden MJ, Santoso A, Cai W. El Niño Southern Oscillation in a Changing Climate. Hoboken, NJ: John Wiley & Sons, 2020.

16. Brenton P, Chemutai V, Pangestu, M. Trade and food security in a climate change-impacted world. Agricultural Economics 2022; 53: 580–591.

17. Kummu M, Kinnunen P, Lehikoinen E, et al. Interplay of trade and food system resilience: Gains on supply diversity over time at the cost of trade independency. Global Food Security 2020; 24: 100360. https://doi.org/10.1016/j.gfs.2020.100360.

18. Burke M, Lobell D. Climate effects on food security: An overview. In D Lobell, M Burke (eds.). Climate Change and Food Security: Adapting Agriculture to a Warmer World. Dordrecht, The Netherlands: Springer Netherlands, 2010, pp. 13–30. Available at: https://web.stanford.edu/~mburke/papers/Chap2_overview.pdf. Accessed November 1, 2022.

19. van Dijk M, Morley T, Rau ML, Saghai Y. A meta-analysis of projected global food demand and population at risk of hunger for the period 2010–2050. Nature Food 2021; 2: 494–501.

20. Global Panel on Agriculture and Food Systems for Nutrition. Food systems and diets: Facing the challenges of the 21st century. September 2016. Available at: https://glopan.org/sites/default/files/ForesightReport.pdf. Accessed November 1, 2022.

21. Headey DD, Martin WJ. The impact of food prices on poverty and food security. Annual Review of Resource Economics 2016; 8: 329–351.

22. Berti PR. Intrahousehold distribution of food: A review of the literature and discussion of the implications for food fortification programs. Food and Nutrition Bulletin 2012; 33: S163–S169.

23. Food and Agriculture Organization of the United Nations, International Fund for Agricultural Development, United Nations International Children's Emergency Fund, World Food Programme, and World Health Organization. The State of Food Security and Nutrition in the World 2021: Transforming Food Systems for Food Security, Improved Nutrition, and Affordable Healthy Diets for All. Rome: FAO, 2021. Available at: https://www.fao.org/documents/card/en/c/cb4474en. Accessed November 1, 2022.

24. Sibakwe CB, Kasambara-Donga T, Njoroge SMC, et al. The role of drought stress on aflatoxin contamination in groundnuts (Arachis hypogea L.) and Aspergillus flavus population in the soil. Modern Agricultural Science and Technology 2017; 3: 22–29.

25. Stewart LD, Elliott CT. The impact of climate change on existing and emerging microbial threats across the food chain: An island of Ireland perspective. Trends in Food Science and Technology 2015; 44: 11–20.

26. King T, Cole M, Farber JM, et al. Food safety for food security: Relationship between global megatrends and developments in food safety. Trends in Food Science and Technology 2017; 68: 160–175.

27. Nicholson CF, Stephens EC, Kopainsky B, et al. Food security outcomes in agricultural systems models: Case examples and priority information needs. Agricultural Systems 2021; 188: 103030. https://doi.org/10.1016/j.agsy.2020.103030.

28. Gregory PJ, Ingram JSI, Brklacich M. Climate change and food security. Philosophical Transactions of the Royal Society. B: Biological Sciences 2005; 360: 2139–2148.

29. Ensor J, Forrester J, Matin N. Bringing rights into resilience: Revealing complexities of climate risks and social conflict. Disasters 2018; 42(Suppl 2): S287–S305.

30. Hawkes C, Haddad L, Udomkesmalee E. & Co-Chairs of the Independent Expert Group of the Global Nutrition Report. The Global Nutrition Report 2015: What we need to do to advance progress in addressing malnutrition in all its forms. Public Health Nutrition 2015; 18: 3067–3069.

31. Kjellström T, Maître N, Saget C, et al. Working on a warmer planet: The effect of heat stress on productivity and decent work. International Labour Office, 2020. Available at: https://www.ilo.org/wcmsp5/groups/public/---dgreports/---dcomm/---publ/documents/publication/wcms_711919.pdf. Accessed November 1, 2022.

32. Tigchelaar M, Battisti DS, Spector JT. Work adaptations insufficient to address growing heat risk for U.S. agricultural workers. Environmental Research Letters 2020; 15: 094035. https://doi.org/10.1088/1748-9326/ab86f4.

33. Murray CJL, Aravkin AY, Zheng P, et al. Global burden of 87 risk factors in 204 countries and territories, 1990–2019: A systematic analysis for the Global Burden of Disease Study 2019. Lancet 2020; 396: 1223–1249.

34. Afshin A, Sur PJ, Fay KA. Health effects of dietary risks in 195 countries, 1990–2017: A systematic analysis for the Global Burden of Disease Study 2017. Lancet 2019; 393: 1958–1972.

35. Swinburn BA, Kraak VI, Allender S, et al. The global syndemic of obesity, undernutrition, and climate change: The Lancet Commission report. Lancet 2019; 393: 791–846.

36. Dicken SJ, Batterham RL. Ultra-processed food: A global problem requiring a global solution. Lancet Diabetes & Endocrinology 2022; 10: 691–694.

37. Springmann M, Mason-D'Cruz D, Robinson S, et al. Global and regional health effects of future food production under climate change: A modelling study. Lancet 2016; 387: 1937–1946.

38. Grace K, Davenport F, Funk C, Lerner AM. Child malnutrition and climate in sub-Saharan Africa: An analysis of recent trends in Kenya. Applied Geography 2012; 35: 405–413.

39. Levy K, Woster AP, Goldstein RS, Carlton EJ. Untangling the impacts of climate change on waterborne diseases: A systematic review of relationships between diarrheal diseases and temperature, rainfall, flooding, and drought. Environmental Science & Technology 2016; 50:4905–4922.

40. United Nations Childrens Fund. Strategy for improved nutrition of children and women in developing countries. Indian Journal of Pediatrics 1991; 58: 13–24.

41. Whitmee S, Haines A, Beyrer C, et al. Safeguarding human health in the Anthropocene epoch: Report of The Rockefeller Foundation-Lancet Commission on planetary health. Lancet 2015; 386: 1973–2028.

42. Springmann M, Mason-D'Cruz D, Robinson S, et al. Mitigation potential and global health impacts from emissions pricing of food commodities. Nature Climate Change 2016; 7: 69–74.

43. Black RE, Allen LH, Bhutta ZA, et al. Maternal and child undernutrition: Global and regional exposures and health consequences. Lancet 2008; 371: 243–260.

44. Frankenberg E, Friedman J, Ingwersen N, Thomas D. Linear child growth after a natural disaster: A longitudinal study of the effects of the 2004 Indian Ocean tsunami. Lancet 2017; 389: 21.

45. UNICEF. Child malnutrition. Available at: https://data.unicef.org/topic/nutrition/malnutrition/. Accessed March 8, 2023.

46. International Food Policy Research Institute. 2017 Global Food Policy Report. Washington, DC: IFPRI, 2017. Available at: https://www.ifpri.org/publication/2017-global-food-policy-report. Accessed November 1, 2022.

47. Our World in Data. Malnutrition: Share of children who are wasted. 2020. Available at: https://ourworldindata.org/grapher/share-of-children-with-a-weight-too-low-for-their-height-wasting. Accessed November 1, 2022.

48. Phalkey RK, Aranda-Jan C, Marx S, et al. Systematic review of current efforts to quantify the impacts of climate change on undernutrition. Proceedings of the National Academy of Sciences of the United States of America 2015; 112: E4522–E4529.

49. Blom S, Ortiz-Bobea A, Hoddinott J. Heat exposure and child nutrition: Evidence from West Africa. Journal of Environmental Economics and Management 2022; 115: 102698. https://doi.org/10.1016/j.jeem.2022.102698.

50. Cooper MW, Brown ME, Hochrainer-Stigler S, Zvoleff A. Mapping the effects of drought on child stunting. Proceedings of the National Academy of Sciences of the United States of America 2019; 116: 17219–17224.

51. Brown ME, Backer D, Billing T, et al. Empirical studies of factors associated with child malnutrition: Highlighting the evidence about climate and conflict shocks. Food Security 2020; 12: 1241–1252.

52. Danysh HE, Gilman RH, Wells JC, et al. El Niño adversely affected childhood stature and lean mass in northern Peru. Climate Change Responses 2014; 1: 1–10.

53. Grace K, Davenport F, Hanson H, et al. Linking climate change and health outcomes: Examining the relationship between temperature, precipitation and birth weight in Africa. Global Environmental Change 2015; 35: 125–137.

54. Hoddinott J, Kinsey B. Child growth in the time of drought. Oxford Bulletin of Economics and Statistics 2001; 63: 409–436.

55. Niles MT, Emery BF, Wiltshire S, et al. Climate impacts associated with reduced diet diversity in children across nineteen countries. Environmental Research Letters 2021; 16: 015010.

56. McMichael AJ. Extreme weather events and infectious disease outbreaks. Virulence 2015; 6: 543–547.

57. Mora C, McKenzie T, Gaw IM, et al. Over half of known human pathogenic diseases can be aggravated by climate change. Nature Climate Change 2022; 12: 869–875.

58. Cabrol J-C. War, drought, malnutrition, measles--A report from Somalia. New England Journal of Medicine 2011; 365: 1856–1858.

59. Schulte I, Bakhtary H, Siantids S, et al. Enhancing NCDs for Food Systems: Recommendations for Decision-Makers. Berlin: WWF Germany & WWF Food Practice, 2020. Available at: https://wwfint.awsassets.panda.org/downloads/wwf_ndc_food_final_low_res.pdf. Accessed October 30, 2023.

60. Global Commission on Adaptation. Adapt now: A global call for leadership on climate resilience. 2019. Available at: https://gca.org/wp-content/uploads/2019/09/GlobalCommission_Report_FINAL.pdf. Accessed November 1, 2022.

61. Fabregas R, Kremer M, Schilbach F. Realizing the potential of digital development: The case of agricultural advice. Science 2019; 366: eaay3038. doi: 10.1126/science.aay3038.

62. Benami E, Jin Z, Carter MR, et al. Uniting remote sensing, crop modelling and economics for agricultural risk management. Nature Reviews Earth & Environment 2021; 2: 140–159.

63. Boucher S, Carter MR, Flatnes JE, et al. Bundling genetic and financial technologies for more resilient and productive small-scale agriculture. National Bureau of Economic Research, 2021 Available at: https://www.nber.org/papers/w29234. Accessed November 1, 2022.

64. Nakalembe C, Becker-Reshef I, Bonifacio R, et al. A review of satellite-based global agricultural monitoring systems available for Africa. Global Food Security 2021; 29: 100543. https://doi.org/10.1016/j.gfs.2021.100543.

65. Gouel C, Laborde D. The crucial role of domestic and international market-mediated adaptation to climate change. Journal of Environmental Economics and Management 2021; 106: 102408. https://doi.org/10.1016/j.jeem.2020.102408.

66. van Berkum S, Ruben R. Exploring a food system index for understanding food system transformation processes. Food Security 2021; 13: 1179–1191.

67. Tenzing JD. Integrating social protection and climate change adaptation: A review. WIREs Climate Change 2020; 11. doi: 10.1002/wcc.626.

68. Ulrichs M, Slater R, Costella C. Building resilience to climate risks through social protection: From individualised models to systemic transformation. Disasters 2019; 43 (Suppl 3): S368–S387.

69. Fanzo J, Davis C, McLaren R, Choufani, J. The effect of climate change across food systems: Implications for nutrition outcomes. Global Food Security 2018: 18: 12–19.

70. Alderman H. Safety nets can help address the risks to nutrition from increasing climate variability. Journal of Nutrition 2010; 140: 148S–152S.

71. Sabates-Wheeler R, Devereux S. Cash transfers and high food prices: Explaining outcomes on Ethiopia's Productive Safety Net Programme. Food Policy 2010; 35: 274–285.

72. Vermeulen SJ, Campbell BM, Ingram JSI. Climate change and food systems. Annual Review of Environment and Resources 2012; 37: 195–222.

73. Miranda MJ, Farrin K. Index insurance for developing countries. Applied Economics Perspectives and Policy 2012; 34: 391–427.

74. Thomson MC, Mason SJ. Climate Information for Public Health Action. London: Routledge, 2019.

75. Gitonga ZM, Visser M, Mulwa C. Can climate information salvage livelihoods in arid and semiarid lands? An evaluation of access, use and impact in Namibia. World Development Perspectives 2020; 20: 100239. doi:10.1016/j.wdp.2020.100239.

Further Reading

Campbell BM, Vermeulen SJ, Aggarwal PK, et al. Reducing risks to food security from climate change. Global Food Security 2016; 11: 34–43.

This paper describes key gaps in existing evidence regarding the impact of climate change on food security and calls for an action-oriented research agenda, including stakeholder-driven priority technology and policy innovation options (for farmers, communities, and countries) as well as focused effort on adaptation for the most vulnerable.

Fanzo JC, Downs SM. Climate change and nutrition-associated diseases. Nature Reviews Disease Primers 2021; 7: 90.

This paper presents a summary of evidence on the pathways (direct and indirect) through which climate change impacts nutrition (malnutrition in all its forms) as well as the contribution of food systems to exacerbating climate change and mitigation potential in the food sector.

Global Panel on Agriculture and Food Systems for Nutrition. Future Food Systems: For People, Our Planet, and Prosperity. London: 2020. Available at: https://www.glopan.org/wp-content/uploads/2020/09/Foresight-2.0_Future-Food-Systems_For-people-our-planet-and-prosperity.pdf. Accessed September 13, 2023.

This report by leading experts aimed at government officials and other decision-makers contains key recommendations to transform food systems to increase health and to decrease their contribution to climate change. Recommendations include making sufficient, nutritious foods that are sustainably produced available for all; optimizing value chain efficiency; empowering consumers to make informed food choices and demand health and sustainability; and ensuring healthy and sustainable diets are affordable for all.

van Dijk M, Morley T, Rau ML, Saghai Y. A meta-analysis of projected global food demand and population at risk of hunger for the period 2010–2050. Nature Food 2021; 2: 494–501.

This systematic review and meta-analysis of the literature on food security projections to 2050 found consistent expectations of increased food demand and divergent estimates of the change in the proportion of people at risk of hunger. It includes a detailed discussion of the challenges in comparing predictions across studies and the role of key assumptions in models and integrated modeling research related to future food security.

Swinburn B, Kraak VI, Allender S, et al. The global syndemic of obesity, undernutrition, and climate change: The Lancet Commission report. Lancet 2019; 393: 791–846.

This report coined the word syndemic, which encompasses three global challenges that share pandemic features: obesity, undernutrition, and climate change. It summarizes evidence on key drivers, promising interventions, and the economic and health rationale for swift action and concludes with clear policy recommendations.

Wheeler T, von Braun J. Climate change impacts on global food security. Science 2013; 341: 508–513.

This influential paper called attention to the threat climate change poses for progress on global food security—specifically via pathways related to food availability, access, and utilization—and recommended attention to "climate-smart" actions within food systems to increase resiliency to climate change.

Willett W, Rockström J, Loken B, et al. Food in the anthropocene: The EAT-Lancet Commission on healthy diets from sustainable agri-food systems. Lancet 2019; 393: 447–492.

This paper presents the evidence of the food system drivers of climate change, food-related drivers of malnutrition and diet-related disease, and recommendations for a dietary pattern that optimizes human health and planetary impact.

9

Mental Health Impacts

Thomas J. Doherty and Amy D. Lykins

After a major hurricane affected the coasts of Connecticut, New York, and New Jersey, many people developed posttraumatic stress disorder (PTSD).

During a prolonged heat wave in the northwestern United States in 2021, many people living elsewhere worried about the health risks associated with climate change.

Because of repeatedly being forced to relocate their villages over the past two decades due to melting permafrost, Native Alaskans have repeatedly suffered from depression.

Mental health impacts of climate change, as these vignettes illustrate, can be placed into the following three categories: (a) direct impacts of disasters, extreme weather events, and a changed environment; (b) indirect vicarious impacts based on observation of global events and concern about future risks; and (c) indirect psychosocial impacts at the community and regional levels (Figure 9-1).[1]

In this chapter, *psychological* refers to basic processes of mental life, relationships, and behavior and to therapeutic activities, such as self-help or counseling, and *psychiatric* refers to mental health disorders and treatment, as diagnosed using the medical model. The psychological consequences of climate change have gone from being treated as a catalog of risks, hypotheses, and probabilities to a set of documented current effects throughout the world.[2-6]

There are several models and disciplinary lenses through which the psychological effects of climate change may be discerned, including

- Personal history: How a person has experienced climate change directly through personal exposure to hazardous events, media images, and/or social contacts
- Observations: How climate change has affected others in a person's family or social groups; whether a person discusses, or avoids discussing, climate change

Thomas J. Doherty and Amy D. Lykins, *Mental Health Impacts* In: *Climate Change and Public Health*. Edited by: Barry S. Levy and Jonathan A. Patz, Oxford University Press. © Oxford University Press 2024. DOI: 10.1093/oso/9780197683293.003.0009

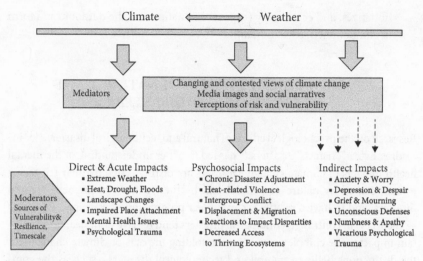

Figure 9-1 Climate change: Differentiating between classes of psychological impacts. (*Source*: Doherty TJ, Clayton S. The psychological impacts of global climate change. American Psychologist 2011; 66: 265–276. doi: 10.1037/a0023141.)

- Subjective impacts: How a person has experienced—or has seen in others' experience—emotions regarding climate change, such as feelings of curiosity, concern, stress, confusion, anger, or anxiety; whether or not a person has experienced heightened reactivity, sleep disturbances, altered mood, or decreased energy.

CONCEPTUAL MODELS

Conceptual models of the relationship of the environment to mental health include those related to the following:

- Influences of earlier climatic eras and associated physical environments—as long as thousands of years ago—on the human genome, physiology, migration, and culture[7]
- Environmental and ecological determinants of mental health, subjective well-being, and psychiatric disorders[8,9]
- Effects of natural landscapes and the built environment on mental health and mental well-being and how individuals develop identities and ethical behaviors in relation to the natural world[10,11]
- Psychotherapeutic approaches to address emotions and despair associated with climate change, such as environmental social work, grassroots

initiatives, and ecotherapy (exposure to nature and the outdoors as a form or component of psychotherapy).[12-13]

PSYCHOLOGICAL EFFECTS OF NATURAL AND TECHNOLOGICAL DISASTERS

Research on the psychological effects of natural and technological disasters,[14,15] including hybrid "natech" disasters,[16] has led to better understanding of the mental health impacts of climate change. Disasters are distinguished by their triggering events (caused by nature or people) and by their pattern of impacts. Natural disasters tend to have a relatively clear, linear progression of warning → impacts → recovery; in contrast, technological disasters tend to have nonlinear and uncertain impacts. The complex causes and unfolding impacts of climate change blur the distinctions between natural and technological disasters—such as the combination of a heat wave and failure of a local power grid causing loss of artificial cooling—and prompt responses associated with both of these types of disasters.[5,17]

Different types of disasters tend to evoke different reactions. Natural disasters tend to evoke altruistic or community-supportive responses; in contrast, technological disasters tend to evoke uncertainty and divisiveness (often exacerbated by preexisting social fissures).[18,19]

Disasters generally have distinct phases and evolving acute or chronic psychological impacts.[14,15,20] Human error and lack of preparedness may contribute to the severity of any disaster. However, technological disasters, given their human causes, have incubation, forewarning, and post-recovery periods that focus on inquiry, social justice, and social and legislative reform.[21]

CAUSAL PATHWAYS

The causal pathways and mechanisms for the direct and indirect psychosocial impacts of climate change are relatively clear. For example, underlying geophysical effects[22] include increased flooding in coastal cities, river delta regions, and inland cities near large rivers—a particular problem in low-lying countries, such as Bangladesh. Hotter and drier weather in the western United States, including Alaska, and in Australia has led to longer wildfire seasons and more widely distributed wildfires (Boxes 9-1 and 9-2). In the Arctic, recession of summer sea ice, which provides coastal protection, leads to more erosion following storms, threatening the displacement and relocation of many communities.

By increasing the frequency, severity, and duration of extreme weather events, climate change:

(*Text continues on page 186*)

Box 9-1 Assessing and Responding to Mental Health Impacts of Wildfires in Alaska

Micah Hahn

In Alaska, wildfires are increasing in intensity and frequency due to rapid climate change.[1] The wildfire season in Alaska is getting longer because earlier snow melt exposes dry grass in the spring, summer temperatures are higher, drought conditions are more extreme, and lightning strikes are becoming more frequent.[1] The frequency of million-acre fire seasons in Alaska in the past 20 years has been unprecedented.[1] A limited road network, vast forests, and dispersed rural populations mean that wildfire management and suppression in Alaska are costly and involve unique logistical challenges.[2]

Very little is known about the health impacts of wildfires in Alaska or possible adaptation strategies to protect the public's health from them. There has been a significant association between wildfire smoke and emergency department visits for cardiorespiratory illnesses,[3] and, by mid-century, it is estimated that exposure to fine particulate matter from wildfires will double in Interior Alaska.[4] In other regions, wildfires have been associated with behavioral changes, anxiety, PTSD, depression, and other mental health problems.[5]

In the summer of 2019, temperature records were broken across the state, reaching 11.2°-16.8°C (about 20°-30°F) above average in some communities. In Anchorage, the temperature reached 32.2°C (90°F) for the first time, compared to an average maximum summer temperature between 1991 and 2020 ranging from 18.1°C to 18.9°C (64.6°F to 66.0°F). Many communities experienced the most extreme drought recorded in Alaska in the past 20 years. In early June, lightning strikes ignited four fires on the Kenai Peninsula, one of which burned more than 167,000 acres over almost 150 days, required more than 3,000 firefighters and support personnel, and cost almost $49 million to suppress.

The almost 700 other wildfires that burned in the state in 2019 brought a long-lasting blanket of smoke across much of Alaska that summer. Anchorage had its first-ever dense smoke advisory at the end of June, and nearly one-third of the days between June and August that summer had 24-hour levels of fine particulate matter that were considered to be at least moderately unhealthy. Most Alaskan homes do not have air conditioning so the dual burden of heat and smoke left many residents unsure how to protect their families.

My colleagues and I determined that it would be useful to perform a local vulnerability and adaptation (V&A) assessment of mental health impacts of wildfires to guide strategies, policies, and interventions to promote resilience and recovery.[6] As a first step toward this V&A assessment, we used interviews and ranking workshops to identify mental health problems, coping strategies, existing supports, and gaps in support in communities that were affected by these wildfires. We interviewed 51 community members and wildland firefighters, mental healthcare providers, community advocates, policymakers, and public health professionals. Many of them also participated in discussion-based workshops to refine descriptions of and rank mental health problems and the supports that had been identified during the interviews. (See Chapter 15.)

As a result of this process, we identified the following categories of mental health issues related to the wildfires:

- Apprehension due to unknown wildfire-related risks and uncertainties, which were exacerbated by a perceived lack of communication from government officials
- Resignation and feelings of being overwhelmed after periods of prolonged wildfire-related stress
- Grief due to loss of natural landscapes, property, normalcy, and associated memories
- Feelings of being trapped due to dense smoke and to limited options for and understanding of evacuation
- Immediate and intense fear and distress related to the imminent threat of death or loss of property
- Lingering, intrusive memories, and anxiety related to previous natural disasters.

Interviewees also identified more than 100 potential interventions to address the mental health problems related to wildfires. They ranked a package of communications-related interventions as being the most practical and useful supports for these problems. This package included ensuring early communication, strategies to support equitable access to information in rural areas, and the use of community-based and participatory approaches to wildfire preparation and management. They also highly ranked leadership that acknowledged the connections between wildfires and mental health, measures to connect community members to formal and informal systems of mental healthcare, measures that enhanced the emergency shelter system, and crisis debriefing during evacuations from wildfires.

While the mental health problems and interventions that were identified may be generalizable, the information obtained in the V&A assessment is most relevant locally and in other Alaskan communities for developing and improving local intervention measures. Storytelling has been used extensively to communicate about people's experiences with wildfire-related mental health problems. And in this V&A assessment, it was an effective way to facilitate conversations among diverse stakeholders regarding assessments of wildfire-related mental health problems and identification of locally relevant adaptation strategies.

Box References

1. Grabinski Z, McFarland H. Alaska's Changing Wildfire Environment [Outreach Booklet]. Alaska Fire Science Consortium, International Arctic Research Center, 2020. Available at: https://uaf-iarc.org/wp-content/uploads/2020/12/Alaskas-Changing-Wildfire-Environment.pdf. Accessed January 23, 2023.

2. Todd SK, Jewkes HA. Wildland Fire in Alaska: A History of Organized Fire Suppression and Management in the Last Frontier. School of Agriculture and Land Resources Management, Agricultural and Forestry Experiment Station 2006. Available at: http://hdl.handle.net/11122/1313. Accessed January 23, 2023.

3. Hahn MB, Kuiper G, O'Dell K, et al. Wildfire smoke is associated with an increased risk of cardiorespiratory emergency department visits in Alaska. GeoHealth 2021; 5. doi: 10.1029/2020GH000349.

4. Woo SHL, Liu JC, Yue X, et al. Air pollution from wildfires and human health vulnerability in Alaskan communities under climate change. Environmental Research Letters 2020; 15: 094019. doi: 10.1088/1748-9326/ab9270.

5. To P, Eboreime E, Agyapong VIO. The impact of wildfires on mental health: A scoping review. Behavioral Sciences 2021, 11: 126. https://doi.org/10.3390/bs11090126.

6. Hahn MB, Michlig GJ, Hansen A, et al. Mental health during wildfires in Southcentral Alaska: An assessment of community-derived mental health categories, interventions, and implementation considerations. PLOS Climate, in press.

Box 9-2 Australia's "Black Summer Bushfire Season"

Amy D. Lykins

For many years, Australia has had a highly variable climate. It is now experiencing disproportionate increases in temperature, which are much greater than global warming worldwide.[1]

Following years of extreme drought, especially in the southeastern part of the country, bushfires started in June 2019 (winter in the Southern

Hemisphere) and continued until March 2020 in what is now referred to as the "Black Summer Bushfire Season"—the most extreme ever recorded, with enormous impacts and costs.[2] The fires burned an estimated 59.3 million acres (24 million hectares); destroyed approximately 6,000 buildings (including more than 3,000 homes); caused the death or displacement of nearly 3 billion mammals, birds, reptiles, and frogs; and killed 33 people. An estimated 429 additional people died as a result of smoke-induced respiratory problems, and more than 4,000 people were admitted to hospitals for respiratory and cardiovascular conditions. Almost half of the country's population was estimated to have been affected by bushfire smoke.

Data collected on people 16-25 years of age from New South Wales in the immediate aftermath of these fires revealed greater prevalence of depression, anxiety, stress, adjustment disorder symptoms, and substance use and lower psychological resilience in individuals who reported direct exposure to the bushfires compared to those who did not.[3] Although causality cannot be inferred from these data, the results are consistent with other reports of mental health sequelae to natural hazard exposure. Young people who reported exposure to the fires also showed higher levels of concern about and distress related to climate change and perceived less distance from the effects of climate change. Given the increasing likelihood of wildfires and related effects on mental health, the healthcare system and society at large need to be better prepared to address these challenges.

Box References

1. Intergovernmental Panel on Climate Change (IPCC). Regional Fact Sheet: Australasia. Sixth Assessment Report, Working Group I-The Physical Science Basis. 2021. Available at: https://www.ipcc.ch/report/ar6/wg1/downloads/factsheets/IPCC_AR6_WGI_Regional_Fact_Sheet_Australasia.pdf. Accessed September 7, 2022.

2. Australian Academy of Science. The Risks to Australia of a 3°C Warmer World. March 2021. Available at: www.science.org.au/warmerworld. Accessed September 7, 2022.

3. Lykins AD, Parsons M, Craig BM, et al. Australian youth mental health and climate change concern after the Black Summer bushfires. EcoHealth 2023; 20, 3–8. doi: 10.1007/s10393-023-01630-1.

- Affects mental health by inflicting more frequent and more severe natural disasters on human settlements, which, in turn, increases the likelihood of serious anxiety-related responses—and, later, the prevalence of chronic and severe psychological problems

- Increases the risk of injuries and other physical health problems, which are causally and reciprocally related to mental health
- Endangers the natural and social environments on which people depend for their livelihoods, health, and well-being.[4]

DIRECT PSYCHOSOCIAL IMPACTS

The environmental impacts of climate change, including extreme weather events, heat waves, droughts, wildfires, and floods, are associated with a wide range of mental health problems.[23–28] These include acute stress disorder and PTSD, somatic disorders, major depression, drug and alcohol abuse, suicide, and domestic violence. Local or immediate consequences, such as stressors resulting from extreme weather events or witnessing rapidly changing or degraded landscapes, may be perceived as both direct personal impacts and community-level manifestations of climate change.[29] Longer and more severe heat waves are associated with increasing mortality, homicide, suicide, physical assault, and spousal abuse in the general population; among people taking psychiatric medications, impaired temperature regulation during heat waves increases the risk of dehydration and heat stroke (Chapters 3, 4, and 10). Increases in the prevalence and severity of mental disorders in affected communities amplify the need for mental health services and disrupt the social, economic, and environmental determinants of mental health.[20,25,30]

TRAUMA AND STRESSOR-RELATED DISORDERS

Acute Stress Disorder and PTSD

Climate change can lead to short-term acute stress disorder (lasting 3 days to 1 month) and PTSD, characterized by significant mental impairment, intrusive memories, negative mood, dissociation, avoidance, and heightened arousal (manifested as sleep disturbance, irritable behavior, and/or hypervigilance). Diagnosis of PTSD requires that a person has experienced a traumatic event, has witnessed it occurring to family members or friends, or has had repeated or extreme exposure to aversive details of events.[31] Traumatic exposure via electronic media is considered a diagnostic criterion if it occurred in a workplace, such as where first-responders work. PTSD symptoms can occur in members of the general public, especially those exposed to repeated media images of traumatic events.[31,32]

Anxiety and Worry

Due its inherent uncertainty, unpredictability, and uncontrollability, climate change can induce anxiety. It is often difficult to differentiate between normative and pathological anxiety and worry about ecological threats, such as climate change. Even in the absence of direct exposure to known consequences of climate change, such as natural disasters and loss of biodiversity, symptoms can arise from autonomic stress-mediated responses and increased sensitization to stimuli (such as news media reports) associated with perceived climate threats. While habitual worrying is correlated with pathological conditions, such as generalized anxiety disorder, worries and apprehensions about ecological issues are correlated with pro-environmental attitudes and behaviors, and with positive personality traits, such as openness and agreeableness.[33] Much of what is considered *eco-anxiety* or *climate change anxiety* can be considered a common experience of anxiety and worry, which generally promotes engagement with climate change issues in the form of pro-environmental behaviors and activism.[34]

In contrast, *pathological eco-anxiety* is characterized by a chronic, strong form of fear and worry,[35] which can be debilitating. It may warrant a psychiatric diagnosis consistent with its prominent presenting features, such as generalized anxiety disorder, specific phobia, or obsessive-compulsive disorder.[36] Pathological eco-anxiety can overlap with normal-range experiences of existential anxiety and dread as people with minimal personal agency to effect significant change wrestle with questions of meaning and problems of daily life.[36] Individuals may also have realistic climate concerns overlaid on a preexisting anxiety disorder or history of anxiety disorders. Therefore, a nuanced assessment of individuals' mental health goals, their current and past functioning, and their subjective perception of climate impacts is warranted when helping to guide them through these stressors.

Depression and Grief

Responses to natural disasters may also be associated with depressive symptoms, such as intense sadness, rumination, insomnia, poor appetite, and weight loss. Those experiencing major impairment lasting 2 weeks or more may meet diagnostic criteria for a major depressive episode. It is important to delineate normal sadness and grief (bereavement) from a concurrent major depressive episode and/or the new diagnosis of *prolonged grief disorder*,[31] especially in highly vulnerable people and those with more severe symptoms or functional impairment. Making an accurate diagnosis requires clinical judgment based on the affected person's history and cultural norms for expressing distress in the context of loss.[31]

Ecological grief, as distinct from eco-anxiety, is characterized by anger, frustration, fear, stress, distress, hopelessness, and helplessness.[34] It can be caused by (a) physical losses, including damage or destruction to ecosystems and/or species; (b) disruption or erosion of environmental and place-based knowledge and identities; and (c) anticipation of these factors.[37] Although not a diagnosable disorder, these experiences are not uncommon among those who are concerned about the impact of climate change on people and the environment.

Adjustment Disorders

Less severe emotional or behavioral symptoms associated with identifiable disaster stressors lasting 3 to 6 months may be due to an adjustment disorder. Diagnostic criteria include marked distress and significant impairment relative to external context and cultural norms. Adjustment disorders may affect a large group of people or, when a natural disaster has occurred, an entire community.[31] Adjustment disorders are common but underdiagnosed due to the common experiences that can trigger them, taxing coping abilities of people but not so severely as in those who develop clinical depression or anxiety or other severe stress-related disorders.

VICARIOUS IMPACTS

Climate change can have mental health impacts on people and communities that do not directly experience its physical impacts. The vicarious mental health impacts of climate change are difficult to measure in many people, given the gradual, cumulative nature of these impacts. Experiences of impacts often occur via virtual media representations of climate change rather than from direct experience of stressors or environmental impacts.[38,39]

Causal mechanisms for vicarious impacts are complex. Ubiquitous social media and instantaneous communications technologies allow visceral and repeated exposure to far-away disaster events, thereby requiring a reconsideration of traditional epidemiological models that gauge the severity of impacts based on physical proximity to a disaster.[32,39] In addition, communications technology has mediating and moderating effects. For example, representations of disasters in the media can amplify the public's sense of threat or moderate public response by encouraging people to seek counseling and providing information about how to do so.

Basic psychological tasks associated with climate change include (a) adjustment to and coping with changing ecosystems and (b) acceptance of human causation of and responsibility for climate change.[40] Climate change provides an overarching story or narrative that connects and frames disparate global events. It can adversely affect people and their expectations for the future, thereby influencing their attitudes and behaviors related to climate change mitigation and adaptation.[38]

Vicarious impacts of climate change have been exacerbated by disinformation concerning fossil fuels,[41] which began in the late 1970s, and politically partisan interests that have created divisions in public opinion about climate science. Fossil fuel disinformation campaigns have included reframing climate issues as risks versus present dangers, shifting responsibility to individuals from corporate and government actors, and rationalizing continued reliance on fossil fuels.[42] (See Chapter 18.) These campaigns have caused a sense of confusion about the systemic causes of and solutions to climate change and created self-blame in some people, especially young people and those with strong environmental ethics. Individuals experiencing a sense of powerlessness concerning the scope of climate-related issues and the limits of their individual actions can feel like "climate hostages."[43]

Emotions Associated with Climate Change

Vicarious psychological impacts of climate change include intense emotions associated with (a) observing at a distance unprecedented global environmental and social changes and (b) uncertainty about future risks to humans and other species.[44] International data have demonstrated that negative emotions related to climate change are associated with insomnia and poor self-rated mental health.[45]

Emotional reactions to climate change differ among people living in different areas. For example, Indigenous Peoples of Arctic areas in Canada have reported that changes in landscapes, ice cover, and weather have made them feel anxious, sad, depressed, fearful, and angry and have lowered their sense of self-worth and perception of health.[46]

Denial and Apathy

Climate change has been likened to a serious progressive disease.[47] While much is known about the diagnosis and its cause, many questions remain about the prognosis and the efficacy and safety of proposed mitigation and adaptation "treatments." These questions have led some people to be skeptical about the seriousness of climate change.[48]

In psychological terms, denial can be understood as either a conscious social process—creating a group culture in which discussion of climate change is discouraged—or as an unconscious psychological defense—an involuntary mental mechanism that distorts perception of internal and external reality to reduce subjective distress.[48,49] Social theorists and interventionists contend that public apathy about climate change is better understood as a paralysis in the face of the magnitude of the issue. Public apathy may be better understood in the context of psychological defense mechanisms, such as splitting, intellectualization, and rationalization—retaining knowledge of reality but divesting it of emotional meaning.[49,50]

INDIRECT PSYCHOSOCIAL IMPACTS

Disorders and Causes

Indirect psychosocial impacts result from ongoing social and economic disruptions, psychological stressors, and physical health impacts of climate disasters and may also include elevated rates of chronic mood disorders and stress; relationship strain and physical violence; and increased suicidal ideation, attempts, and deaths by suicide.[44] Indirect psychosocial impacts of climate change also include distress associated with chronic, unfolding environmental disruptions over years or decades, such as drought, sea-level rise, large-scale environmental degradation, shortage of resources, and forced migration.[28,35,46] These impacts are best understood at the community and regional levels.[2]

Climate change is most concretely represented in the public mind as "global warming."[51] In addition to impacting health, heat and drought impact human behavior and well-being, with rural residents, members of farming families, or others with emotional and lifestyle connections to land being at greater risk of the social and emotional impacts of drought.[52-54] Indigenous Peoples may be especially vulnerable because the degradation or destruction of their lands may compound the loss of cultural traditions, leading to increased substance abuse, psychiatric problems, and suicide.[53] Children and young people may also be at especially high risk of mental health consequences, both from the direct effects of drought and from the effects of growing up in homes with caregivers suffering mental illness, such as those whose parents are farmers living in drought conditions.[55] Laboratory-based studies and comparisons of crime rates associated with seasonal and regional temperature differences indicate that uncomfortably hot temperatures increase the likelihood of physical aggression and violence.[56] Drought has also been associated with an increased suicide rate among adult males.[57]

Intergroup Conflict

Some climate impacts are indirect results of changes in how people use and occupy territory. Diminishing natural resources, such as freshwater, can lead to intergroup conflict. Ecological degradation can force one group to impinge upon another group's territory.[58,59] Some evidence suggests that prolonged severe drought may have contributed to the start of the Syrian Civil War.[59] (See Chapter 10.)

Displacement and Relocation

It has been projected that there could be 150 million *climate refugees* by 2050.[60] Communities in low-lying coastal areas, especially those in Small Island Developing States in the Pacific Ocean, are already being forced to relocate because of current or anticipated climate changes. Such forced relocations can involve a severing of emotional ties to place, disruption of social networks, and attempts to maintain cultural integrity despite relocation.[58,59] These disruptions of geographic and social connections may lead to grief, anxiety, and a sense of loss, especially among those with a strong sense of place or national identity.[61] (See Box 10-1 in Chapter 10.)

Diminished Access to Thriving Ecosystems

Even when secure in one's location, the degradation of familiar landscapes can lead to an emotional sense of loss and diminishment (*solastalgia*).[62] As ecosystems decline and the resources required to maintain them increase, climate change may adversely affect public parks and other green spaces in cities (Chapter 17).[63] Effects of climate change on local animal and plant species are likely to profoundly affect people living nearby. Rural residents who rely on natural resources for their economic well-being, such as those working in farming, forestry, and eco-tourism, and people whose identities are tied to a specific conception of place, such as Indigenous Peoples, are likely to be seriously affected by changes to existing ecosystems.[30,53,54] People who rely on subsistence living, such as Indigenous Peoples living in the Arctic, are especially sensitive to these impacts.[46]

MEDIATORS AND MODERATORS OF PSYCHOLOGICAL IMPACTS

The severity of climate change impacts is due not only to extreme events, but also to the interaction between human systems and these events. For example, direct impacts and indirect psychosocial impacts are moderated by structural and personal resilience or vulnerability.[2] Indirect vicarious psychological impacts can be moderated by social and cognitive factors.[45,48,49] Some people may perceive current danger and harm, while others may perceive that society will be able to adapt to the adverse effects of climate change. Personal concern about climate change is often mediated by cognitive appraisals, such as estimates of personal risk and attributions of responsibility, and moderated by the responses of people in one's social circle.[39]

SOCIAL JUSTICE IMPLICATIONS

The effects of climate change fall disproportionately on people with low socioeconomic status or people who are otherwise underprivileged, which raises questions of human rights and environmental justice. Some regions are disproportionately affected by the impacts of climate change, and societal vulnerability to these impacts is highly uneven due to varying sensitivity to risks and varying ability to cope and adapt.[64,65] Generally, those people and communities who face the highest risks related to climate change are those who are the least prepared and have inadequate access to mental health services.[66] And, in the aftermath of a disaster, mental health services may be further diminished.[67] (See Box 1-1 in Chapter 1 and also Chapter 20.)

Heat

Increasing heat exposure among those with relatively few resources can reduce income, disrupt daily social and family contacts, and create psychological distress.[4] After the "heat dome" over the northwestern United States occurred in 2021, there was increased climate change anxiety.[68] (See Chapters 3 and 4.)

Drought

In addition to impacts on the natural environment, drought results in financial hardships for residents of rural areas, especially for those who farm: lost income, debt, and damage to property. When livelihoods and identities are placed at risk, there is despair, which may lead to suicide, especially among those most vulnerable geographically and socioeconomically. Individuals living in rural areas are likely to suffer from multiple sources of disadvantage, placing them at increased risk of mental illness.[30,54,69] The severity and distribution of these mental health problems can be moderated by community cohesion, resilience, and external supports.

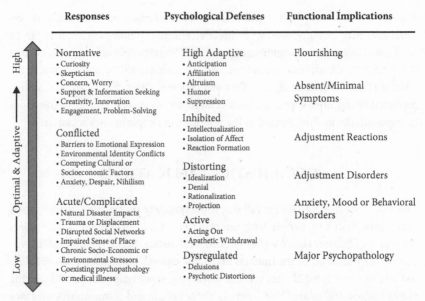

Figure 9-2 Responses to global climate change: A provisional psychological framework. (*Source*: Doherty TJ, Clayton S. The psychological impacts of global climate change. American Psychologist 2011; 66: 265–276. doi: 10.1037/a0023141.)

AN INTEGRATIVE FRAMEWORK

Understanding individual psychological responses to the range of mental health impacts potentially associated with climate change—and the adaptive benefits of those responses—is contingent on contextual factors, such as time frame, pre-existing health status, individual and cultural differences, social influences, and community resources. An integrative framework for categorizing responses to the mental health impacts of climate change, associated psychological defense mechanisms, and functional implications is shown in Figure 9-2.[2] This framework contains a range of healthy or adaptive responses, including:

- Emotions such as curiosity, concern, skepticism, and creativity
- Impulses toward conservation behaviors or other pro-social interests or causes
- High adaptive psychological defenses, such as anticipation, altruism, humor, and suppression, which provide an optimal balance between a conscious awareness of threatening issues and a sense of control and agency.

At the other extreme, maladaptive, acute, and disordered responses include:

- Trauma or displacement associated with environmental disasters or regional conflicts
- Debilitating stress
- Emergent psychopathology, such as anxiety or depressive disorders, and/or increased comorbidity with preexisting psychopathology or medical illness
- Avoidant coping, such as psychological minimization and substance abuse
- Psychotic or highly dysregulated psychological defenses, such as delusions, gross denial, or acting out.

Between these extremes of adaptive and maladaptive responses lies the potential for psychological distress, inhibition, and internal conflicts, especially in the face of vicarious impacts. These may include feelings of hopelessness or nihilism; distress regarding competing motivations related to personal, social, or socioeconomic factors; and compromise-level defenses, such as intellectualization, rationalization, or projection, which serve to keep potentially threatening events, ideas, or feelings out of awareness or misattributed to external causes.

PREVENTIVE MEASURES

To plan measures to prevent the mental health impacts of climate change, health professionals need to understand individual and community risks; sources of vulnerability and resilience in the people they are serving, especially in high-risk groups; and possibilities for intervention and transformation.[44] They also need to be aware of the tradeoffs and co-benefits of potential preventive measures.[35,70]

Health professionals should focus on the linkages between adaptation and mitigation. Promoting the health of affected people through effective adaptation measures can better equip them to understand the personal impacts of climate change and increase their capacity for coping, hope, and personal and collective agency. This increased capacity is associated with behaviors to promote mitigation of climate change.[71,72]

Vulnerability and Resilience

Proximity to the physical impacts of climate change, such as extreme weather events or sea-level rise, combined with other social vulnerability factors, such as population density, racial and ethnic minority status, and low socioeconomic status, can increase the severity of mental health impacts of climate change.[23,25,30,64] However, severity of the mental health impacts of climate change can be moderated by vulnerability and resilience factors. Processes that put some people in harm's way, such as preexisting medical conditions, living in marginal areas, or having precarious livelihoods, limit their options for mitigation and adaptation. In contrast, community resilience and potential to adapt to climate change is supported by uniform levels of mental health and functioning that provide the energy, creativity, and social cohesion needed to recover from stressors and disasters.[70,73] Community resilience is also associated with adaptive capacities for economic development, reduction of risk and resource inequalities, and engagement of a comprehensive range of community stakeholders in disaster preparedness and response.[72,73]

High-Risk Groups

Groups likely to be at greater psychosocial risk from climate change include women, children, older people, poor people, racial and ethnic minorities, migrants, people with poor family or social support, those with a history of emotional disability, people in low-income countries, and people otherwise marginalized.[74] People with preexisting mental illness are especially vulnerable to heat-related morbidity and mortality due to risk factors, such as use of psychotropic medications, preexisting respiratory and cardiovascular disease, substance misuse, and poor-quality housing.[28]

Young people have heightened vulnerability to the mental health impacts of climate change because of their relative inability to effect political change, their less developed abilities to cope and to find meaning in life, and the likelihood that they will experience continued exposure to these impacts over the course of their lives.[74] Data collected on an international sample of 10,000 people 16 to 25 years of age revealed that 84% were moderately or extremely worried about climate change and almost 50% reported that their feelings about climate change affected their daily lives.[75]

Even the most healthy and resilient persons may develop fatigue, strain, anxiety, and mood or behavioral problems when subjected to extreme stressors, such as those associated with climate-related disasters. Those who face overwhelming stressors could be at risk for developing psychiatric disorders.[76–78]

Tradeoffs and Co-Benefits

Mitigation and adaptation measures to address climate change have tradeoffs and co-benefits regarding mental health. For example, reduced automobile use by some people would reduce greenhouse gas emissions and potentially increase their physical activity, creating co-benefits of reducing risks for obesity and cardiovascular disease, as well as improvements in mood.[63,70] However, alternative travel by public transport, walking, or bicycling may increase travel time, possibly reducing access to work opportunities and human services and limiting supportive contact with family members and friends. (See Chapter 13.)

A *caring-for-country approach* recognizes that Indigenous Peoples derive strength and meaning from their deep knowledge and active stewardship of their ancestral lands. This approach promotes climate change adaptation for these highly vulnerable people and contributes to their social capital, psychological resilience, and community well-being.[70]

INTERVENTION STRATEGIES

Risk Factors

Traumatic mental reactions may be superimposed on preexisting anxiety, mood, or psychotic disorders or physical injuries. Trauma- and stressor-related psychological disorders are also associated with a variety of other risk factors, including preexisting socioeconomic disadvantage, age, and having directly experienced personal injury or the death or injury of a loved one. Negative appraisals, maladaptive coping strategies, exposure to upsetting reminders of the events and subsequent adverse life events, and financial or other trauma-related losses also increase risk.[72,76-78] Knowledge of risk factors and protective factors, such as social support and optimistic coping strategies, can help provide more accurate acute and chronic prognoses.[76,79]

Disaster Response

In the aftermath of disasters, early interventions to assist with basic needs and functional recovery have profound benefits.[80-82] Measures that, at the personal level, are attentive to family context and affective and emotional factors, and, at the community level, promote interagency cooperation and coordination are most effective. Intervention principles to guide early to mid-stage intervention and prevention measures include promoting safety, calming, a sense of self-efficacy and community efficacy, connectedness, and hope.[81]

Coping with Vicarious and Indirect Impacts of Climate Change

Some people believe that climate change caused by human behavior will ultimately cause widespread extinction of species, a belief that can be associated with profound feelings of loss. Treatment models have been designed to help people adjust to the reality of climate change and to move through grief and mourning toward a reinvestment of emotional energy in more ecologically sustainable life choices.

ROLES OF PUBLIC HEALTH PROFESSIONALS

Public health professionals can provide guidance to communities, governments, and nongovernmental and grassroots organizations on healthy coping with the mental health impacts of climate change. In the wake of disasters, they can intervene with short-term, practical coping assistance and identification of those who need more intensive support (psychological first aid) to limit unnecessary pain and suffering, reduce or prevent the development of psychiatric disorders among community members, and support coping and readjustment. Public health professionals can teach that optimal coping with the mental health impacts of climate change depends on accurate recognition of risks, effective management of emotions, a focus on pro-social outcomes, and engagement in mitigation and adaptation measures.

Public health professionals need to understand that reducing the mental health impacts of disasters requires, at the local level, first-responder interventions and effective triage and preparedness planning to reduce vulnerabilities, especially for high-risk groups. They also need to understand, at the societal and global levels, how natural systems and human activities interact to cause climate change and what can be done to mitigate and adapt to climate change.[83–85]

FROM ADAPTATION TO EMPOWERMENT

When contemplating the threat of mental health impacts associated with climate change, one can adopt a solution-focused perspective that highlights the benefits of engaging in conservation and experiencing healthy natural settings (Chapter 17). There are intrinsic benefits to be gained from pro-environmental behaviors, including a sense of participation and competence. Some people respond to the threat of climate change with social engagement, leading to a sense of agency and purpose and other positive emotions.

Promoting adaptive and effective coping and empowerment by people who are minimally affected by climate change but are nevertheless experiencing vicarious stressors and traumas can have benefits beyond emotional relief; it can lead to an *adaptation-mitigation cascade*, empowering people to take action to promote mitigation. Emotional responses to climate change and feelings of personal responsibility and efficacy are intertwined aspects of a psychological response to the threat of climate change. Emotional processes, in turn, mediate the extent of individual and collective engagement in environmentally significant behaviors, such as those that contribute to mitigation of greenhouse gas emissions. Therefore, adaptation to indirect vicarious impacts can promote mitigation, such as people reducing their carbon footprints.[85]

SUMMARY POINTS

- Mental health impacts of climate change include direct impacts, vicarious impacts, and indirect psychosocial impacts.
- Direct impacts of climate change on mental health include stress disorders, anxiety and worry, depression and grief (including ecological grief), and adjustment disorders.
- Vicarious impacts of climate change include intense emotions associated with unprecedented environmental and social changes and with uncertainty about the future.
- Indirect psychosocial impacts include substance abuse, violent behavior, and suicide.
- Preventive measures should focus on addressing risk factors for psychological disorders, preparedness and response to climate-related disasters, and helping people cope with the vicarious impacts of climate change.

References

1. Bourque F, Cunsolo Willox A. Climate change: The next challenge for public mental health? International Review of Psychiatry 2014; 26: 415–422.
2. National Research Council. Understanding and Responding to Climate Change. Washington, DC: National Research Council/The National Academies, 2009.
3. Doherty TJ, Clayton S. The psychological impacts of global climate change. American Psychologist 2011: 66: 265–276.
4. Jay A, Reidmiller DR, Avery CW, et al. Overview. In DR Reidmiller, CW Avery, DR Easterling, et al. (eds.). Fourth National Climate Assessment Volume II: Impacts, Risks, and Adaptation in the United States. Washington, DC: U.S. Global Change Research Program, 2018, pp. 33–71. Available at: https://nca2018.globalchange.gov/chapter/1/. Accessed September 12, 2022.
5. Cianconi P, Betro S, Janiri L, et al. The impact of climate change on mental health: A systematic descriptive review. Frontiers in Psychiatry 2020; 11: 74. https://doi.org/10.3389%2Ffpsyt.2020.00074.
6. Obradovich, N, Migliorini, R, Paulus MP, Iyad Rahwan I. Empirical evidence of mental health risks posed by climate change. Proceedings of the National Academy of Sciences 2018; 115: 10953–10958.

7. Burroughs WJ. Climate Change in Prehistory. Cambridge: Cambridge University Press, 2005.
8. Frumkin H. Environmental Health: From Global to Local. San Francisco: Jossey Bass, 2010.
9. Gatrell AC, Elliot SJ. Geographies of Health: An Introduction. Malden, MA: Wiley-Blackwell, 2009.
10. Clayton S. Oxford Handbook of Environmental and Conservation Psychology. New York: Oxford University Press, 2012.
11. Kahn PH, Hasbach PH. (eds.). Ecopsychology: Science, Totems, and the Technological Species. Cambridge, MA: MIT Press, 2012.
12. Gray M, Coates J, Hetherington T. Environmental Social Work. New York: Routledge, 2012.
13. Randall R. Loss and climate change: The cost of parallel narratives. Ecopsychology 2009; 1: 118–129.
14. Halpern J, Tramontin M. Disaster Mental Health: Theory and Practice. Belmont, CA: Thomson Brooks/Cole, 2007.
15. Reyes G, Jacobs GA (eds.). Handbook of International Disaster Psychology (Vol.1–4). Westport, CT: Praeger, 2006.
16. Cruz AM, Steinberg LJ, Vetere-Arellano AL. Emerging issues for natech disaster risk management in Europe. Journal of Risk Research 2006; 9: 483–501.
17. Marshall BK, Picou JS. Postnormal science, precautionary principle, and worst cases: The challenge of twenty-first century catastrophes. Sociological Inquiry 2008; 78: 230–247.
18. Gill DA. Secondary trauma or secondary disaster? Insights from Hurricane Katrina. Sociological Spectrum 2007; 27: 613–632.
19. Baum A, Fleming I. Implications of psychological research on stress and technological accidents. American Psychologist 1993; 48: 665–672.
20. Norris FH, Friedman MJ, Watson PJ. 60,000 disaster victims speak: Part II. Summary and implications of the disaster mental health research. Psychiatry: Interpersonal and Biological Processes 2002; 65: 240–260.
21. Aini MS, Fakhrul-Razi A. The development of socio-technical disaster model. Safety Science 2010; 48: 1286–1295.
22. Melillo JM, Richmond TC, Yohe GW (eds.). Climate Change Impacts in the United States: The Third National Climate Assessment. Washington, DC: U.S. Global Change Research Program, 2014. doi:10.7930/J0Z31WJ2.
23. Braun-Lewensohn O. Coping resources and stress reactions among three cultural groups one year after a natural disaster. Clinical Social Work Journal, October 2013. doi: 10.1007/s10615-013-0463-0.
24. Mills MA, Edmondson D, Park CL. Trauma and stress response among Hurricane Katrina evacuees. American Journal of Public Health 2007; 97: S116–S123.
25. Davidson JRT, McFarlane AC. The extent and impact of mental health problems after disaster. Journal of Clinical Psychiatry 2006; 67: 9–14.
26. Ahern M, Kovats RS, Wilkinson P, et al. Global health impacts of floods: Epidemiologic evidence. Epidemiologic Reviews 2005; 27: 36–46.
27. Bulbena A, Sperry L, Cunillera J. Psychiatric effects of heat waves. Psychiatric Services 2006; 57: 1519.
28. Stöllberger C, Lutz W, Finsterer J. Heat-related side-effects of neurological and non-neurological medication may increase heatwave fatalities. European Journal of Neurology 2009; 16: 879–882.

29. Furgal M, Martin D, Gosselin P. Climate Change and Health in Nunavik and Labrador: Lessons from Inuit Knowledge. In I Krupnik, D Jolly (eds.). The Earth is Faster Now: Indigenous Observations of Arctic Environmental Change (2nd ed.). Fairbanks, AK: Arctic Research Consortium of the United States 2010, pp. 266–299.
30. Morrissey SA, Reser JP. Natural disasters, climate change and mental health considerations for rural Australia. Australian Journal of Rural Health 2007; 15: 120–125.
31. American Psychiatric Association. Diagnostic and Statistical Manual of Mental Disorders (5th ed.). Arlington, VA: American Psychiatric Publishing, 2022.
32. Marshall RD, Bryant RA, Amsel L. et al. The psychology of ongoing threat: Relative risk appraisal, the September 11 attacks, and terrorism-related fears. American Psychologist 2007; 62: 304–316.
33. Verplanken B, Roy D. "My worries are rational, climate change is not": Habitual ecological worrying is an adaptive response. PLoS One 2013; 8: e74708.
34. Ojala M, Cunsolo A, Ogunbode CA, Middleton J. Anxiety, worry, and grief in a time of environmental and climate crisis: A narrative review. Annual Review of Environment and Resources 2021; 46: 35–58.
35. Clayton S, Manning CM, Speiser M, Hill AN. Mental Health and our Changing Climate: Impacts, Inequities, Responses. Washington, DC: American Psychological Association, and ecoAmerica, 2021.
36. Pihkala P. Anxiety and the ecological crisis: An analysis of eco-anxiety and climate anxiety. Sustainability 2020; 12: 7836. https://doi.org/10.3390/su12197836.
37. Cunsolo A, Ellis NR. Ecological grief as a mental health response to climate change-related loss. Nature Climate Change 2018; 8: 275–281.
38. Reser JP, Swim JK. Adapting to and coping with the threat and impacts of climate change. American Psychologist 2011; 66: 277–289.
39. Stokols D, Misra S, Runnerstrom MG, Hipp JA. Psychology in an age of ecological crisis: From personal angst to collective action. American Psychologist 2009; 64: 181–193.
40. Kolbert E. The Sixth Extinction: An Unnatural History. New York: Henry Holt and Co, 2014.
41. Franta B. Early oil industry disinformation on global warming. Environmental Politics 2021; 30: 663–668.
42. Supran G, Oreskes N. Rhetoric and frame analysis of ExxonMobil's climate change communications. One Earth 2021; 4. https://doi.org/10.1016/j.oneear.2021.04.014.
43. O'Hara D. Thomas Doherty works at the intersection of psychology and environmental science. American Psychological Association, June 7, 2018. Available at: https://ww.apa.org/members/content/doherty-psychology-environmental-science. Accessed September 12, 2022.
44. Fritze JG, Blashki GA, Burke S, Wiseman J. Hope, despair and transformation: Climate change and the promotion of mental health and wellbeing. International Journal of Mental Health Systems 2008; 2: 13.
45. Ogunbode CA, Pallesen S, Bohm G, et al. Negative emotions about climate change are related to insomnia symptoms and mental health: Cross-sectional evidence from 25 countries. Current Psychology 2021; https://doi.org/10.1007/s12144-021-01385-4.
46. Willox AC, Harper S. Edge V, et al. "The land enriches the soul": On environmental change, affect, and emotional health and well-being in Nunatsiavut, Canada. Emotion, Space and Society 2013 6: 14–24. doi:10.1016/j.emospa.2011.08.005.

47. Stern PC. Thinking appropriately about climate change. Behavioral Scientist, February 4, 2014. Available at: https://behavioralscientist.org/thinking-appropriat ely-about-climate-change/. Accessed September 12, 2022.
48. Norgaard KM. The Social Organization of Climate Denial: Emotions, Culture and Political Economy. In JS Dryzek, RB Norgaard, D Schlosberg (eds.). The Oxford Handbook of Climate Change and Society. Oxford: Oxford University Press, 2011.
49. Vaillant GE. Adaptive mental mechanisms: Their role in a positive psychology. American Psychologist 2000; 55: 89–98.
50. Weintrobe S. Engaging with Climate Change. New York: Routledge, 2013.
51. Meehl G, Tebaldi C More intense, more frequent, and longer lasting heat waves in the 21st century. Science 2004; 305: 994–997.
52. Dean JG, Stain HJ. Mental health impact for adolescents living with prolonged drought. Australian Journal of Rural Health 2010; 18: 32–37.
53. Rigby CW, Rosen A, Berry HL, Hart CR. If the land's sick, we're sick: The impact of prolonged drought on the social and emotional wellbeing of Aboriginal communities in rural New South Wales. Australian Journal of Rural Health 2011; 19: 249–254.
54. Berry HL, Hogan A, Owen J, et al. Climate change and farmers' mental health: Risks and responses. Asia-Pacific Journal of Public Health 2011; 23: 119S–132S.
55. UNICEF. In Their Own Words: The Hidden Impact of Prolonged Drought on Children and Young People. 2019. Available at: https://www.unicef.org.au/Upload/UNICEF/ Media/Documents/Drought-Report-2019.pdf. Accessed September 12, 2022.
56. Anderson CA, DeLisi M. Implications of Global Climate Change for Violence in Developed and Developing Countries. In J Forgas, A Kruglanski, K Williams (eds.). Social Conflict and Aggression. London: Psychology Press, 2011.
57. Hanigan IC, Butler CD, Kokic PN, Hutchinson MF. Suicide and drought in New South Wales, Australia, 1970-2007. Proceedings of the National Academy of Sciences 2012; 109: 13950–13955.
58. Reuveny R. Ecomigration and violent conflict: Case studies and public policy implications. Human Ecology 2008; 36: 1–13.
59. Abel GJ, Brottager M, Crespo Cuaresma J, Muttarak R. Climate, conflict and forced migration. Global Environmental Change 2019; 54: 239–249.
60. Rigaud KK, de Sherbinin A, Jones B, et al. Groundswell: Preparing for Internal Climate Migration. World Bank 2018. Available at: https://www.worldbank.org/en/ news/infographic/2018/03/19/groundswell---preparing-for-internal-climate-migrat ion. Accessed September 13, 2023.
61. Nelson DR, West CT, Finan TJ. Introduction to "In focus: Global change and adapta- tion in local places." American Anthropologist 2009; 111: 271–274.
62. Albrecht G, Sartore, G-M Connor L, et al. Solastalgia: The distress caused by environ- mental change. Australasian Psychiatry 2007; 15: S95–S98.
63. Younger M, Morrow-Almeida HR, Vindigni SM, Dannenberg AL. The built environ- ment, climate change, and health: Opportunities for co - benefits. American Journal of Preventive Medicine 2008; 35: 517–552.
64. McMichael AJ, Friel S, Nyong A, Corvalan C. Global environmental change and health: Impact, inequalities, and the health sector. British Medical Journal 2008; 336: 191–194.
65. Brouwer R, Akter S, Brander L, Haque E. Socioeconomic vulnerability and adaptation to environmental risk: A case study of climate change and flooding in Bangladesh. Risk Analysis 2007; 27: 313–326.
66. Jacob K, Sharan P, Mirza I, et al. Mental health systems in countries: Where are we now? Lancet 2007; 370: 1061–1077.

67. Jones L, Asare J, El Masri M, et al. Severe mental disorders in complex emergencies. Lancet 2009; 374: 654–661.
68. Bratu A, Card KG, Closson K, et al. The 2021 Western North American heat dome increased climate change anxiety among British Columbians: Results from a natural experiment. Journal of Climate Change and Health 2022; 6: 100116. https://doi.org/10.1016/j.joclim.2022.100116.
69. Stain HJ, Kelly B, Carr VJ, et al. The psychological impact of chronic environmental adversity: Responding to prolonged drought. Social Science & Medicine 2011; 73: 1593–1599.
70. Berry HL, Butler JRA, Burgess CP, et al. Mind, body, spirit: Co-benefits for mental health from climate change adaptation and caring for country in remote aboriginal Australian communities. NSW Public Health Bulletin 2010; 21: 139–145.
71. Koerth J, Vafeidis A, Hinkel J, Sterr H. What motivates coastal households to adapt pro-actively to sea-level rise and increasing flood risk? Regional Environmental Change 2013; 13: 897–909.
72. Norris FH, Stevens SP, Pfefferbaum B, et al. Community resilience as a metaphor, theory, set of capacities, and strategy for disaster readiness. American Journal of Community Psychology 2008; 41: 127–150.
73. Ebi KL, Semenza JC. Community-based adaptation to the health impacts of climate change. American Journal of Preventive Medicine 2008; 35: 501–507.
74. Crandon TJ, Scott JG, Charlson FJ, Thomas HJ. A social-ecological perspective on climate anxiety in children and adolescents. Nature Climate Change 2022; 12: 123–131.
75. Hickman C, Marks E, Pihkala P, et al. Climate anxiety in children and young people and their beliefs about government responses to climate change: A global survey. Lancet Planet Health 2021; 5: e863–e873.
76. Cutter SL, Finch C. Temporal and spatial changes in social vulnerability to natural hazards. Proceedings of the National Academy of Sciences 2008; 105: 2301–2306.
77. Haskett ME, Scott SS, Nears K, Grimmett MA. Lessons from Katrina: Disaster mental health service in the Gulf Coast region. Professional Psychology: Research and Practice 2008; 39: 93–99.
78. Neria Y, Shultz JM. Mental health effects of Hurricane Sandy: Characteristics, potential aftermath, and response. JAMA 2012; 308: 2572.
79. Doherty TJ, Lykins A, Piotrowski N, et al. Clinical Psychology Responses to the Climate Crisis. In GJG Asmundson (ed.). Comprehensive Clinical Psychology (2nd ed). Amsterdam: Elsevier, 2022, volume 11, pp. 167–183.
80. Vernberg EM, Steinberg AM, Jacobs AK, et al. Innovations in disaster mental health: Psychological first aid. Professional Psychology: Research and Practice 2008; 39: 381–388.
81. Hobfoll SE, Watson PB, Carl C, et al. Five essential elements of immediate and mid-term mass trauma intervention: Empirical evidence. Psychiatry: Interpersonal and Biological Processes 2007; 70: 283–315.
82. James LE, Welton-Mitchell C, Noel JR, James AS. Integrating mental health and disaster preparedness in intervention: A randomized controlled trial with earthquake and flood-affected communities in Haiti. Psychological Medicine 2019; 50: 342–352.
83. Frumkin H. Beyond toxicity: Human health and the natural environment. American Journal of Preventive Medicine 2001; 20: 234–240.
84. Weber EU, Stern PC. Public understanding of climate change in the United States. American Psychologist 2011; 66: 315–328.
85. Milfont TL. The interplay between knowledge, perceived efficacy and concern about global warming and climate change: A one-year longitudinal study. Risk Analysis 2012; 32: 1003–1020.

Further Reading

Doherty TJ, Lykins AD, Piotrowski NA, et al. Clinical Psychology Responses to the Climate Crisis. In: GJG Asmundson (ed.). Comprehensive Clinical Psychology (2nd ed). Amsterdam: Elsevier, 2022, Volume 11, pp. 167–183. https://dx.doi.org/10.1016/B978-0-12-818 697-8.00236-3.

> This chapter shows how a clinical psychology lens can be applied to the mental health impacts of climate change. It addresses a range of issues from therapeutic approaches to social justice.

Hayes K, Berry P, Ebi KL. Factors influencing the mental health consequences of climate change in Canada. International Journal of Environmental Research and Public Health 2019; 16: 1583. doi: 10.3390/ijerph16091583.

> This paper identified key factors that influenced the capacity to adapt to climate change, including social capital, sense of community, government assistance, access to resources, community preparedness, intersectoral/transdisciplinary collaboration, vulnerability and adaptation assessments, communication and outreach, mental health literacy, and culturally relevant resources.

Charlson F, Ali S, Benmarhnia T, et al. Climate change and mental health: A scoping review. International Journal of Environmental Research and Public Health 2021; 18: 4486. doi:10.3390/ijerph18094486.

> This scoping review of 120 original studies published between 2001 and 2020 found that most were quantitative, cross-sectional, conducted in high-income countries, and focused on the first of the World Health Organization global resource priorities: assessing mental health risks related to climate change.

van Nieuwenhuizen A, Hudson K, Chen X, Hwong AR. The effects of climate change on child and adolescent mental health: Clinical considerations. Current Psychiatry Reports 2021; 23: 88. https://doi.org/10.1007/s11920-021-01296-y.

> This paper reviewed the impacts of climate change on child and adolescent mental health and recommended that mental health clinicians address research gaps, obtain relevant clinical training, educate their communities, and support young people in their advocacy activities.

Berry HL, Bowen K, Kjellstrom T. Climate change and mental health: A causal pathways framework. International Journal of Public Health 2010; 55: 123–132.

> This paper provides a causal pathways framework for better understanding the relationship between climate change and mental health.

Australian Psychological Society. Climate Change. Available at: https://psychology.org.au/for-the-public/psychology-topics/climate-change-psychology.

> A description of one of the most long-standing climate initiatives among psychological associations globally.

Climate Change and Happiness Podcast. Available at: https://climatechangeandhappiness.com/.

> An example of a public-facing podcast that addresses research and therapy findings on the emotional impacts of climate change and coping activities in accessible language for a range of listeners.

10

Violence

Barry S. Levy

> After a severe hurricane that damaged their home, a husband and wife get into a heated argument during which he physically attacks her.

> After intense flooding that destroys much of a community, teenage gangs engage in gun violence.

> After a prolonged drought in a rural area, farmers and their families migrate to a large city, where their presence contributes to social and political instability and, ultimately, to a civil war.

At first glance, each of these three scenarios may appear to be unrelated to climate change. However, increasingly, heat waves, extreme weather events, floods, droughts, and other adverse consequences of climate change are contributing to the occurrence of violence. The relationships between climate change and violence are often not well understood. However, there is increasingly strong evidence that climate change can cause or contribute to many forms of violence.

In 2014, the Intergovernmental Panel on Climate Change stated: "Climate change can indirectly increase risks of violent conflicts by amplifying well-documented drivers of these conflicts such as poverty and economic shocks."[1] This chapter reviews some of the available scientific evidence, discusses needs for further research, and presents ways in which violence related to climate change can be prevented.

Violence, which is increasingly recognized as a major public health problem, is "the intentional use of physical force or power, threatened or actual, against oneself, another person, or against a group or community that either results in or has a high likelihood of resulting in injury, death, psychological harm, maldevelopment or deprivation."[2] Violence can be divided into three categories: self-inflicted, interpersonal, and collective violence. The following sections describe some of the evidence of associations between climate change and each of these categories of violence.

Barry S. Levy, *Violence* In: *Climate Change and Public Health.* Edited by: Barry S. Levy and Jonathan A. Patz, Oxford University Press. © Oxford University Press 2024. DOI: 10.1093/oso/9780197683293.003.0010

SELF-INFLICTED VIOLENCE

Two systematic reviews have found an association between climate change and self-inflicted violence. In 2018, a systematic review was conducted of 35 studies concerning associations between high temperatures and heat waves with mental health outcomes. Of the 17 studies that described the relationship between temperature and suicide, 15 found a significant association between increasing temperature and suicide frequency. This systematic review also found that three of four studies that examined the difference between male and female suicide risk at higher temperatures found a higher risk in males than females. In addition, this systematic review found increased risks of hospital admissions and emergency department visits related to mental health at higher temperatures.[3]

In 2019, a systematic review and meta-analysis of 14 studies in 10 countries on the association between temperature and suicide found that, for each 1.0° C (1.8°F) increase in temperature, there was a significant association with a 1% increased incidence of suicide. The authors concluded that their pooled results "provide strong evidence that rising temperature has a positive relationship with increased risk of suicide, especially completed suicide."[4]

INTERPERSONAL VIOLENCE

In 1989, an extensive review of studies on temperature and aggression found ubiquitous effects of heat on the occurrence of violence. The review concluded that hot temperatures increase aggressive motives and tendencies. Analyses of hotter years, quarters of years, seasons, months, and days have all found higher rates of aggressive behaviors, such as murders, rapes, assaults, and riots. Field studies yielded clear evidence that uncomfortably hot temperatures increase aggressive motives and behaviors.[5]

Experimental studies strongly support the conclusion that hot temperatures can and do increase aggression in many contexts.[6] For example, an individual-level, longitudinal study that investigated the relationship between ambient temperature and externalizing behaviors, such as aggression and delinquency, among urban-dwelling adolescents found that aggressive behaviors significantly increased with rising temperatures. This association was greatly decreased in neighborhoods that had more greenspace.[7]

A study of 38 police officers in the Netherlands who were assigned to a training session found that officers in uncomfortably hot conditions (27°C, or 81°F) were about four times more likely to draw their weapons and shoot targets than those in comfortable temperature conditions (21°C, or 70°F).[8]

Gender-Based Violence

Studies from many different countries have found an association between gender-based violence and both warmer temperatures and extreme weather events. For example, in Kenya, a study found a 60% increase in reported intimate partner violence (physical or sexual violence by a partner or husband) affecting women in counties that experienced a severe weather event.[9] In Bangladesh, a case study demonstrated that gender-based violence increased during and after a cyclone; in the immediate aftermath of the cyclone, women became vulnerable not only in their damaged homes, but also when they moved from place to place to receive healthcare or other services, usually without a male chaperone to accompany them.[10] In Brisbane, Australia, a study found that domestic violence occurred more frequently on hotter days.[11]

A systematic review of 41 studies found that gender-based violence increased during and after extreme weather and climate events, including storms, floods, droughts, heat waves, and wildfires. Most of these studies showed an increase in one or several forms of gender-based violence during or after extreme events, often associated with economic instability, food insecurity, mental stress, disrupted infrastructure, increased exposure to men, and exacerbated gender inequality.[12]

COVID-19 may further intensify the association between climate change and gender-based violence. Helen Clark, administrator of the United Nations Development Program, and colleagues have stated that conflict, climate change, and COVID-19 together create a breeding ground for sexual violence and gender-based violence.[13]

Community Violence

Several studies in the United States and other countries have found an association between ambient heat and violent crime. For example, studies have found that hotter U.S. cities and regions have higher violent crime rates than cooler ones, even after controlling for various factors.[5,14] A study in Minneapolis found an association between heat and rate of physical assault.[15] In the aftermath of Hurricane Katrina in 2005, which displaced more than 1 million people, homicide rates increased in cities to which migrants had moved, such as Houston.[16]

Similar findings have been found in other countries. For example, a study in Australia, which investigated the relationship between temperature and crime using an 11-year dataset, found that assault and theft were significantly higher in summer than in winter. In 96% of local government areas in New South Wales, the assault rate during summer was higher than the rate during winter.

Controlling for seasonality, they found that the numbers of daily assaults significantly increased with rising temperature.[17] Another study demonstrated that changes in rainfall patterns and frequent droughts in three African countries threatened the livelihood of pastoralists, precipitating violent conflict as they were forced to share dwindling resources and pastures.[18]

A review of 16 studies on a possible relationship between temperature and homicide rate found that in 9 of these studies there was a significant increase in the homicide rate with a rise in temperature and that all of the 7 other studies found an increase, although it was not statistically significant. The authors conservatively estimated that, for each 1.0°C (1.8°F) increase in temperature, there would be a 4–5% increase in homocides.[19]

A study which examined the association between climate change and neighborhood levels of violence over 20 years in St. Louis found that neighborhoods with higher levels of social disadvantage are very likely to experience higher levels of violence as a result of anomalously warm temperatures. The study's authors predicted that 20% of the most disadvantaged neighborhoods in St. Louis would experience more than half of the climate change-related increase in cases of violence.[20]

COLLECTIVE VIOLENCE

Collective violence—"the instrumental use of violence by people who identify themselves as members of a group . . . against another group or set of individuals, in order to achieve political, economic, or social objectives"[21]—includes war and other forms of armed conflict, state-sponsored violence (such as genocide and torture), and organized violent crime (such as gang warfare). This section focuses on armed conflict, which causes morbidity and mortality not only directly, but also indirectly by damaging the health-supporting infrastructure of society, forcibly displacing people, diverting human and financial resources, and leading to more violence. Armed conflict and other forms of collective violence also violate human rights and adversely affect the environment—and release large amounts of greenhouse gases (GHGs).[22]

Studies in widely varying geographical locations and time periods demonstrate that climate change contributes to armed conflict along with other factors, such as extreme poverty and socioeconomic inequities, militarism and availability of weapons, poor governance, intergroup animosity, and environmental stress.[22] Indeed, climate change is worsening extreme poverty and the socioeconomic inequities that contribute to armed conflict.[23] There is also evidence that climate-related disasters in ethnically fractionalized countries increase the risk of armed conflict.[24]

Starting in the 1990s, Thomas Homer-Dixon and colleagues warned that growing scarcities of renewable resources can contribute to social instability and violent conflict. Scarcity of important environmental resources, mainly cropland, forests, river water, and fish, can contribute to violence, often by generating severe social stresses that lead, within countries, to urban unrest, clashes among ethnic and cultural groups, and insurgency campaigns.[25-27]

Scarcity of renewable resources is not the only factor that contributes to violent conflict. For example, Solomon Hsiang and colleagues have asserted that there are eight nonexclusive pathways that may explain the association between climate change and armed conflict (and other forms of collective violence): government capacity, labor markets, inequality, food prices, altering of logistical constraints, misattribution of the causes of random events (such as to the performance of governments), psychological responses, and migration and urbanization.[28]

Complex linkages exist between factors in the climate system, natural resources (such as land, ecosystems, and biodiversity), human security (such as food, health, and energy), and societal stability (such as violence and conflict).[29,30]

Eleven climate and conflict experts from diverse disciplines have agreed that "climate has affected organized armed conflict within countries. However, other drivers, such as low socioeconomic development and low capabilities of the state, have been judged to be substantially more influential, and the mechanisms of climate-conflict linkages are not fully understood. Intensifying climate change is estimated to increase future risks of conflict."[31]

HISTORICAL STUDIES ON THE ASSOCIATION BETWEEN CLIMATE CHANGE AND VIOLENCE

Research has provided strong evidence of the association between climate change and violence. For example, from 1500 to 1800 in the Northern Hemisphere, climate change was the major driver of armed conflict and other large-scale humanitarian crises—and social mechanisms failed to prevent these crises. Falling ambient temperatures decreased agricultural production, which led to war and other major social problems, including inflation, famine, and population decline.[32] From 1560 to 1660 in Europe, cooling of the climate was the ultimate cause of successive agro-ecological, socioeconomic, and demographic catastrophes.[33] In eastern China over the past millennium, the frequency of warfare was significantly associated with Northern Hemisphere temperature oscillations, especially cooling phases that significantly decreased agricultural production.[34] In Europe over the past millennium, conflict was more intense when the climate was colder—although this association weakened in the Industrial Era.[35]

MODERN-ERA RESEARCH ON THE ASSOCIATION BETWEEN CLIMATE CHANGE AND VIOLENCE

Influence of Temperature

From 1950 to 2004, in the tropics, the probability of new civil conflicts increased 3–6% during El Niño years (when much of the continental tropics become substantially warmer and drier). Although changes were usually less extreme in more temperate latitudes, the El Niño-Southern Oscillation (ENSO) may have influenced the development of 21% of civil conflicts since 1950.[36]

From 1981 to 2002, in sub-Saharan Africa, there was a strong association between warmer temperature and civil war.[37] A 54% increase in armed conflict was predicted in Africa by 2030, with 393,000 battle-related deaths.[5] (However, during the past 30 years, while temperature in most of Africa has *increased*, the incidence of civil war has *decreased*.[38])

From 1990 to 2009, in East Africa, temperatures much warmer than normal were significantly associated with increased violence, but warmer temperatures were only a modest predictor of violence compared to political, economic, and physical geographic predictors.[39]

From 1997 to 2009, in Somalia, drought fueled conflict through shocks in the price of livestock. If the average temperature in East Africa rises approximately 3.2°C (5.8°F) by 2100, it is predicted that cattle prices will *decrease* 4%, but violent conflict will *increase* by about 58%.[40]

Influence of Rainfall

Much research has demonstrated that a substantial decrease in rainfall and associated drought results in increased conflict:

- When rainfall was significantly decreased, there was a significantly increased probability of internal conflict starting in the following year.[41]
- In sub-Saharan Africa, from 1990 to 2008, rainfall much above historic norms was associated with communal conflict.[42]
- In Africa, from 1997 to 2011, negative climate shocks during the growing seasons of main crops had a sizeable and persistent impact on increasing

conflict at the subnational level, especially violence against civilians; it was predicted that severe climate shocks during growing seasons would become 2.5 times as frequent during the next two decades, leading, in turn, to a 7% rise in the average incidence of conflict.[43]

- In India, from 1970 to 1995, marked decreases in rainfall were associated with conflict and *excessive* rainfall decreased the likelihood of conflict by almost 8%.[44]
- In Asia and Africa, from 1989 to 2014, for agriculturally dependent groups and politically excluded groups in very poor countries, a local drought was associated with increased likelihood of sustained violence.[45]

Some analysts argue that the causation of drought needs to be perceived in a broader and historical context. For example, it has been asserted that the drought in northern Uganda in 2016 and 2017 was erroneously understood as a globally induced natural hazard affecting local vulnerability. Rather, it is argued that "it should be seen as embedded with ongoing, longstanding, multiscalar processes of environmental devastation and generated by equally multiscalar histories of human engagement with the material processes of the planet."[46]

Less rainfall as a result of climate change will worsen freshwater shortages, especially in water-stressed countries in the Middle East, North Africa, and South Asia. These shortages are already worsening due to high population growth rates, rapid urbanization, industrialization, and modernization.

There is much evidence that water shortages increase conflict. In recent decades, intrastate and interstate conflicts over water have been increasing substantially, with more than two-thirds of them occurring in the context of violence—although generally low-level violence. The Water Conflict Chronology

Table 10-1 Water-related conflicts, 1800–2021, by time period

Time period	Number	Annual average
1800–1899	15	0.15
1900–1999	181	1.81
2000–2009	220	22.00
2010–2019	629	62.90
2020–2021	195	97.50

Source: Gleick PH. The Water Conflict Chronology. Pacific Institute. Available at: https://www.wor ldwater.org/conflict/list/. Accessed January 4, 2023.

of the Pacific Institute has demonstrated an especially sharp increase in the number of water-related conflicts during the past two decades (Table 10-1).

Climate change may have contributed to the start of the civil war in Syria. A drought there, from 2006 to 2010, transformed almost 60% of the land into desert. By 2009, the drought may have killed 80% of cattle in Syria. Hundreds of thousands of farmers migrated to cities, seeking work, and many perceived that they were mistreated by the government. Both the dislocation and difficult conditions of the farmers were among the complex set of factors that contributed to the start of the civil war.[47,48] A reassessment, however, argued that there was little merit to the claim that prewar drought and resultant large-scale migration and socioeconomic stresses contributed to the start of the war.[49] In addition, over the duration of the war, there has been evidence that dry conditions during growing seasons led to an increase in government-initiated attacks.[50]

In Mali, since the mid-1990s, droughts have occurred more often, placing increased stress on vulnerable people and a fragile environment in a country with weak political institutions as well as religious and ethnic tensions. In 2012, rebels in northern Mali began an anti-government uprising, which led to a declaration of an independent Islamic state, which was eventually overtaken by French and West African military forces.[47]

Results of some research in Africa, however, have *not* been consistent with an association between conflict and decreased rainfall and/or drought:

- In Africa, from 1960 to 2004, no direct association was found between drought and civil war, and it was concluded that the main causes of intrastate armed conflict were political.[51]
- In Kenya, from 1989 to 2004, years with below-average rainfall tended to have a peaceful influence on the following year.[52]
- In pastoral areas in Kenya, climate change was found not to explain violent conflict.[53]
- In pastoral areas in drought-prone northwestern Kenya, violent conflicts were found to have resulted from a complex interaction of sociocultural, economic, and political factors.[54]
- In the Sahel in West Africa, factors other than those directly related to the environment and resource scarcity, such as government corruption and obstructed mobility of livestock and herders, were found to have represented the most plausible explanations for violent conflict.[55]
- Another study found that civil wars in Africa can be best explained by ethno-political exclusion, poor national economies, and the collapse of the Cold War system.[56]
- A study found that, globally, from 1979 to 2006, *abundance* of water correlated with political violence.[57]

- Another study found, in sub-Saharan Africa from 1991 to 2007, that rainfall was associated with civil war and insurgency, and extreme deviations in rainfall—especially abundant rainfall—were strongly associated with violent events.[58]

META-ANALYSES OF STUDIES ON CLIMATE CHANGE AND VIOLENCE

Meta-analyses provide the strongest evidence of a causal association between climate change and violence. The most comprehensive investigation on climate change and human conflict has been a meta-analysis based on 60 longitudinal studies.[59] It found that deviations from normal precipitation—either below or above normal—and mild temperatures significantly increased conflict, especially in poorer populations. It estimated that each standard deviation in climate toward more *extreme* rainfall or warmer temperatures (equivalent to about a 3.0°C [5.4°F] rise above average in New York City temperatures) increased the frequency of intergroup conflict overall by 14%—and in some places by more than 50%. This meta-analysis concluded that, with rising temperatures over future decades, there could be substantial increases in conflict.[59] Critics have suggested that this meta-analysis suffers from selection bias and conflates climate with weather.[60]

Another meta-analysis and review, based on 50 quantitative studies of the relationship of climatological variables on violent conflict and sociopolitical instability, led to similar conclusions.[28] It demonstrated that, when temperatures are hot and precipitation is extreme, both conflict and sociopolitical instability increase. It found that studies that were best designed to determine causation overwhelmingly found strong associations between climatic anomalies and conflict/social instability and that climatic events influence many different types of conflict on a wide range of spatial scales.

INFLUENCE OF SEA-LEVEL RISE ON VIOLENCE

By 2100, sea level is projected to rise by 40–100 centimeters (15.7–39.4 inches).[61] How much sea level will *actually* rise depends on the degree of global warming, the amount of ice melting in polar regions (excluding sea ice), and other factors.

Sea-level rise will have its greatest impact on people living in coastal areas, where one-fifth of the world's population resides. This impact will be most profound on people living in low-lying Small Island Developing States and in densely populated coastal areas of countries like Bangladesh. Some of these island nations may disappear and some coastal regions may become uninhabitable, forcing many people to migrate within their own countries or to neighboring countries. In addition, sea-level rise will likely increase saltwater incursion into river deltas and coastal groundwater aquifers, damaging cropland, causing food and water shortages, and forcing people to migrate. These displacements are likely to cause social, economic, and political upheavals associated with violence as individuals and groups compete for control of land and other resources.

INFLUENCE ON VIOLENCE OF CLIMATE-INDUCED MIGRATION

Gradual climate change as well as extreme weather events are leading to an increased number of *climate refugees* (Box 10-1). Many people, especially in low- and middle-income countries, have been forcibly displaced because of higher temperatures accompanied by drought and extreme weather events accompanied by flooding.

PREVENTION OF VIOLENCE DUE TO CLIMATE CHANGE

In addition to mitigation of GHG emissions and implementation of a range of adaptation measures, violence due to climate change can be reduced by preventing conflicts becoming violent, addressing underlying factors that contribute to violence, and building an infrastructure for peace. A potential pathway from climate change to war is shown in Figure 10-1, including opportunities for interventions.

Prevention

There are no simple solutions to preventing collective violence due to climate change and the environmental scarcity caused by climate change, which is often a cause of this collective violence. In addition to measures to mitigate climate change, adaptation measures will need to be developed and implemented to reduce the risk of collective violence. These measures must include preventing or reducing the socioeconomic inequalities, violations of human rights, and other forms of social injustice that are often the underlying causes of violence. They also need to include implementing nonviolent approaches to resolving conflict, such as mediation.

Box 10-1 Migration Due to Climate Change

Barry S. Levy

As of early 2023, there were about 103 million people who had been forced to flee their homes, including about 33 million refugees and about 53 million internally displaced persons.[1] When people are forced to leave their homes and communities, their health and human rights are threatened.

The leading root causes of this forced migration are political instability, economic tensions, ethnic conflict, and environmental degradation. Except for ethnic conflict, these root causes can result from climate change. In addition, migration is increasingly recognized as an adaptation response to climate change.[2]

For thousands of years, *environmental migrants* have migrated because of drought, soil erosion, desertification, deforestation, and other environmental problems, often in association with population pressure and poverty.[3] (The term *environmental refugees* can be misleading because it may erroneously imply that these people always receive official refugee status in their host countries and because it does not include internally displaced persons, who often face greater threats to their health and human rights and are less likely than refugees to have access to safe food and water, sanitation, and health and social services.) At present, there are many *climate migrants* who have been forced to migrate because of higher temperatures, sea-level rise, droughts, flooding, extreme weather events, and the indirect consequences of climate change.

It has been projected that hundreds of millions of people may become climate migrants by 2050. The Institute for Economics and Peace, in 2020, projected that as many as 1.2 billion people in 31 countries that are not as sufficiently resilient to withstand ecological threats could be forced to migrate by mid-century.[4]

Migration due to short-term extreme weather events is usually temporary, generally occurs over short distances within countries, and follows established channels of movement.[5,6] These migrants generally return to their homes eventually to rebuild their lives. Migration due to long-term consequences of climate change, such as persistent drought and sea-level rise, is more likely to be permanent. These migrants are unlikely to ever return to their homes.

Risk of becoming a climate migrant varies depending on temporal position, geographic location, social position, and the type of society in which people live. Many people who are at high risk of becoming climate migrants have contributed negligibly, or not at all, to climate change. Climate migration raises important issues concerning human responsibility and the duty to provide humanitarian assistance, jurisdictional responsibility, assignment of liabilities and costs, political responsibility, and the need to change societal structures and conditions so that people are not forced to migrate.

Policy recommendations concerning climate migration have included the following:

- Avoiding promotion of migration as an adaptive response to climate hazards
- Preserving cultural and social ties of migrants
- Enabling and promoting participation of migrants in decision-making in places where they are relocated and resettled
- Reducing barriers for migrants' access to healthcare.[7]

Box References

1. UNHCR: The UN Refugee Agency. Refugee data finder: Key indicators. Available at: https://www.unhcr.org/refugee-statistics/. Accessed March 17, 2023.

2. Balsari S, Dresser C, Leaning J. Climate change, migration, and civil strife. Current Environment Health Reports 2020; 7: 404–414.

3. Myers N. Environmental refugees: A growing phenomenon of the 21st century. Philosophical Transactions of the Royal Society B: Biological Sciences 2002; 357: 609–613.

4. Henley J. Climate crisis could displace 1.2bn people by 2050, report warns. The Guardian September 9, 2020. Available at: https://www.theguard ian.com/environment/2020/sep/09/climate-crisis-could-displace-12bn-peo ple-by-2050-report-warns. Accessed March 17, 2023.

5. McMichael C, Barnett J, McMichael AJ. An ill wind? Climate change, migration, and health. Environmental Health Perspectives 2012; 12; 646–654.

6. Bardsley DK, Hugo G. Migration and climate change: Examining thresholds of change to guide effective adaptation decision-making. Population and Environment 2010; 32: 238–262.

7. Schwerdtle PN, Stockemer J, Bowen KJ, et al. A meta-synthesis of policy recommendations regarding human mobility in the context of climate change. International Journal of Environmental Research and Public Health 2020; 17: 9342. doi: 10.3390/ijerph17249342/.

Climate change is most likely to cause collective violence in situations and locations where manifestations of social injustice are prevalent, especially in LMICs and among socioeconomically disadvantaged people in high-income countries. Therefore, measures need to be developed and implemented to address the needs of these populations.

High temperature and drought

1 ↓

Crop failure and damage to farmland

2 ↓

Food shortage, loss of income, and distress migration

3 ↓

Socioeconomic and political instability

4 ↓

War

Figure 10-1 A potential pathway from climate change to war. Opportunities for interrupting this pathway, and therefore preventing war, include: At 1, providing technology assistance, irrigation, and resistant seeds; at 2, providing food, aid, and/or income support and building individual and community resilience; at 3, strengthening governance and civil society; and at 4, restricting arms and resolving conflicts without violence. (Diagram by Barry S. Levy.)

A better understanding of the complex pathways between climate change and violence may help to prevent violence related to climate change. Some models have been proposed to improve this understanding, such as the Climate, Aggression, and Self-Control in Humans (CLASH) model, which proposes that aggression and violence increase as climates become hotter and seasonal variation becomes smaller by influencing time orientation and self-control.[62]

Research

Future studies on the relationship between climate change and violence needs to strike a balance concerning the size of the geographic areas and populations studied. These studies need to be large enough to gather a sufficient amount of data but small enough so that the effects of subnational or local climatic conditions on the occurrence of violence are not diluted. In addition, they need to be based on adequate timeframes so that the effects of short-term weather events can be differentiated from the impacts of longer-term climate change.

In recent years, research has helped to improve our understanding of the impacts of climate change on social unrest; the complexity of the relationships among climate, migration, and social unrest; and how agricultural production patterns influence the risk of conflict. Important remaining gaps in research include the long-term implications of gradual climate change and the potential for violent conflict as well as the effectiveness of various interventions to reduce the risk of climate-related conflict.[63]

Psychological research may also improve our understanding of the relationship between climate change and conflict and the likely effectiveness of various interventions. Specifically, psychological research on intergroup threat, climate change mitigation and adaptation, and culture may provide valuable insights into understanding and responding to climate-induced conflict.[64]

Legal Obligations of Countries

Because of their legal and moral obligations to protect human rights, countries must work to prevent collective violence and protect human rights that are threatened by collective violence due to climate change. They have legal and moral obligations to mitigate climate change and thereby reduce the risk of its adverse consequences to health and human rights. And they have legal and moral obligations to promote and support adaptation to climate change.[65]

SUMMARY POINTS

- The ways in which climate change contributes to violence are generally not recognized.
- There is increasing evidence that climate change contributes to the occurrence of self-inflicted violence, interpersonal violence, and collective violence, including armed conflict.
- Concerning interpersonal violence, higher temperatures and extreme weather events have been associated with gender-based violence and community violence.
- There is substantial evidence that climate change, especially warmer temperature and extremes of precipitation, can contribute to the cause of armed conflict, although climate change is often less of a causative factor than social and political factors.
- In addition to mitigation of and adaptation to climate change, measures can be taken to prevent conflicts from becoming violent, address the underlying causes of violence, and build a stronger infrastructure for peace.

ACKNOWLEDGMENT

This chapter is based, in part, on the chapter on "Collective Violence," written by Barry S. Levy and Victor W. Sidel, in the first edition of this book.

References

1. Intergovernmental Panel on Climate Change. Climate Change 2014: Synthesis Report. Geneva: IPCC, 2015, p. 16. Available at: https://www.ipcc.ch/site/assets/uplo ads/2018/05/SYR_AR5_FINAL_full_wcover.pdf. Accessed January 5, 2023.
2. WHO Global Consultation on Violence and Health. Violence: A Public Health Priority (WHO/EHA/SPI.POA.2). Geneva: World Health Organization, 1996.
3. Thompson R, Hornigold R, Page L, Waite T. Associations between high temperatures and heat waves with mental health outcomes: A systematic review. Public Health 2018; 161: 171–191.
4. Gao J, Cheng Q, Duan J, et al. Ambient temperature, sunlight duration, and suicide: A systematic review and meta-analysis. Science of the Total Environment 2019; 646: 1021–1029.
5. Anderson CA. Temperature and aggression: Ubiquitous effects of heat on occurrence of human violence. Psychological Bulletin 1989; 106; 74–96.
6. Anderson CA, Anderson KB, Door N, DeNeve K. Temperature and aggression. Advances in Experimental Social Psychology 2000; 32: 63–133.
7. Younan D, Li L, Tuvlad C, et al. Long-term ambient temperature and externalizing behaviors in adolescents. American Journal of Epidemiology 2018; 187: 1931–1941.
8. Vrij A, Van der Steen J, Koopelaar L. Aggression of police officers as a function of temperature: An experiment with the fire arms training system. Journal of Community & Applied Social Psychology 1994; 4: 365–370.
9. Allen EM, Munala L, Henderson JR. Kenyan women bearing the cost of climate change. International Journal of Environmental Research and Public Health 2021; 18: 12697. https://doi.org/10.3390/ijerph182312697.
10. Rezwana N, Pain R. Gender-based violence before, during, and after cyclones: Slow violence and layered disasters. Disasters 2020; 45: 741–761.
11. Auliciems A, DiBartolo L. Domestic violence in a subtropical environment: Police calls and weather in Brisbane. International Journal of Biometeorology 1995; 39: 34–39.
12. van Daalen KR, Kallesøe SS, Davey F, et al. Extreme events and gender-based violence: A mixed-methods systematic review. Lancet 2022; 6: e504-e523.
13. Clark H, Bachelet M, Albares JM. Conflict, climate change, and COVID-19 combine to create a breeding ground for sexual and gender based violence. British Medical Journal 2022; 378:o2093. doi:10.1136/bmj.o2093.
14. Anderson CA, Anderson KB. Violent crime rate studies in philosophical context: A destructive testing approach to heat and southern culture of violence effects. Journal of Personality and Social Psychology 1996; 70: 740–756.
15. Bushman BJ, Wang MC, Anderson CA. Is the curve relating temperature to aggression linear or curvilinear? Assaults and temperature in Minneapolis reexamined. Journal of Personality and Social Psychology 2005; 89: 62–66.
16. Reuveny R. Ecomigration and violent conflict: Case studies and public policy implications. Human Ecology 2008; 36: 1–13.

17. Stevens HR, Beggs PJ, Graham PL, Chang H-C. Hot and bothered? Associations between temperature and crime in Australia. International Journal of Biometeorology 2019; 63: 747–77.

18. Leff J. Pastoralists at war: Violence and security in the Kenya-Sudan-Uganda border region. International Journal of Conflict and Violence 2009; 3: 188–203.

19. Chersich MF, Swift CP, Edelstein I, et al. Violence in hot weather: Will climate change exacerbate rates of violence in South Africa? South African Medical Journal 2019; 109: 447–449.

20. Mares D. Climate change and levels of violence in socially disadvantaged neighborhood groups. Journal of Urban Health 2013; 90: 768–783.

21. Krug EG, Dahlberg LL, Mercy JA, et al. Collective Violence, in World Report on Violence and Health. Geneva: World Health Organization, 2002, p. 215. Available at: https://apps.who.int/iris/handle/10665/42495. Accessed March 20, 2023.

22. Levy BS. From Horror to Hope: Recognizing and Preventing the Health Impacts of War. New York: Oxford University Press, 2022.

23. Stott R. Climate change, poverty, and war. Journal of the Royal Society of Medicine 2007; 100: 399–402.

24. Schleussner C-F, Donges JF, Donner RV, Schellnhuber HJ. Armed-conflict risks enhanced by climate-related disasters in ethnically fractionalized countries. Proceedings of the National Academy of Sciences USA 2016; 113: 9216–9221.

25. Homer-Dixon TF, Boutwell JH, Rathjens GW. Environmental change and violent conflict. Scientific American 1993; 268; 38–45.

26. Homer-Dixon TF. Environmental scarcities and violent conflict: Evidence from cases. International Security 1994; 19: 5–40.

27. Homer-Dixon TF. Environment, Scarcity, and Violence. Princeton: Princeton University Press, 1999.

28. Hsiang SM, Burke M. Climate, conflict, and social stability: What does the evidence say? Climatic Change 2014; 123: 39–55.

29. Scheffran J, Link PM, Schilling J. Theories and Models of Climate-security Interaction: Framework and Application to a Climate Hot Spot in North Africa. In J. Scheffran, M. Brzoska, HG Brauch, et al. (eds.). Climate Change, Human Security and Violent Conflict: Challenges for Societal Stability. Hexagon Series on Human and Environmental Security at Peace, Volume 8. Heidelberg: Springer-Verlag, 2012, pp. 91–131.

30. Scheffran J, Brzoska M, Kominek J, et al. Climate change and violent conflict. Science 2012; 336: 869–871.

31. Mach KJ, Kraan CM, Adger WN, et al. Climate as a risk factor for armed conflict. Nature 2019; 571: 193–197.

32. Zhang DD, Brecke P, Lee HF, et al. Global climate change, war, and population decline in recent human history. Proceedings of the National Academy of Sciences 2007; 104: 19214–19219.

33. Zhang DD, Lee HF, Wang C, et al. The causality analysis of climate change and large-scale human crisis. Proceedings of the National Academy of Sciences 2011; 108: 17296–17301.

34. Zhang DD, Zhang J, Lee HF, He Y. Climate change and war frequency in eastern China over the last millennium. Human Ecology 2007; 35: 403–414.

35. Tol RSJ, Wagner S. Climate change and violent conflict in Europe over the last millennium. Climatic Change 2010; 99: 65–79.

36. Hsiang SM, Meng KC, Cane MA. Civil conflicts are associated with the global climate. Nature 2011; 476: 438–440.

37. Burke MB, Miguel E, Satyanath S, et al. Warming increases the risk of civil war in Africa. Proceedings of the National Academy of Sciences 2009; 106: 20670–20674.
38. Buhaug H. Reply to Burke et al: Bias and climate war research. Proceedings of the National Academy of Sciences 2010; 107: E186–E187.
39. O'Loughlin J, Witmer FD, Linke AM, et al. Climate variability and conflict risk in East Africa, 1990–2009. Proceedings of the National Academy of Sciences 2012; 109: 18344–18349.
40. Maystadt JF, Ecker O, Mabiso A. Extreme Weather and Civil War in Somalia: Does Drought Fuel Conflict Through Livestock Price Shocks? International Food Policy Research Institute (IFPRI Discussion Paper 01243), 2013. Available at: http://www. ifpri.org/sites/default/files/publications/ifpridp01243.pdf. Accessed January 3, 2023.
41. Levy MA, Thorkelson C, Vörösmarty C, et al. Freshwater Availability Anomalies and Outbreak of Internal War: Results from a Global Spatial Time Series Analysis. Human Security and Climate Change: An International Workshop, June 2005. Available at: http://www.ciesin.org/pdf/waterconflict.pdf. Accessed January 3, 2023.
42. Fjelde H, von Uexkull N. Climate triggers: Rainfall anomalies, vulnerability and communal conflict in sub-Saharan Africa. Political Geography 2012; 31: 444–453.
43. Harari M, La Ferrara E. Conflict, Climate and Cells: A Disaggregated Analysis. Discussion Paper Series No. 9277. Centre for Economic Policy Research 2013. Available at http://www.cepr.org/pubs/dps/DP9277.asp. Accessed July 17, 2014.
44. Sarsons H. Rainfall and conflict: A cautionary tale. Journal of Development Economics 2015; 115: 62–72.
45. von Uexkull N, Croicu M, Fjelde H, Buhaug H. Civil conflict sensitivity to growing-season drought. Proceedings of the National Academy of Sciences USA 2016; 113: 12391–12396.
46. Branch A. From disaster to devastation: Drought as war in northern Uganda. Disasters 2018; 42: S306–S327.
47. NPR staff. How could a drought spark a civil war? Available at: http://www.npr.org/ 2013/09/08/220438728/how-could-a-drought-spark-a-civil-war. Accessed January 3, 2023.
48. Gleick PH. Water, drought, climate change and conflict in Syria. Weather, Climate, and Society 2014; 6: 331–340.
49. Selby J, Dahi OS, Fröhlich C, Hulme M. Climate change and the Syrian civil war revisited. Political Geography 2017; 60: 232–244.
50. Linke AM, Ruether B. Weather, wheat, and war: Security implications of climate variability for conflict in Syria. Journal of Peace Research 2021; 58: 114–131.
51. Theisen OM, Holtermann H, Buhaug H. Climate wars? Assessing the claim that drought breeds conflict. International Security 2011; 36: 79–106.
52. Theisen OM. Climate clashes? Weather variability, land pressure, and organized violence in Kenya, 1989-2004. Journal of Peace Research 2012; 49: 81–96.
53. Adano WR, Dietz T, Witsenburg K, Zaal F. Climate change, violent conflict and local institutions in Kenya's drylands. Journal of Peace Research 2012; 49: 65–80.
54. Opiyo FEO, Wasonga OV, Schilling J, Mureithi SM. Resource-based conflicts in drought-prone Northwestern Kenya; The drivers and mitigation mechanisms. Wudpecker Journal of Agricultural Research 2012; 1: 442–453.
55. Benjaminsen TA, Alinon K, Buhaug H, Buseth JT. Does climate change drive land-use conflicts in the Sahel? Journal of Peace Research 2012; 49: 97–111.
56. Buhaug H. Climate not to blame for African civil wars. Proceedings of the National Academy of Sciences 2010; 107: 16477–16482.

57. Salehyan I, Hendrix CS. Climate Shocks and Political Violence. Presentation at the Annual Convention of the International Studies Association, San Diego, CA, April 1, 2012. Available at: http://cshendrix.files.wordpress.com/2007/03/s_h_climateshocks_forweb.pdf. Accessed January 3, 2023.

58. Hendrix CS, Salehyan I. Climate change, rainfall, and social conflict in Africa. Journal of Peace Research 2012; 49: 35–50.

59. Hsiang SM, Burke M, Miguel E. Quantifying the influence of climate on human conflict. Science 2013; 341: doi. 10.1126/science.1235367.

60. Bohannon J. Study links climate change and violence, battle ensues. Science 2013; 341: 444–445.

61. Sweet WV, Hamlington DB, Kopp RE, et al. Global and Regional Sea Level Rise Scenarios for the United States: Updated Mean Projections and Extreme Water Level Probabilities Along U.S. Coastlines. NOAA Technical Report NOS 01. Silver Spring, MD: National Oceanic and Atmospheric Administration, National Ocean Service. Available at: https://aambpublicoceanservice.blob.core.windows.net/oceanserv iceprod/hazards/sealevelrise/noaa-nos-techrpt01-global-regional-SLR-scenarios-US.pdf. Accessed January 16, 2023.

62. Rinderu MI, Bushman BJ, Van Lange PAM. Climate, aggression, and violence (CLASH): A cultural-evolutionary approach. Current Opinion in Psychology 2018; 19: 113–118.

63. von Uexkull N. Security implications of climate change: A decade of scientific progress. Journal of Peace Research 2021; 58: 3–17.

64. Suh SM, Chapman DA, Lickel B. The role of psychological research in understanding and responding to links between climate change and conflict. Current Opinion in Psychology 2021; 42: 43–48.

65. Hall MJ, Weiss DC. Avoiding adaptation apartheid: Climate change adaptation and human rights law. The Yale Journal of International Law 2012; 37: 309–366.

Further Reading

Homer-Dixon TF. Environment, Scarcity, and Violence. Princeton: Princeton University Press, 1999.

This book asserts that environmental scarcities will have profound social consequences, such as contributing to insurrections, ethnic clashes, urban unrest, and other forms of civil violence, especially in LMICs.

Scheffran J, Brzoska M, Brauch HG, et al. (eds.). Climate Change, Human Security, and Violent Conflict: Challenges for Societal Stability. Heidelberg: Springer-Verlag, 2012.

This book explores associations among climate change, environmental degradation, human security, societal stability, and violent conflict.

Crawford NC. The Pentagon, Climate Change, and War: Chartering the Rise and Fall of U.S. Military Emissions. Cambridge, MA: MIT Press, 2022.

This book demonstrates how the U.S. economy and military have created a deep and long-term cycle of economic growth, fossil fuel use, and dependency. It also shows that, even as the U.S. military acknowledged and adapted to human-caused climate change, it resisted reporting its own GHG emissions. The author states that the most effective way of reducing military emissions would be for the United States to reduce the size and operations of its military.

PART III

DEVELOPING AND IMPLEMENTING POLICIES FOR MITIGATION

11

The Public Policymaking Process and the Power of Participation

Kathleen M. Rest

"Never doubt that a small group of thoughtful, committed [people] can change the world. Indeed, it is the only thing that ever has."
—Margaret Mead

"Change will not come if we wait for some other person or some other time. We are the ones we've been waiting for. We are the change that we seek."
—Barack Obama

This chapter will enable you to understand and use the public policymaking process to make change—change that creates broad-based and enduring improvements, interventions, and solutions to the issues and problems we face as a society, including the existential threat of climate change. Developing or improving public policy is critical for mitigating greenhouse gas (GHG) emissions and adapting to climate change. We, the people, have a vital role to play in this process. While this chapter focuses on public policymaking in the United States, the broad outlines of the process are similar to those in other democracies.

A BRIEF PRIMER ON PUBLIC POLICY

Public policy, defined here as the official or unofficial actions and decisions made by federal, state, and local governmental institutions, affects almost every aspect of our daily lives—the quality of the air, food, and water that sustain us; the safety of the products that we use in our homes; the codes and regulations for the buildings, roads, and other infrastructure that we use; the energy that powers our society; and how, when, and where we vote. Nongovernmental institutions also make policies that affect the public and their own members, customers, clients, and constituents, including private-sector businesses and civic, community, and

Kathleen M. Rest, *The Public Policymaking Process and the Power of Participation* In: *Climate Change and Public Health*. Edited by: Barry S. Levy and Jonathan A. Patz, Oxford University Press. © Oxford University Press 2024.
DOI: 10.1093/oso/9780197683293.003.0011

faith-based organizations. These groups are frequent stakeholders in the governmental public policy process.

Participants ("actors") in public policymaking are present in all three branches of government at the federal, state, and local levels. In the legislative branch, they are present in Congress and state and local legislative bodies, where they enact laws and pass budgets. In the executive branch, they are present in the office of the chief executive (such as the President, governor, or mayor) and in the agencies that implement and enforce the laws by developing rules and regulations (terms often used interchangeably) as well as related policies and programs. In the judicial branch, they are present in the courts that interpret and apply laws to individual cases brought to the courts by parties who believe they are adversely affected by a law or agency action. Some of these actors are elected by the voting public; others are appointed. (USA.Gov provides an accessible tutorial on the executive, legislative, and judicial branches of the U.S. government.[1])

Official public policy can take many forms: a law, such as the Clean Air Act[2]; a regulation, such as the Environmental Protection Agency (EPA) ozone air quality standard[3]; a local ordinance, such as the City of Boston Building Emissions Reduction and Disclosure Ordinance (BERDO)[4]; or a formal statement of policy by a government agency or official, such as the EPA's Climate Adaptation Policy Statement.[5] (The annual Congressional Research Service summary of climate change policy efforts in the United States includes examples of these various mechanisms.[6]) (See Box 11-1.)

Unofficial policies can also exist in governing bodies and other entities in the form of unspoken, unwritten, and culturally embedded attitudes, views, opinions, preferences, and historic ways of doing things. They include, for example, a presidential administration's anti-regulatory culture or disregard of science[7] and the "glass ceiling," which has created barriers for women and minorities seeking career advancement and movement into positions of power. Such unofficial policies can powerfully influence agency action or inaction.[8]

THE PUBLIC POLICYMAKING PROCESS

The public policymaking process is generally described as a cycle with a series of stages or steps. The general public and other stakeholders can have influence throughout. Conceptual models and frameworks vary slightly in the number and description of the steps in the process,[9,10] but they all basically involve:

- Identifying the issue or problem and getting it on policymakers' agenda
- Formulating and considering policy options
- Choosing and further developing a specific policy option
- Formally adopting the policy
- Implementing the policy
- Evaluating results of policy implementation.

Box 11-1 Examples of Existing Federal Laws, Regulations, and Other Policies Related to Climate Change

Kathleen M. Rest

- In 2022, Congress passed and the President signed the Inflation Reduction Act, a historic piece of legislation, which addresses the climate crisis and energy security. The Act included provisions that lower energy costs for households and create jobs in the solar panel, wind turbine, and electric vehicle manufacturing sectors.[1]
- The EPA has issued numerous regulations related to climate change, such as those that set and enforce air pollution levels, regulate vehicle emissions and fuel use, and phase down production and consumption of hydrofluorocarbons, which are potent GHGs.[2] For example, in 2021, it proposed a rule to sharply reduce methane and other air pollutants from both new and existing sources in the oil and gas industry, and then updated and strengthened its proposal in 2022.[3] Public hearings were held in January 2023, and the public comment period closed on February 13, 2023. EPA received over 500,000 public comments on the 2022 supplemental proposal on top of the over 400,000 received on the 2021 proposal. While the rule is not final yet (as of November 2023), the EPA has publicly stated the rule will be finalized in 2023. (Its efforts to regulate GHG emissions from power plants were ruled as unconstitutional by the Supreme Court.[4])
- The U.S. Department of Agriculture has supported a diverse range of farmers, ranchers, and private forest landowners through its Partnerships for Climate-Smart Commodities, investing more than $3.1 billion for projects that require meaningful involvement of small and underserved producers.[5]

Box References

1. Blue Green Alliance. A User Guide to the Inflation Reduction Act. Available at: https://www.bluegreenalliance.org/site/a-user-guide-to-the-inflation-reduction-act/. Accessed August 15, 2023.

2. U.S. Environmental Protection Agency. Climate Change Regulatory Actions and Initiatives. Available at: https:www.epa.gov/climate-change/climate-change-regulatory-actions-and-initiatives. Accessed August 15, 2023.

3. U.S. Environmental Protection Agency. EPA issues supplemental proposal to reduce methane and other harmful pollution from oil and natural gas operations. Available at: https://www.epa.gov/controlling-air-pollution-oil-and-natural-gas-industry/epa-issues-supplemental-proposal-reduce. Accessed August 15, 2023.

4. The National Law Review. U.S. Supreme Court case limits the authority of the EPA in regulating air emissions. Available at: https://www.natlawrev iew.com/article/us-supreme-court-case-limits-authority-epa-regulating-air-emissions. Accessed August 15, 2023.

5. U.S. Department of Agriculture. Partnerships for Climate-Smart Commodities. Available at: https://www.usda.gov/climate-solutions/climate-smart-commodities. Accessed January 2, 2023.

Steps in the policymaking process are not necessarily linear or sequential in na-ture; they are interrelated and can have multiple feedback loops. The process can be initiated by a government agency, members of the public, or other groups and organizational stakeholders. As policymakers and stakeholders participate or as situations change, earlier steps in the process may be revisited to clarify, up-date, or otherwise gather additional input for decisions. In addition, the entire policymaking process is influenced by many other factors, including:

- Real-world situations, such as a pandemic, a war, an extreme weather event, a building collapse, a fire, the state of the economy, or social unrest
- Existing laws and regulations, institutional structures and relationships, and other administrative and social processes
- Political power and sociopolitical ideologies
- Financial incentives and disincentives
- Values, visions, and goals of stakeholders
- Public pressure and participation.

NUTS AND BOLTS

The following sections describe the details of policymaking at the federal, state, and local levels in the United States, as well as opportunities for individuals and organizations to participate in the process. (The Appendix to this chapter defines some terms frequently used in the policymaking process.)

At the Federal Level

The House of Representatives and Senate develop and vote on proposed laws (bills) (Figure 11-1).[11] Any elected member of Congress can propose a bill, often inspired by or in response to constituents' concerns, ideas, and requests. Bills proposed and passed in one chamber are then considered and voted on in the

Figure 11-1 The House of Representatives and the Senate (collectively the U.S. Congress), which comprise the legislative branch of the federal government, meet at the U.S. Capitol. (Photograph by MH Anderson Photography/Shutterstock.com.)

other. Any differences between bills passed by the House and Senate are resolved in bicameral conference committees. Both chambers then vote on the identical bill, which, if approved, moves to the President's desk. When signed by the President, the bill becomes law—and is subsequently called an *act* or a *statute*. If the President vetoes the bill, Congress can override the veto by a two-thirds vote. Aggrieved parties may ask federal courts—and ultimately the Supreme Court— to review a law, often on constitutional grounds. If the Court finds the law unconstitutional, it can invalidate the law (Figure 11-2).

The President can also issue executive orders that direct executive branch agencies to take various actions. Executive orders can have the force of law, but they can be revoked by a subsequent President. (The National Archives has an online database of executive orders issued from 1937 to January 2017[12]; subsequent executive orders can be found in the *Federal Register*.[13]) President Joe Biden used this authority to issue many executive orders related to climate change.[14]

Executive branch agencies implement the laws passed by Congress and signed by the President (Figure 11-3). They issue rules and regulations to implement them, describe what is and is not legal under the law, and detail specific compliance requirements for individuals, businesses, state and local governments, and other organizations and stakeholders. The process of agency rulemaking is governed by the enabling legislation and by the Administrative Procedures Act

Figure 11-2 The U.S. Supreme Court is part of the judicial branch of the federal government. (Photograph by Heidi Besen/Shutterstock.com.)

Figure 11-3 The headquarters of the U.S. Environmental Protection Agency, part of the executive branch of the federal government, in the William Jefferson Clinton Federal Building in Washington, DC. (Photograph by Rob Crandall/Shutterstock.com.)

(APA).[15,16] It requires these agencies to publish notices of proposed and final rulemaking in the *Federal Register* and provides opportunities for public participation in the rulemaking process. The APA also establishes a standard for judicial review (unless another is specified in the enabling legislation) and requires most rules and regulations to have an effective date delayed by a minimum of 30 days.

In developing rules and regulations, executive agency experts and advisors research pertinent information. The agency *may* publish an Advanced Notice of Proposed Rulemaking (ANPR) or a Request for Information (RFI) to seek input on the need for a rule and possible alternatives to rulemaking, as well as to collect relevant qualitative and quantitative information. The Small Business Regulatory Enforcement and Fairness Act requires some agencies, such as the Occupational Safety and Health Administration (OSHA) and the Environmental Protection Agency (EPA), to review and issue a report on the potential impact of the proposed regulation on small businesses.

A Notice of Proposed Rulemaking (NPRM), including the proposed regulatory text and preamble, is required by the APA and is published by the relevant executive branch agency in the *Federal Register*, along with details for how the public can comment on the proposal by submission of written comments and/ or participation in public hearings. Based on this input, the agency can issue a final regulation, which is reviewed and approved by the White House and then published in the Code of Federal Regulations (CFR). Alternatively, the agency can decide to rescind it or propose a different regulation. Subsequent lawsuits may delay implementation of the rule/regulation until its legitimacy is resolved in the courts.

In addition to its normal lengthy regulatory process,[17] the Occupational Safety and Health Act allows OSHA to issue emergency temporary standards (ETSs) when it determines that workers are in grave danger and immediate action is needed. ETSs usually take effect immediately, stay in effect for no longer than 6 months or until a permanent standard is issued, and can be challenged in court. OSHA has used this authority to issue an ETS to protect healthcare workers from COVID-19.[18] It *could* use its authority to issue a heat standard that would apply across all its jurisdictions. Oregon, as an OSHA state plan state, did exactly that in 2021.[19]

Congress gave itself power to overturn an agency's final rule (within certain time limits) by using the Congressional Review Act (CRA), a little known but powerful oversight tool.[20,21] It is not subject to filibuster in the Senate. If both chambers approve a resolution to overturn an agency's final rule, it goes to the President's desk for signature. Once a regulation is overturned under the CRA, the agency cannot issue another regulation that is "substantially the same" unless authorized by new legislation.

At the State Level

Each state has its own rules, processes, and procedures for enacting laws and writing and implementing regulations. State courts can review these laws and invalidate them if they determine they violate the state's constitution. The Tenth Amendment of the Constitution defines the principle of federalism and states' rights. Powers that are neither delegated to the federal government by this amendment nor prohibited to the states by the amendment are reserved for the states. However, federal laws can preempt (invalidate) conflicting state laws. (The Library of Congress provides an online guide to lawmaking in each U.S. state and territory.[22])

At the Local Level

Counties, cities, towns, and other municipalities have their own processes and procedures for enacting laws and ordinances, covering such local matters as zoning, development, renting, parking, local taxes, employment, public works, public health and safety, and budgets for local police and fire departments and other local service units. In *home rule states,* localities have the authority to govern themselves unless the state government specifically prohibits them from doing so.[23] In other states, local governments have only the limited authority granted to them by the state legislature.

Localities may also include *special districts—usually* single-purpose government units authorized by state law that operate independently and separately from local government. Special districts may perform a variety of functions, such as those related to schools, airports, highways, hospitals, libraries, parks, cemeteries, sewage, and conservation.[24] For example, school districts may hire and dismiss staff, select curricular materials, and manage budgets. Park districts may oversee and manage public parks and related cultural facilities, such as monuments and zoos. Conservation districts focus on conserving the natural environment and protecting local ecosystems, habitats, and natural resources, including land, water, soil, and forests.

PUBLIC PARTICIPATION IN POLICYMAKING

Public policies affect some, if not most, aspects of our daily lives, and participating in their development and implementation is integral to the principles and values of a democratic society. Public participation helps ensure that governmental institutions are responsive and accountable to its citizens. It creates opportunities for individuals, groups, and organizations to bring their issues, problems, or needs to the attention of policymakers, get them on the policymakers' action agendas, and influence the development of laws and regulations. It enhances mutual learning and understanding and facilitates the contribution of essential community-based knowledge, information, experience, and insight that can often be lacking in technically driven expert processes. Therefore, public participation builds community and government capacity to address societal issues and problems while providing transparency and building trust, validity, and legitimacy of governmental decisions, policies, and policymaking processes.

Public participation is a *process* that engages the public in governmental decision-making and policymaking—from getting the issue on the policy agenda in the first place to influencing the outcome.[25,26] Policymaking is not a single, check-the-box event but rather a process of communication, information exchange, interaction, consultation, and some degree of shared decision-making between government policymakers and affected members of the public—that is, stakeholders who are directly or indirectly impacted by a problem and potential policies to address it.

The individuals and groups most affected are often the least powerful, most vulnerable, and most socially and economically disadvantaged members of a community. They are often not adequately represented in policymaking processes. This inequity requires the policymaking entity to give them special attention, and engage them creatively and effectively.

Tools, Mechanisms, Processes, and Venues for Public Participation

Governments have many tools and options for engaging the public, each with advantages and disadvantages as well as varying time and resource commitments.[27] These options include public meetings, notice and comment periods on proposed rules and policies, negotiated rulemaking processes, educational and technical workshops, forums, task forces, advisory committees, surveys, and polls. Digital and web-based opportunities for input can be useful, especially if they facilitate debate and dialogue; however, they often exclude people who lack access to the technology.[28] There is no single best option for engaging the public; ideally, a variety of measures and tools are used to ensure reach, inclusivity, and ample opportunity for members of the public to participate in meaningful dialogue and deliberation and to influence both the process and the outcome.

When selecting and designing participation measures, other considerations include the following:

- Logistics: Time, location, and frequency
- Substance: Determining level of technical information and addressing public concerns

- Cultural and interpersonal aspects: Existing relationships, comfort in public speaking, and tolerance of debate and conflict
- Socioeconomic barriers: Including higher-priority concerns and obligations.

Individuals and groups have many ways to participate in and have influence on the policymaking process. They can meet, write, or call their elected representatives and urge them to take up an issue or to provide comment on a policy already in the making. Resources that provide contact information for elected officials are available.[29] Engaging friends, neighbors, colleagues, and fellow members of groups, clubs, and organizations can create a groundswell of interest that gets the attention of policymakers. Use of print, broadcast, and social media can be powerful visibility and organizing tools. These tools include letters to the editor, op-eds and opinion pieces, interviews, announcements, posts on listservs and social media platform, and engagement with bloggers and podcasters.

Generating or signing petitions can demonstrate public support for action and exert pressure on policymakers. Many national and international organizations, coalitions, religious institutions, and student groups use petitions to inform, engage, and call for action on climate change. Petitions have called for the declaration of a climate emergency[30]; demands for climate justice[31]; and increased use of renewable energy and the phasing out of fossil fuels.[32]

Demonstrations, protests, marches, sit-ins, strikes, and acts of civil disobedience are more direct and highly visible forms of action, often garnering attention from the news media.[33-35] From the Global Climate March in 2015[36] to the many student sit-ins calling on their educational institutions to divest from fossil fuels,[37] these forms of direct action can exert significant pressure on policymakers, public officials, and organizational leaders.

While the ability and willingness to use these various mechanisms may reasonably vary, everyone can have some influence on public policy and policymakers, such as by voting and otherwise supporting candidates and elected officials who understand and are committed to taking action on the issues of importance to them. Climate change will grow in priority as an issue of grave public concern, given its visible and costly impacts and the growing activism of young people throughout the world (Chapters 19 and 20).

Roles and Responsibilities

All parties in the public participation process can contribute to its success. In government-initiated efforts, the responsibility is primarily on the sponsoring executive branch agency because it generally has the financial, human, and technical resources as well as established mechanisms to engage the public—along with the accountability for policy decisions. The agency may view its role as facilitator, mediator, or arbiter of competing interests and values. In this role, it may weigh and be guided by majority views in making decisions—a *utilitarian approach*. When the voices and views of marginalized, disenfranchised, and otherwise socially and economically disadvantaged communities are underrepresented or totally absent, it has been argued that the agency should act as a trustee for the health and well-being of those who are relatively worse off in order to ensure distributive and social justice—a *communitarian approach*.[38]

Community members and other stakeholders can initiate a policymaking process. Their provision of information, pressure, and advocacy can put a spotlight on a problem, facilitate government attention and action, and generate a participatory policymaking process. Their experiences, perspectives, ideas, and opinions are essential to policymaking; so are their willingness to share their views and listen to the view of others. Both are integral to deliberation and mutual learning. Even when meetings and conversations become heated, participants can help the process by creating and maintaining a climate of respect and integrity throughout.

Elements for Success

Case studies and experiences have identified some important and necessary elements for successful public participation.[38,39] At the highest level, success requires trust and trustworthiness, mutual respect, transparency, integrity, accountability, access to information, sustained commitment, and recognition of the value and importance of technical and experiential knowledge. Operationally, public participation requires the government agency to clearly state the goals of the process and who is accountable for the outcome. It also requires:

- Communication, dialogue, interaction, and mutual learning
- Broad representation and diversity of views
- Sufficient financial, human, and technical resources
- Capacity building to develop and strengthen the skills, abilities, and resources of community participants to engage in the process. (Agency staff can also benefit from building their own capacity to effectively engage with the public.)
- Participation by agency staff members who have the authority to make policy decisions.

Community groups have used a variety of mechanisms to successfully engage in the policymaking process. They have participated in citizen advisory committees at all levels of government; examples include the EPA Clean Air Advisory Committee,[40] the Minnesota Governor's Advisory Council on Climate

Change,[41] and the Concord (Massachusetts) Climate Action Advisory Board.[42] They have participated in designing research and collecting data in partnership with researchers in academic institutions and nonprofit organizations. They have applied for and been awarded environmental justice grants and collaborative problem-solving cooperative agreements from the EPA.[43] Some community organizations, like the Little Village Environmental Justice Organization[44] in Chicago and the Boys & Girls Clubs of America,[45] have conducted youth leadership programs to help young people acquire the skills to engage in their communities and the larger democratic process.

At the Federal Level

Public concern and the involvement of labor unions, community groups, and nonprofit and civil society organizations have resulted in the establishment of critical policymaking agencies. And they have resulted in the enactment of laws and regulations that protect public health and the environment and promote the health, safety, and well-being of workers, consumers, and communities. Some historical examples have included

- The 1969 Federal Coal Mine Safety and Health Act and subsequent Black Lung Benefits Program.[46,47] Both were prompted by the deaths of 78 miners in the 1969 Farmington (West Virginia) mine explosion, the deaths of 533 miners in coal mine disasters in the previous 2 years, and the countless other miners disabled by coal workers' pneumoconiosis ("black lung"). The United Mine Workers of America, individual miners, and leading medical experts testified on the need to strengthen safety protections for coal workers and provide healthcare and benefits to those who develop pneumoconiosis as a result of their exposure to coal mine dust.
- The establishment of the EPA in 1970 in response to heightened public concern about air and water pollution, degraded natural landscapes, and two environmental disasters in the previous year—the offshore oil rig spill in the Santa Barbara (California) channel and the oil-slick fire on the Cuyahoga River in Cleveland.[48] Images of the Earth taken in 1968 by Apollo astronauts also helped spark public appreciation and concern for the beauty of the Earth, as did the creation of the first Earth Day in 1970—now a global event occurring annually on April 22 (Chapter 19).[49]
- The Clean Air Act of 1970 established the basis for the National Ambient Air Quality Standards for both stationary (industrial) and mobile sources, along with statutory deadlines for compliance.[50] Its enactment was a deliberative response to the notable failure of the previous state-based approach to, and growing public concern about, air pollution. Subsequent amendments have

strengthened the original Act. The 1977 amendments established major permit review requirements. The 1990 amendments authorized new regulatory programs to control acid rain and 189 toxic air pollutants and established a program to phase out the use of chemicals that deplete the ozone layer.

- The 1986 Emergency Planning and Community Right to Know Act[51] was enacted to help communities know about, prepare for, and respond to environmental, health, and safety hazards posed by the storage and handling of toxic chemicals—an act prompted by the 1984 chemical release disaster in Bhopal, India.
- OSHA's 1991 Bloodborne Pathogen Standard was promulgated in response to the significant health risks associated with exposure to blood and other potentially infectious materials in healthcare and related occupations. The issue was not even on OSHA's regulatory agenda until unions filed petitions and public health advocates and organizations supported the standard.[52]
- OSHA's 2017 final rule on occupational exposure to beryllium in construction and shipyard sectors was similar to the model jointly proposed by industry and labor.[53] Labor unions, healthcare providers, and Public Citizen Health Research Group had earlier petitioned OSHA for an emergency temporary standard.
- In 2021, the EPA finalized a rule to establish nationwide monitoring for 29 per- and polyfluoroalkyl substances (PFAS) along with lithium in drinking water.[54] This rule was a significant milestone in the agency's PFAS Strategic Roadmap to address human health and environmental impacts of these toxic chemicals. State governments, tribes, public health advocates, and environmental justice and agricultural communities had all called for federal action on these biopersistent chemicals.

At the Local Level

Boxes 11-2 and 11-3 present two case studies that describe how local groups have successfully used participatory research, strategic partnerships, policy advocacy, and a science-to-action mindset to address multiple climate-related threats, environmental injustices, and preferred solutions in their communities. With a focus on extreme heat, GreenRoots in Chelsea, Massachusetts, worked with university-based researchers to co-design a study, engage local youth in data collection, and develop a community-led intervention for heat island mitigation. In Highland Park, Illinois, Souladarity partnered with researchers in a science-based advocacy organization to focus on energy democracy, community resilience, and a just transition to clean energy. Both are stellar examples of strong community engagement, participatory research, and the power of local advocacy for climate change mitigation and adaptation.

Box 11-2 Local Community Takes Action on Deadly Heat

Roseann Bongiovanni

Climate change has unleashed record-breaking heat on communities throughout the world, and low-income communities, particularly those in densely populated urban areas, are especially vulnerable to its dangerous impacts. Residents often struggle to pay utility bills or purchase cooling equipment and may not have access to local cooling centers even when living within sweltering urban heat islands. (See Chapter 4.)

Community groups and coalitions have been organizing and taking action, sometimes in partnership with academic researchers, policymakers, nonprofit organizations, and local experts. The Chelsea and East Boston Heat Study (the C-HEAT Project) is an illustrative example of successful public participation, collaboration, and intervention.[1]

Comprising just 1.8 square miles on a peninsula, Chelsea is the smallest and second most densely populated city in Massachusetts, with a population of approximately 40,800—two-thirds of whom are Hispanic or Latinx. Chelsea is home to GreenRoots,[1] a community-based organization dedicated to achieving environmental and climate justice in Chelsea and surrounding communities. In 2020, with the support of a private foundation, GreenRoots began a collaborative research project with the Boston University School of Public Health (BUSPH) aimed at building community capacity to respond to extreme heat, thereby enhancing the urban environment and public health.

Specifically, the research was co-designed to assess location and population risk factors for heat-related illness, analyze personal and home exposures to heat, and engage city officials, residents, and housing and community development stakeholders to raise their awareness of the impacts of extreme heat on Chelsea and nearby communities.

The C-HEAT team, which has consisted of BUSPH researchers and GreenRoots community leaders, surveyed English- and Spanish-speaking residents of Chelsea and East Boston in various neighborhoods and types of housing and provided them with temperature monitors to place in their homes and Fitbit wrist devices to monitor their health metrics. They also placed temperature sensors in city trees and on rooftops throughout Chelsea. The GreenRoots youth group, Environmental Chelsea Organizers (ECO), worked with the C-HEAT team to measure surface temperature in parks and green spaces and to make observations at city hydration stations.

Although all of Chelsea is considered an urban heat island, some blocks are hotter than others. Heat monitoring data collected by the BUSPH researchers provided GreenRoots, the City of Chelsea, and other partners the basis for an innovative and community-led project to mitigate the heat island effect. This project, called "Cool Block," consists of planting trees, installing sidewalk planters, transforming vacant lots into green space, reducing asphalt, and installing a white roof on a boys and girls club.[2,3] The state's Municipal Vulnerability Preparedness grant program helped to finance the project. The BUSPH researchers plan to monitor temperature changes during warm seasons for the next few years to measure the impact of the project.

The C-HEAT Project is an illustrative example of how participatory research, strong community leadership and engagement, and a science-to-action mindset can address the heat impacts of climate change on community health and well-being. The project has increased awareness of climate change and the urgent need for policies and actions to help local communities adapt to warmer temperature.

Box References

1. GreenRoots, Inc. C-Heat Project. Available at: http://www.greenroots chelsea.org/c-heat. Accessed August 15, 2023.

2. O'Donnell N. Activists work to cool down sweltering 'heat island' near Boston. NBC Bay Area, April 22, 2022. Available at: https://www.nbcbaya rea.com/news/national-international/activists-work-to-cool-down-swelter ing-heat-island-chelsea-on-the-boston-harbor/2871233/. Accessed August 15, 2023.

3. Bebinger M. A Massachusetts city keeps cool in a hotter climate. WBUR Here and Now May 25, 2022. Available at: https://www.wbur.org/hereand now/2022/05/25/climate-change-summer-heat. Accessed August 15, 2023.

Box 11-3 Approaching Climate Action Through the Lens of Local Needs

Shimekia Nichols and James Gignac

Highland Park, a 3-square-mile city in southeast Michigan with 10,000 residents, is among the 8% of U.S. communities most adversely affected by environmental injustice in Michigan.[1] The median household income is about $20,000, and almost half the population lives at or below the poverty level. Residents struggle with the impacts of food insecurity, aging and substandard housing, and inadequate Internet access. Residents are especially vulnerable to the adverse effects of extreme heat, flooding, and other impacts of climate change.

Despite the historic economic and environmental forces stacked against them, Highland Park residents have organized for transformative and forward-thinking changes in areas such as food, education, transportation, and energy. They are claiming a seat at the table in state and federal policy conversations on several important topics and using self-determination to address community challenges. For example, when the local electric utility repossessed and removed two-thirds of the city's streetlights, residents responded by forming Soulardarity, a nonprofit organization that has since installed numerous community-owned, solar-powered streetlights and has become a strong advocate for energy democracy—the idea that the people who are most affected by energy decisions should have the greatest say in shaping them. In addition, two residents launched sustainable neighborhood projects, Avalon Village and Parker Village, which are advancing community-centered sustainable development visions through solar power, green infrastructure, and net-zero homes, as well as garden, community, and educational spaces.[2,3]

In 2021, Soulardarity partnered with the Union of Concerned Scientists (UCS) to analyze local energy usage and how solar power, energy efficiency, and locally owned resources could meet the community's electricity needs. (Local clean energy has been defined as community-owned and community-led energy efficiency and development of solar power.) A subsequent case study focused on options for a solar- and battery-powered microgrid for Parker Village.[4] The resultant report by UCS, Soulardarity, and Highland Park residents, entitled *Let Communities Choose: Clean Energy Sovereignty in Highland Park, Michigan*, illustrated how Soulardarity's approach to climate action has sought to address the overlapping challenges of underinvestment, economic disparity, and environmental injustice facing its community.[5] Increasing local clean energy results in more wealth staying in the community while reducing residents' energy costs, fostering local economic growth, and reducing air pollutants and GHG emissions from fossil fuel plants. In addition, microgrids in neighborhoods, such as Parker Village, can provide residents more control over electricity reliability.

A key part of the collaboration between Soulardarity and UCS included the development of several policy recommendations, which they also included in the *Let Communities Choose* report. These state- and local-level policies would help make the just transition to clean energy not only possible but affordable—thereby improving the quality of life for consumers paying up to or more than 30% of their household income on energy bills. Leveraging its years of relationship-building locally, Soulardarity is using the analysis and policy recommendations to advocate that the city council and city administration implement measures such as a local ordinance to clarify guidelines for solar installations, formation of a city sustainability committee, and research into alternatives to traditional utility service.

The Department of Energy selected Highland Park as one of 23 communities in the United States to participate in a new program called Communities LEAP and to receive technical assistance in pursuing a path toward 100% local, renewable energy—another illustration of how Soulardarity increases community resilience through strategic partnerships and practical programming, such as weatherization and solar bulk purchasing efforts to reduce costs for residents to install rooftop solar panels. At the state level, Soulardarity and UCS are using their research to support stronger policies for equitable deployment of local, customer-owned solar power.

While *Let Communities Choose* focused on Highland Park and state policies in Michigan, the vision articulated in the report, the analytical methodology, and the community's strong advocacy for policy change are applicable elsewhere. The recognition that people should have the ability to choose how their electricity is provided—the right to energy sovereignty— and to choose locally generated clean energy are the core principles of the better energy system we need—not only to reduce climate change, but also to help address the persistent economic and social disparities faced by many communities.

Box References

1. School for Environment and Sustainability, University of Michigan. Screening Tool for Environmental Justice in Michigan. Available at: https:// www.arcgis.com/apps/webappviewer/index.html?id=dc4f0647dda349599 63488d3f519fd24. Accessed August 12, 2022.

2. The Avalon Village. Available at: http://theavalonvillage.org. Accessed August 15, 2023.

3. Parker Village. About Us. Available at: http://www.parkervillagehp.com. Accessed August 15, 2023.

4. Baek Y, Gignac J, Shannon J. Designing a Neighborhood Microgrid. Union of Concerned Scientists March 2022. Available at: https://www.ucs usa.org/sites/default/files/2022-03/designing-neighborhood-microgrid.pdf. Accessed August 12, 2022.

5. Gignac J, Baek Y, Koeppel J, et al. Let Communities Choose: Clean Energy Sovereignty in Highland Park, Michigan. Union of Concerned Scientists and Soulardarity 2021. Available at: www.ucsusa.org/sites/default/files/2021-10/ Let-Communities-Choose-10-12-21.pdf. Accessed October 31, 2023.

PEOPLE, POWER, AND PARTICIPATION

The U.S. Constitution opens with three important words: "We the People." A representative democracy, a form of government in which power resides in the governed, is a precious gift and a responsibility. It gives people a voice in the public policies that affect nearly every aspect of their daily lives.

Power can be a positive or negative force for change.[55] It is not evenly distributed and too often varies by gender, race, ethnic group, class, wealth, and national, geographical, and religious identities. Power can be coercive and domineering, exerting undue and disproportionate control and influence in society—having *power over*. But the power of ordinary people can also be developed, harnessed, and utilized as a force for positive change—having *power to*.

Making and changing public policy can be a complex, confusing, frustrating, and taxing process. But participation can be energizing and empowering, with rewards that are enormous and long-lasting. The Center for Community Health and Development at the University of Kansas has developed a "community toolbox," an online resource for those who want to build healthy communities and create social change. It suggests "Eight Ps" that can aid communities and the public in changing public policies[56]:

1. Planning: Engage with others to determine if a new policy or a policy change is actually needed and to develop an overall strategy for the effort.
2. Preparation: Conduct the research and inquiry necessary to learn as much as possible about the issue and current related policies. Consider who your allies and opponents will be.
3. Personal contact: Personal relationships are the key to effective advocacy. Establish contact with policymakers, local opinion leaders, and individuals and groups that support—or oppose—the desired change.
4. Pulse of the community: Learn what community members think about the issue and any proposed policy change—and what might engender their support.
5. Positivism: Accentuate the positive elements of the policy change.
6. Participation: Engage as many people as possible in the planning and advocacy effort. Broad participation will enhance its credibility.
7. Publicity: Do all you can to keep people informed and aware of the effort. Use the media, your social networks and relationships, and other creative tactics to give the issue a high profile, such as displaying signs, organizing events, and distributing information at local gathering places.
8. Persistence: Policy change can take a long time. Keep at it. Do not give up.

SUMMARY POINTS

- Public policy—the decisions and actions taken by governmental entities—is a powerful tool for addressing societal issues and improving the health and well-being of the public.
- Public participation in the policymaking process is integral to the principles and values of a democratic society; it informs and builds trust, legitimacy, and accountability of governmental decisions.
- The policymaking process provides opportunities for the public to initiate, participate in, and influence the focus and outcome.
- Governmental agencies and civil society organizations provide and use a variety of mechanisms, tools, and resources to facilitate individual, group, and stakeholder participation in the policymaking process.
- Often, those most affected by a problem are the most vulnerable and socioeconomically disadvantaged members of a community—a maldistribution of access and power that places a special responsibility on policymaking institutions and those advocating for change.

References

1. USA.gov. Branches of the U.S. Government. Available at: https:www.usa.gov/branches-of-government#item-214500. Accessed August 15, 2023.
2. Environmental Protection Agency. Overview of the Clean Air Act and Air Pollution. Available at: https://www.epa.gov/clean-air-act-overview. Accessed August 15, 2023.
3. Environmental Protection Agency. Ozone (O_3) Air Quality Standards. Available at https://www.epa.gov/naaqs/ozone-o3-air-quality-standards. Accessed August 15, 2023.
4. City of Boston. Building Emission Reduction and Disclosure Ordinance (BERDO). Available at: https://www.boston.gov/departments/environment/building-emissions-reduction-and-disclosure. Accessed August 15, 2023.
5. Environmental Protection Agency. Policy Statement on Climate Change Adaptation. Available at: https://www.epa.gov/system/files/documents/2021-09/epa-climate-adaptation-plan-pdf-version.pdf#page=3. Accessed August 15, 2023.
6. Congressional Research Service. U.S. Climate Change Policy. October 28, 2021. Available at: https://crsreports.congress.gov/product/pdf/R/R46947/4. Accessed August 15, 2023.
7. Webb RM, Kurtz L. Politics v. science: How President Trump's war on science impacted public health and environmental regulation. Progress in Molecular Biology and Translational Science 2022; 188(1): 65–80. https://pubmed.ncbi.nlm.nih.gov/35168747/. Accessed August 15, 2023.
8. Kagan J. The glass ceiling: Definition, history, effects, and examples. Investopedia October 19, 2022. Available at: https://www.investopedia.com/terms/g/glass-ceiling.asp.AccessedAugust 15,2023.

9. Goodman D. What are the steps in the public policy process? HistoricalIndex.org December 15, 2022. Available at: https://www.historicalindex.org/what-are-the-steps-in-the-public-policy-process.htm. Accessed August 15, 2023.

10. Schito M. Public Policy 101: The stages of the policy process. Arcadia March 12, 2022. Available at: https://www.byarcadia.org/post/public-policy-101-the-stages-of-the-policy-process. Accessed August 15, 2023.

11. USA.gov. How Laws are Made. Available at: https://www.usa.gov/how-laws-are-made. Accessed August 15, 2023.

12. U.S. National Archives. Executive Orders Disposition Tables Historical Index. Available at: https://www.archives.gov/federal-register/executive-orders/disposition. Accessed August 15, 2023.

13. Federal Register. Executive Orders. Available at: https://www.federalregister.gov//presidential-documents/executive-orders. Accessed August 15, 2023.

14. Melilo G. A look at Biden's past executive orders on climate change. The Hill, August 16, 2022. Available at: https://thehill.com/changing-america/sustainability/climate-change/3603947-a-look-at-bidens-past-executive-orders-on-climate-change/ Accessed August 15,2023.

15. U. S. National Archives. Administrative Procedures Act. Available at: https://www.archives.gov/federal-register/laws/administrative-procedure. Accessed August 15, 2023.

16. Office of the Federal Register. A Guide to the Rulemaking Process. Available at: https://www.federalregister.gov/uploads/2011/01/the_rulemaking_process.pdf. Accessed August 15, 2023.

17. Occupational Safety and Health Administration. OSHA Standards Development. Available at: https://www.osha.gov/laws-regs/standards-development. Accessed August 15, 2023.

18. Occupational Safety and Health Administration. COVID-19 Healthcare ETS (Emergency Temporary Standard). December 27, 2021. Available at: https:www.osha.gov/coronavirus/ets. Accessed August 15, 2023.

19. Oregon.gov. Oregon OSHA adopts emergency rule bolstering protections for workers against the hazards of high and extreme heat. Available at: https://osha.oregon.gov/news/2021/pages/nr2021-26.aspx. Accessed August 15, 2023.

20. Congressional Research Service. The Congressional Review Act (CRA): A Brief Overview. Available at: https://crsreports.congress.gov/product/pdf/IF/IF10023. Accessed August 15, 2023.

21. Congressional Research Service. The Congressional Review Act: Frequently Asked Questions. Available at: https://crsreports.congress.gov/product/pdf/R/R43992. Accessed August 15, 2023.

22. Library of Congress. Guide to Law Online: U.S. States and Territories. Available at: https://guides.loc.gov/us-states-territories. Accessed August 15, 2023.

23. Center for the Study of Federalism. Home Rule. Available at: https://encyclopedia.federalism.org/index.php/Home_Rule. Accessed August 15, 2023.

24. California Special District Association. Learn About Districts. Available at: https://www.csda.net/special-districts/learn-about. Accessed August 15, 2023.

25. Ashford NA, Rest KM. Public Participation in Contaminated Communities. March 1999. Available at: http://ashford.mit.edu/public-participation-contaminated-communities. Accessed August 15, 2023.

26. U.S. Environmental Protection Agency. Public Participation Guide. Available at: https://www.epa.gov/international-cooperation/public-participation-guide. Accessed December 23, 2022.

27. U.S. Environmental Protection Agency. Public Participation Guide: Tools. Available at: https://www.epa.gov/international-cooperation/public-participation-guide-tools. Accessed August 15, 2023.

28. Coehlo TR, Pozzebon M, Cuhna MA. Citizens influencing public policy-making: Resourcing as source of relational power in e-participation platforms. Information System Journal 2022; 32: 344–376.

29. Common Cause. Find Your Representatives. Available at: https://www.commonca use.org/find-your-representative/addr/. Accessed August 15, 2023.

30. Climate Emergency Coalition. Sign Petition: Declare a Global Warming State of Emergency. Available at: https://www.cecoalition.org/emergency_petition. Accessed August 15, 2023.

31. Change.Org. Climate Justice!! Not Climate Change! Available at: https://www.cha nge.org/p/climate-justice-not-climate-change. Accessed August 15, 2023.

32. Reed K. Students petition for renewable energy on campus. The Villanovan January 31, 2022. Available at: https://villanovan.com/19505/news/students-pen-petition-for-renewable-energy-on-campus/. Accessed August 15, 2023.

33. Lowery T, Banjo F. The 2023 Global Climate Strike is Almost Here. Here's Everything to Know. Global Citizen, February 22, 2023. Available at: htps://www.globalcitizen. org/en/content/global-climate-strike-fridays-future-what-to-know/.   Accessed September 14, 2023.

34. Gayle D. Scientists call on colleagues to protest climate crisis with civil disobedience. The Guardian August 29, 2022. Available at: https://www.theguardian.com/envi ronment/2022/aug/29/scientists-call-on-colleagues-to-protest-climate-crisis-with-civil-disobedience. Accessed August 15, 2023.

35. Yamin F. This is the only way to tackle the climate emergency. Time June 14, 2019. Available at: https://time.com/5607152/extinction-rebellion-farhana-yamin/. Accessed August 15, 2023.

36. 350.org. We Sent a Message to Paris. Available at: https://350.org/global-climate-march/. Accessed August 15, 2023.

37. Hirji Z. Standing up by sitting down: Student sit-ins demand divestment. Inside Climate News April 27, 2015. Available at: https://insideclimatenews.org/news/ 27042015/standing-sitting-down-student-sit-ins-demand-divestment/. Accessed August 15, 2023.

38. Ashford NA, Rest KM. Public participation in contaminated communities. March 1999. Available at: http://ashford.mit.edu/public-participation-contaminated-comm unities. Accessed August 15, 2023.

39. U.S. Environmental Protection Agency. Public Participation Guide: Foundational Skills, Knowledge, and Behaviors. Available at: https://www.epa.gov/international-cooperation/public-participation-guide-foundational-skills-knowledge-and-behavi ors. Accessed August 15, 2023.

40. U.S. Environmental Protection Agency. Clean Air Act Advisory Committee (CAAAC). Available at: https://www.epa.gov/caaac. Accessed August 15, 2023.

41. Our Minnesota Climate. Governor's Advisory Council on Climate Change. Available at: https://climate.state.mn.us/advisory-council. Accessed August 15, 2023.

42. The Town of Concord, Massachusetts. Climate Action Advisory Board. Available at: https://concordma.gov/DocumentCenter/View/19827/Climate-Action-Advisory-Board-Charge---06-04-18. Accessed August 15, 2023.

43. U.S. Environmental Protection Agency. EPA awards 2021 environmental justice small grants to local communities in the Southeast. Available at: https://www.epa.gov/newsreleases/epa-awards-2021-environmental-justice-small-grants-local-communities-southeast. Accessed August 15, 2023.

44. Little Village Environmental Justice Organization. Available at: www.lvejo.org. Accessed August 15, 2023.

45. Boys and Girls Club of America. Available at: https://www.bgca.org. Accessed August 15, 2023.

46. Mine Safety and Health Administration. Federal Coal Mine and Safety Act of 1969. Available at: https://www.msha.gov/federal-coal-mine-and-safety-act-1969. Accessed August 15, 2023.

47. Social Security Administration Office of Retirement and Disability Policy. Black Lung Benefits Program Description and Legislative History. Available at: https://www.ssa.gov/policy/docs/statcomps/supplement/2016/blacklung.html#:~:text=The%20Black%20Lung%20benefit%20program,pneumoconiosis%2C%20and%20to%20their%20dependents. Accessed August 15, 2023.

48. U.S. Environmental Protection Agency. The Origins of EPA. Available at: https://www.epa.gov/history/origins-epa. Accessed August 15, 2023.

49. U.S. Environmental Protection Agency. EPA History: Earth Day. Available at: https://www.epa.gov/history/epa-history-earth. Accessed August 15, 2023.

50. U.S. Environmental Protection Agency. EPA History: The Clean Air Act of 1970. Available at: https://www.epa.gov/archive/epa/aboutepa/epa-history-clean-air-act-1970.html. Accessed August 15, 2023.

51. U.S. Environmental Protection Agency. What is EPCRA? Available at: https://www.epa.gov/epcra/what-epcra. Accessed August 15, 2023.

52. U.S. Department of Labor/OSHA. Final Rule on Occupational Exposure to Bloodborne Pathogens. Available at: https://www.osha.gov/laws-regs/federalregister/1991-12-06. Accessed August 15, 2023.

53. Federal Register. Occupational Exposure to Beryllium and Beryllium Compounds in Construction and Shipyard Sectors. Available at: https://www.federalregister.gov/documents/2017/06/27/2017-12871/occupational-exposure-to-beryllium-and-beryllium-compounds-in-construction-and-shipyard-sectors. Accessed August 15, 2023.

54. U.S. Environmental Protection Agency. EPA announced nationwide monitoring effort to better understand extent of PFAS in drinking water. Available at: https://www.epa.gov/newsreleases/epa-announces-nationwide-monitoring-effort-better-understand-extent-pfas-drinking. Accessed August 15, 2023.

55. PowerCube. Understanding analysis for social change. Available at: https://www.powercube.net/an-introduction-to-power-analysis/. Accessed August 15, 2023.

56. KU Center for Community Health and Development. (2023). Chapter 25, Section 1: Changing Policies: An Overview. Lawrence, Kansas: University of Kansas. Available at: https://ctb.ku.edu/en/table-of-contents/implement/changing-policies/overview/checklist. Accessed August 15, 2023.

Further Reading

Pohlmann A, Walz K, Engels A, et al. It's not enough to be right! The climate crisis, power, and the climate movement. GAIA 2021; 30: 4: 231–236.

A concise discussion of social factors and obstacles to climate action as well as advice to the climate movement.

Harris P, Baum F, Friel S, et al. A glossary of theories for understanding power and policy for health equity. Journal of Epidemiology and Community Health 2020; 74: 548–552.

An overview of different theories of and perspectives on power aimed at researchers and advocates involved in public health policy.

Ashford NA, Rest KM. Public Participation in Contaminated Communities. March 1999. Available at: http://ashford.mit.edu/public-participation-contaminated-comm unities. Accessed August 15, 2023.

A case study-based compilation of many dimensions of successful public participation.

U.S. Environmental Protection Agency. Public Participation Guide: Internet Resources on Public Participation – Public Participation Ethics, Values, and Principles. Available at: https://www.epa.gov/international-cooperation/public-participation-guide-inter net-resources-public-participation. Accessed August 15, 2023.

A useful guide and compilation of online resources for public participation.

Center for Community Health and Development, University of Kansas. Community Toolbox: Tools to Change the World. Available at: https://ctb.ku.edu/en. Accessed August 15, 2023.

A voluminous collection of practical and user-friendly information, tips, and tools for taking action to create social change.

National Collaborating Centre for Healthy Public Policy. Public Policy Models and Their Usefulness in Public Health: The Stages Model. 2013. Available at: https://www. ncchpp.ca/docs/ModeleEtapesPolPubliques_EN.pdf. Accessed August 15, 2023.

Although dated, this publication remains a useful articulation of policymaking models and information that public health actors can contribute at the various stages of the policymaking process.

Appendix: Glossary of Some Policymaking Terms

Kathleen M. Rest

Amendment: A proposed change to a pending text (such as a bill, resolution, or another amendment)

Bill: The primary form of legislative measure used to propose a law (Depending on the chamber of origin, bills begin with a designation of either "H.R." or "S.")

Calendar: Lists of measures, motions, and matters that are (or soon will become) eligible for consideration on the chamber floor; also, the official document that contains these lists and other information about the status of legislation and other matters

Capacity building: Activities designed to develop and strengthen the skills, abilities, and effectiveness of people, groups, and organizations to achieve their mission and goals

Cloture: The method by which a supermajority (typically three-fifths) of the Senate may agree to limit further debate and consideration of a question, such as a bill amendment or other matter

Conference committee: Temporary joint committee created to resolve differences between House-passed and Senate-passed versions of a bill

The Congressional Record: The official record of the proceedings and debates of the U.S. Congress

Enabling legislation: A law that gives government officials or bodies the authority to create the rules/policies to implement or enforce a statute or law

Filibuster: In the Senate, the use of dilatory or obstructive tactics to block passage of a measure by preventing it from coming to a vote

Hold: A request by a Senator to his or her party leader to delay floor action on a measure (such as a bill) or matter (such as a nomination), to be consulted on its disposition, and/or an indication that he/she would object to a unanimous consent request to consider said item of business or otherwise delay or obstruct consideration

Markup: Meeting by a committee or subcommittee during which committee members offer, debate, and vote on amendments to a measure

Motion to table: A non-debatable motion in the House and Senate (and in their committees) by which a simple majority may agree to negatively and permanently dispose of a question (such as an amendment)

Policy agenda: A set of issues, problems, or subjects that are viewed as important by people or groups involved in making or influencing policy, such as government officials, other decision-makers, political parties, nongovernmental organizations, and interest groups

Point of order: A member's statement to the presiding officer that the chamber (or committee) is taking action contrary to the rules or precedents, and a demand that they be enforced

Ranking member: The most senior (although not necessarily the longest serving) member of the minority party on a committee or subcommittee

Recess appointment: A temporary presidential appointment, during a recess of the Senate, of an individual to a federal government position, where such appointment usually requires the advice and consent of the Senate

Stakeholder: An individual or group that is involved in, has an interest in, or is affected by a decision or course of action

Supermajority: A term sometimes used for a vote on a matter that requires approval by more than a simple majority of those members present and voting with a quorum being present

Suspension of the rules: In the House, a procedure that streamlines consideration of a measure with wide support by prohibiting floor amendments, limiting debate to 40 minutes, and requiring a two-thirds majority for passage

Veto: Presidential disapproval of a bill passed by Congress (If a President vetoes a bill, it can become law only if the House and Senate separately vote, by two-thirds, to override the veto. A less common form of presidential veto—a *pocket veto*—occurs if Congress has adjourned without the possibility of returning for the remainder of the legislative session and the President does not sign the measure within the required 10-day period.)

For definitions of additional policymaking terms, see glossary of legislative terms at: https:/www.congress.gov/help/legislative-glossary and https://www.ncsl.org/research/about-state-legislatures/glossary-of-legislative-terms.

12

Energy Policy

Nova M. Tebbe, Nicholas A. Mailloux, and Gregory F. Nemet

A priority goal of energy policy is reducing greenhouse gas (GHG) emissions, primarily by decreasing—and ultimately eliminating—the burning of fossil fuels. Since fossil fuel combustion emits not only GHGs but also harmful air pollutants, such as particulate matter, sulfur dioxide, and oxides of nitrogen, reducing GHG emissions provides substantial health co-benefits.

The primary focus of GHG emission reductions is human-generated carbon dioxide, about 86% of which is derived from fossil fuel combustion globally.[1] The current concentration of atmospheric carbon dioxide is about 420 ppm, the highest that it has been in 800,000 years. Ambient levels of carbon dioxide have increased by approximately 50% since 1750.[2] Of all carbon dioxide emitted since 1850, 43% has been released since 1990.[3] Although ambient carbon dioxide levels are continuing to rise, the annual rate of increase of GHG emissions from energy supply has slowed; between 2000 and 2009 they rose 2.3% per year, but between 2010 and 2019 they rose only 1.0% per year.[4]

Although carbon dioxide can remain in the atmosphere for decades, other GHGs, such as methane, nitrous oxide, and fluorinated gases, remain in the atmosphere for shorter periods but are more potent than carbon dioxide in causing global warming (Table 1-1 in Chapter 1). These other GHGs together account for about 25% of global warming.[5]

The scale of transformations of the energy system needed to limit global warming by 2100 to 2.0°C (3.6°F) above preindustrial levels—the goal set in the 2015 Paris Agreement—will require unprecedented policies and actions. It will require achieving *net-zero emissions* globally—the condition in which the amount of emissions released equals the amount removed from the atmosphere. Policymakers need to consider the potential benefits and the adverse effects of proposed policies to achieve ambitious emissions reductions. For example, reducing GHG emissions in the energy system can, by improving air quality, provide substantial health co-benefits (Box 12-1). However, adverse effects can occur when biofuels compete for land use with food production, thereby increasing the cost of food and food insecurity (Chapters 8 and 14).

Nova M. Tebbe, Nicholas A. Mailloux, and Gregory F. Nemet, *Energy Policy* In: *Climate Change and Public Health*. Edited by: Barry S. Levy and Jonathan A. Patz, Oxford University Press. © Oxford University Press 2024.
DOI: 10.1093/oso/9780197683293.003.0012

Box 12-1 Air Quality and Health Co-Benefits of Mitigation Policies

Nicholas A. Mailloux

In addition to emitting GHGs, fossil fuel combustion releases air pollutants that adversely affect health. The U.S. government regulates six criteria air pollutants: particulate matter, ozone, sulfur dioxide, nitrogen dioxide, carbon monoxide, and lead. Policies to reduce GHG emissions can provide substantial health co-benefits from improved air quality while mitigating climate change.

Most harmful is fine particulate matter ($PM_{2.5}$), which consists of solid and liquid particles with an aerodynamic diameter equal to or less than 2.5 microns. Estimates of deaths due to $PM_{2.5}$ vary. A study found that more than 4.1 million people died prematurely from ambient (outdoor) $PM_{2.5}$ exposure in 2019.[1] Another study found that about 8.8 million people died prematurely because of $PM_{2.5}$ exposure in 2015; an estimated 3.6 million of these deaths were due to $PM_{2.5}$ from fossil fuel emissions.[2] $PM_{2.5}$ exposure is associated with death from ischemic heart disease, stroke, lung cancer, diabetes, chronic obstructive pulmonary disease, and infections of the lower respiratory tract.[1]

An additional 2.3 million people die prematurely each year from household (indoor) air pollution, especially in India and sub-Saharan African countries, where people rely heavily on solid fuels for heating and cooking.[1] About 3 billion people rely on solid fuels—including charcoal, coal, crop waste, dung, and wood—for cooking and heating. Using cleaner cookstoves that rely on electricity or cleaner liquefied petroleum gas products, such as propane, can greatly reduce disease from household air pollution. (See Box 5-2 in Chapter 5.)

Some studies have estimated health co-benefits resulting from shifts to renewable energy production, including the following:

- Deployment of wind and solar power in the United States from 2007 to 2015 prevented between 3,000 and 12,700 premature deaths from reduced exposure to $PM_{2.5}$.[3]
- Between 260,000 and 410,000 premature deaths could be avoided in the United States from 2020 to 2050 by reductions in ambient $PM_{2.5}$ exposure.[4]
- China's Air Clean Plan prevented 230,000 premature deaths between 2014 and 2020.[5]
- Eliminating $PM_{2.5}$ emissions from electricity generation, transportation, buildings, and industry in the United States could prevent 53,200 premature deaths each year and provide $608 billion in health co-benefits,

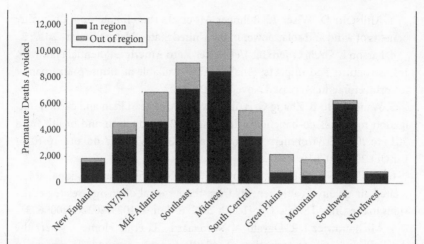

Figure 12-1 Avoided premature deaths by elimination of energy-related emissions, by region. (*Source*: Adapted from Mailloux NA, Abel DW, Holloway T, et al. Nationwide and regional $PM_{2.5}$-related air quality health benefits from the removal of energy-related emissions in the United States. GeoHealth 2022; 6. doi: 10.1029/2022GH000603.)

mainly from eliminating on-road vehicle emissions and decreasing fuel use in the residential and commercial sectors and for electricity generation.[6]

Air pollutants do not necessarily remain where they were emitted; prevailing winds can carry pollutants thousands of miles from their source. The last of these studies also found that between 32% and 95% of health co-benefits would occur in regions other than where emissions were eliminated (Figure 12-1).[6]

Natural gas stoves release into homes harmful pollutants, such as nitrogen dioxide and volatile organic compounds, including benzene, hexane, and toluene.[7,8] Electrifying end uses, such as by replacing gas stoves with induction ranges, can greatly reduce fossil fuel use for cooking. Residential fossil fuel use is a major source of GHG emissions, especially methane, which is the primary component of natural gas.[8]

Box References

1. Murray CJL, Avarkin A, Zheng P, et al. Global burden of 87 risk factors in 204 countries and territories, 1990–2019: A systematic analysis for the Global Burden of Disease Study 2019. Lancet 2020; 396: 1223–1249.

2. Lelieveld J, Klingmüller K, Pozzer A, et al. Effects of fossil fuel and total anthropogenic emission removal on public health and climate. Proceedings of the National Academy of Science 2019; 116: 7192–7197.

3. Millstein D, Wiser R, Bolinger M, et al. The climate and air-quality benefits of wind and solar power in the United States. Nature Energy 2017; 2.

4. Larson E, Greig C, Jenkins J, et al. Net-Zero America: Potential Pathways, Infrastructure, and Impacts: Final Report. Available at: https://netzeroamer ica.princeton.edu. Accessed November 21, 2022.

5. Wang Z, Hu B, Zhang C, et al. How the Air Clean Plan and carbon mitigation measures co-benefited China in PM2.5 reduction and health from 2014 to 2020. Environment International 2022; 169. https://doi.org/10.1016/ j.envint.2022.107510.

6. Mailloux NA, Abel DW, Holloway T, et al. Nationwide and regional $PM_{2.5}$-related air quality health benefits from the removal of energy-related emissions in the United States. GeoHealth 2022; 6. doi: 10.1029/2022GH000603.

7. Michanowicz DR, Dayalu A, Nordgaard CL, et al. Home is where the pipeline ends: Characterization of volatile organic compounds present in natural gas at the point of the residential end user. Environmental Science & Technology 2022; 56: 10258–10268.

8. Lebel ED, Finnegan CJ, Ouyang Z, et al. Methane and NO_x emissions from natural gas stoves, cooktops, and ovens in residential homes. Environmental Science & Technology 2022; 56: 2529–2539.

ENERGY SOURCES

Primary energy resources, such as crude oil and solar radiation, can be converted into energy carriers, such as gasoline and electricity, which can be used to provide useful end-use energy services, such as transportation and home heating. The *energy sector* includes generation of electric power, space and water heating, lighting, and other applications in transportation, agriculture, and other sectors (Chapters 13 and 14). This chapter focuses mainly on the generation and use of electricity.

Global electricity demand has increased steadily over the past several decades, driven by economic development and population growth.[6] Electricity demand in the United States is primarily met by fossil fuels, especially natural gas (38%) and coal (22%). *Low-carbon energy sources* include (a) nuclear power and (b) renewable energy sources: wind, solar, geothermal, and hydroelectric; biomass; and tidal and wave power. *Renewable energy* refers to energy that can replenish itself at the same rate at which it is used.

In 2021, low-carbon energy sources accounted for more than 33% of electricity generation globally and about 40% in the United States.[7,8] Nuclear power is the largest source of low-carbon energy in the United States, accounting for about 20% of electricity generation. Renewables comprise the remainder of low-carbon electricity generation in the United States: wind power, 9.2%; hydro-electric power, 6.2%; solar power, 2.8%; wood and waste, 1.3%; and geothermal power, 0.4%. Renewable energy sources also include tidal power and wave power, which use the difference in height between tides or waves to spin turbines to generate electricity.

Wind and Solar Power

The generating capacity of wind and solar power in the United States has greatly increased in recent years (Figure 12-2). Wind power supplied almost 67 times more electricity in the United States in 2021 than in 2000. The amount of electricity generated from solar power in the United States was more than 230 times higher in 2021 than it was in 2000.

Wind power is generated by large-scale wind farms on land or offshore (Figure 12-3). Wind passing over the blades of a wind turbine induces lift to spin a rotor connected to an electromagnetic generator, which converts mechanical energy into electrical energy, which is then distributed to end-users via a power grid.

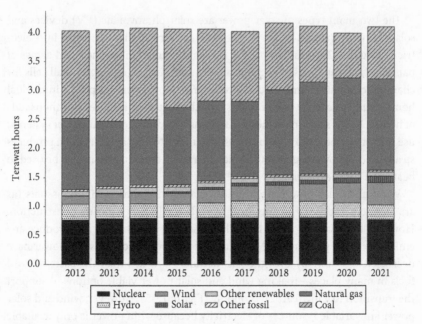

Figure 12-2 U.S. electricity generation, by source and year, 2012–2021. (*Source*: U.S. Energy Information Administration.)

Figure 12-3 Wind power has grown substantially in the United States in recent decades, producing almost 67 times as much electricity in 2021 as it did in 2000. (Copyright Associated Press.)

The two main types of solar power are solar photovoltaic (PV) devices and solar-thermal systems. Solar PV devices convert sunlight directly into electricity (Figure 12-4). Individual PV cells are grouped into panels and arrays of panels, which can be used in many applications, including single small cells for charging calculators and watch batteries, rooftop systems to power individual homes, and large-scale power plants, which can power hundreds or thousands of homes. Large-scale solar-thermal systems (concentrated solar power systems) use mirrors or lenses to concentrate solar energy to heat a fluid, which produces steam used to power a generator. Solar-thermal technologies can also be used to heat water for use in homes.

Wind and solar power are both associated with some GHG emissions (as are all energy sources) in the form of embodied energy during construction. However, they have very low lifecycle emissions since they do not produce any emissions while generating electricity. The costs of wind and solar power have decreased substantially over the past decade—now less than the costs of fossil fuels in many places—making wind and solar power viable options to support the transition to a low-carbon economy (Figure 12-5). However, wind and solar power are variable resources of electricity because wind power is only available when it is windy and solar power can only produce electricity when it is sunny. Energy storage technologies, such as batteries or pumped hydropower, help align

Figure 12-4 Solar power plants can generate enough electricity to power hundreds, or even thousands, of homes. (Photograph by Dennis Schroeder, NREL 17842.)

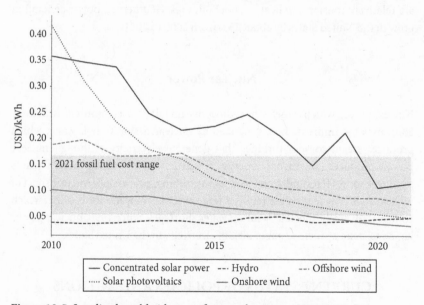

Figure 12-5 Levelized worldwide cost of energy, by source, 2010–2021. In 2010, the highest cost of energy in these categories was from solar photovoltaics, followed by concentrated solar power, offshore wind, onshore wind, and hydro. (*Source*: International Renewable Energy Agency.)

supply and demand so that energy produced from wind and solar power can be available when needed.

Hydroelectric and Geothermal Power

Hydroelectric power (hydropower) uses the energy of running water—either flowing naturally or forced through a dam by the weight of water in a reservoir—to spin turbines and electromagnetic generators to produce electricity. The flexibility and storage capacity of hydroelectric power plants enable them to be used in combination with intermittent energy sources, such as wind and solar power. However, there is growing concern about the ability of hydroelectric power plants to produce energy during water shortages due to climate-driven droughts. Hydroelectric power has provided a fairly constant amount of electricity in the United States since 2000.

Geothermal power, which uses heat generated in the Earth's core, is harnessed by digging wells and pumping to the surface steam, which spins a turbine that generates electricity. Geothermal power is also harnesssed in geothermal heat pumps ("ground-source" heat pumps), in which water or another fluid is circulated through pipes just feet below the Earth's surface (where temperatures are relatively constant) to heat or cool buildings. Geothermal power generation grew in the United States by about 15% from 2000 to 2021.

Nuclear Power

Nuclear power, which is fueled by radioactive metals (most commonly uranium-235), uses the immense heat generated by nuclear fission to boil water, which produces steam to spin turbines that generate electricity. Although nuclear power produces low-carbon energy, construction of nuclear power plants produces some GHG emissions, as does the mining, processing, and transport of nuclear fuel.[9] Nuclear power requires large amounts of water for cooling, which could be adversely affected by future water shortages. Since 1990, nuclear power has consistently provided about 20% of electricity in the United States.[7]

CURRENT STATUS OF POLICIES AND ACTIONS

In designing and implementing energy policies, governments must address sometimes conflicting social demands for energy systems that provide a reliable supply of energy at low cost while protecting human health and the environment.

Climate change adds an additional set of challenges for governments to transform the energy system—reducing GHG emissions while meeting increasing demands for energy. The challenges and opportunities for energy policy are affected by (a) the scale of the transformation required, (b) the need for innovation-oriented policy, (c) the long timeframes required for making major changes in energy systems, and (d) the need for persistence and credibility in policymaking.

In 2015, during the 21st United Nations Framework Convention on Climate Change (UNFCCC) Conference of the Parties (COP21), the Paris Agreement was agreed to by 196 parties, including the United States. This international treaty requires parties to submit nationally determined contributions (NDCs), which are designed to reduce their GHG emissions to limit global warming to well below 2.0°C (3.6°F) above preindustrial levels by 2100. (See Box 1-2 in Chapter 1.)

The Paris Agreement requested the Intergovernmental Panel on Climate Change (IPCC) to issue a report on the impacts of global warming of 1.5°C (2.7°F) since many countries considered the warming target of 2.0°C (3.6°F) to be inadequate. This resulted, in 2018, in the *IPCC Special Report on Global Warming of 1.5°C*, which concluded that (a) limiting global warming to 1.5°C would result in lower climate-related risks than if warming were limited to 2.0°C and (b) limiting global warming to less than 1.5°C would require reaching net-zero emissions by 2050.

Many countries set net-zero emissions targets in their NDCs. But achieving net-zero targets will not completely eliminate GHG emissions. Net-zero targets also do not account for previous GHG emissions, which remain in the atmosphere and have led to current global warming of about 1.1°C (2.0°F) above preindustrial levels. Therefore, while net-zero targets help limit warming to 1.5°C (2.7°F), further reductions of GHG emissions may be needed to reduce ambient temperatures.

Recent Energy Mix in the United States

The mix of energy sources in the U.S. electricity sector has changed over time. Coal use for electric power was 52% lower in 2021 than at its historical high in 2007.[4] This decline occurred while there was a 75% increase in use of natural gas for electricity, starting in 2007, when hydraulic fracturing (fracking) rapidly expanded. Between 2012 and 2021, while energy generated from nuclear and hydroelectric power remained fairly constant, energy from wind power nearly tripled and solar power production increased more than 25-fold. Carbon dioxide emissions from electricity generation, which peaked in 2007, declined 36% by 2021 because of (a) the switch from coal to natural gas, which is less carbon-intensive than coal; (b) the rapid growth of wind and solar power; and (c) widespread adoption of energy efficiency measures.

Energy Policy in the United States

In 2015, during the Barack Obama administration, the United States submitted its NDC, which aimed to reduce GHG emissions 26–28% below 2005 levels by 2025. This target was based on ongoing work by the Obama administration to address climate change, including proposing, in 2014, the Clean Power Plan, which sought to reduce electric power sector emissions 30% below 2005 levels by 2030. However, in 2022, the U.S. Supreme Court found the Clean Power Plan to be unconstitutional and invalidated it in *West Virginia vs. EPA*.

The Donald Trump administration was openly hostile toward climate and clean-energy policy. In 2017, President Trump said that the United States would withdraw from the Paris Agreement, a decision that took effect in late 2020. During the Trump administration, many states, cities, businesses, universities, and other entities promised to increase their efforts to reduce GHG emissions because of the lapse in federal leadership on climate change. Despite setbacks during the Trump administration, by 2020, the United States had reduced GHG emissions by 21% below 2005 levels—exceeding the interim Obama administration goal by 4%.[10]

When Joe Biden became President in 2021, the United States rejoined the Paris Agreement and submitted an updated NDC, setting a target of reducing GHG emissions 50–52% below 2005 levels by 2030. In addition, President Biden set goals of a carbon-free electricity sector by 2035 and net-zero emissions for the entire economy by 2050.

In 2022, President Biden signed into law the Inflation Reduction Act (IRA), which then represented the most comprehensive climate and clean energy policy of the U.S. government. The law allocated $369 billion for climate change and energy, including investments in the electricity sector to expand tax credits for wind and solar power and to create a new tax credit for battery storage. Implementation of the law might, by 2030, reduce GHG emissions by approximately 42%, compared to their peak in 2005—well below the estimated 27% reduction that would have occurred without the law.[11-13] However, the law also included provisions that could bolster fossil fuels, such as making new offshore oil and gas leases available. Of concern, it also contains more "carrots," such as tax credits, rebates, and grant and loan programs, than "sticks," such as regulations and mandates. In addition, its fossil fuel provisions could exacerbate environmental harm in communities predominantly comprised of low-income people, people of color, and Indigenous Peoples, who have disproprotionately suffered from the adverse impacts of extractive industries; for example, the law included provisions to reinstate cancelled oil and gas lease sales, thereby opening harmful oil and gas developments in communities near the Gulf of Mexico and in Alaska.

In the United States, there are barriers to sustained implementation of public policies. Policy priorities change over time. Politically powerful interest groups often oppose measures to address climate change in order to maintain the status quo. While policies like the IRA may be less subject to policy change because they focus on changes to the tax code and are included in legislation passed by both houses of Congress, other policies created by executive orders or through federal agency rulemaking can be more vulnerable to changing policy priorities. (See Chapter 11.)

Even without federal action, state and local governments have many opportunities to address climate change. Since 2015, more than 20 states, including California, Colorado, and Hawaii, have committed to policies for achieving net-zero emissions in their electricity sectors. Some of these policies use Renewable Portfolio Standards (RPSs) or Clean Electricity Standards (CESs), which mandate that electric utilities generate a specific percentage of electricity from renewable sources to achieve emission reduction targets. Other policies include nonbinding goals that signal the intent of state legislatures or governors to address climate change.

Energy Policy in Other Countries

Many countries have been global leaders in setting climate targets and transitioning to low-carbon energy systems. Other countries with emerging economies are high emitters of GHGs because they are building energy-intensive infrastructure and developing modern energy systems based on fossil fuels to support economic growth and alleviate poverty. Still other countries, which have not built substantial energy systems based on fossil fuels, may be able to "leap-frog" directly to energy systems based on renewable energy sources.[14]

At least 18 countries have maintained GHG emissions reductions for at least 10 years since 2005. Some countries are generating almost all of their electricity from renewable sources (Figure 12-6). Scotland has one of the most ambitious climate targets—a legally binding target of net-zero emission by 2045.[15] It is investing in and building renewable energy systems. In 2020, renewable energy sources, especially onshore and offshore wind power, accounted for 97% of its energy generation. In Norway, 98% of electricity comes from renewable energy sources, mainly hydropower and also some wind and geothermal power.[16] Costa Rica, by 2018, generated 98% of its electricity from renewable energy sources, mainly hydropower and geothermal power.[17]

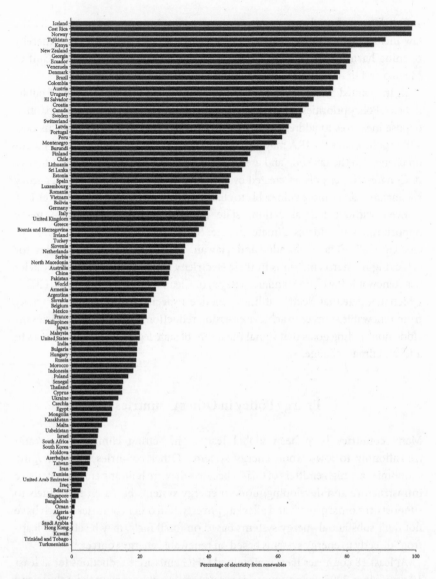

Figure 12-6 Share of electricity from renewable energy sources, by country, 2021. (*Source*: BP Statistical Review of World Energy.)

Some countries with emerging economies that are focused on growth and building infrastructure for a modern energy system based around fossil fuels, such as India and China, emit large amounts of GHGs. These countries are building fossil fuel infrastructure to ensure sufficient reliable energy while also investing in renewable energy infrastructure. China, the country with the most GHG emissions, is one of the largest investors and manufacturers of solar panels and wind turbines. India has substantially increased its investment in local renewable energy infrastructure. In the future, China, India, and other countries may have to grapple with stranded assets because the fossil fuel facilities they are now building will yield diminishing returns on investment as these countries transition toward increased reliance on renewable energy.

Countries with limited modern energy services may be able to industrialize using renewable energy sources while making electricity available to all their citizens. These countries could "leapfrog" to an electricity system based on renewable energy if the knowledge and technology to do so is made available. Kenya, the first African country to install geothermal power on a large scale, has started this process by installing a vast solar power infrastructure. Myanmar, although only 35% of its residents have access to electricity, uses hydropower to supply about half of its electricity.[18,19] International investment or financial mechanisms could motivate other countries to leapfrog to renewable energy sources.

2050 VISION: NET-ZERO EMISSIONS

Achieving net-zero emissions in the energy sector requires a wholesale transformation of the production and consumption of energy. The changes to energy systems needed to achieve ambitious emission reduction goals include:

- <u>Decarbonizing electricity sources</u>: Shifting sources of electricity generation from fossil fuels to low- and zero-carbon sources
- <u>Electrifying end-users</u>: Replacing technologies in vehicles, buildings, and factories that rely on fossil fuels with electric alternatives, such as electric vehicles and heat pumps
- <u>Increasing energy efficiency</u>: Reducing overall energy demand by increasing the efficiency of energy technologies and, by weatherization and use of smart devices, improving the energy efficiency of buildings
- <u>Building out infrastructure</u>: Constructing critical infrastructure, such as an expanded and more interconnected electric transmission grid, which could connect low-carbon power sources to population centers
- <u>Investing in innovation</u>: Expanding research, development, and demonstration for nascent technologies, such as clean hydrogen and carbon-removal technologies, which could help to mitigate emissions from "hard-to-abate" sectors, such as aviation and manufacturing.[20]

POLICY MECHANISMS TO ACHIEVE NET-ZERO GOALS

The four main categories of policies that governments can use to reduce GHG emissions are investments and subsidies, justice-based mechanisms, standards and regulations, and market-based mechanisms. Each category has tradeoffs that policymakers must consider when designing and implementing energy policy.

Investments and Subsidies

Governments can help reduce financing to fossil fuel industries and promote new renewable energy infrastructure with investments and subsidies. Policy mechanisms, such as tax credits, grant programs, loan programs, and rebates, can encourage the use and building of renewable energy infrastructure. Incentives can also be removed from fossil fuel industries in order to reduce reliance on them and reduce GHG emissions. Governments can reduce or eliminate subsidies for fossil fuels, which were $5.2 trillion globally in 2019 and are estimatd to be $20.5 billion in the United States per year (approximately 72% from the federal government and 28% from state governments).[21] Governments and businesses can also completely divest from fossil fuels by removing financial assets, such as stocks and bonds.

The scale of the technological change required for energy transition makes government incentives a central component of energy policy. Tax incentives and credits for both electric utilities and consumers using and supporting alternative energy sources can accelerate market adoption, create jobs, reduce pollution, and encourage research. Incentivizing production through "up-front" rebates and performance-based incentives can help ensure strong development of the market in alternative energy sources. Direct investments in research, development, and demonstration projects are also needed to encourage innovation and help exhibit a new technology. Government support and promotion of solar power has helped accelerate adoption of this technology. The European Union, by implementing financial incentives and promoting public–private partnerships that support innovation, encourages continuous improvement on and new development of "green" products.[22]

Justice-Based Mechanisms

Environmental justice is the fair treatment and meaningful involvement of all people regardless of race, color, national origin, or income in the development, implementation, and enforcement of environmental laws, regulations, and policies.[23] *Equity* recognizes each person's circumstances and allocates necessary resources and opportunities to each to achieve an equal outcome.[24] Policies ensuring environmental justice and equity are critically important on the path to decarbonization. Some ways to ensure environmental justice while facilitating decarbonization include reversing policies that had previously been inequitable, improving the power of communities to make their own decisions, and establishing an equitable framework for energy transitions. (See Chapter 20.)

Fossil fuel industries have been built and operated mainly in low-income communities and communities of Black people, Indigenous Peoples, and people of color. Therefore, decarbonization and the transition to renewable energy must ensure that these communities participate in the energy decisions that affect them. Community input and engagement are important elements to ensure projects and new ideas are relevant to community needs. Equitable policies for transition to renewable energy also need to include opportunities and incentives for fossil fuel industry workers to learn new knowledge and skills, such as for work in the renewable energy sector.

Requiring all new energy projects to have community input and engagement— with prior notice, dissemination of information, and equitable representation and opportunity to speak—allows communities to help determine what is brought into and built in their communities. Enabling communities to participate in the planning of development can help revitalize communities, promote local economies, and provide other co-benefits. However, two issues should be noted. First, communities—even those who support renewable energy in general—may oppose renewable energy projects in their own communities, a phenomenon known as *NIMBYism* (for "Not in My Back Yard").[25,26] This opposition could slow the transition to renewable energy. Second, fossil fuel companies and other interest groups may use practices such as *astroturfing*, in which they falsely present themselves as grassroots organizations to create the appearance of broad support or opposition to policies where little exists.[27,28]

Standards and Regulations

Government regulations (also known as standards or rules), which are the means by which laws are implemented, represent a basic mechanism to institute energy policy (Chapter 11). For example, in 2015, the Obama administration developed the Clean Power Plan to regulate GHG emissions from power plants. Another example is the use of RPSs or CESs to achieve emission reduction targets, as previously described.

Another regulatory approach to reduce GHG emissions is the use of emissions penalties. For example, the Clean Air Act provides the basis for setting emission standards for stationary and mobile sources of air pollution, such as motor vehicles, power plants, and industrial facilities. When states do not meet these standards, the Environmental Protection Agency requires them to implement a set of policies or programs to meet these standards within a specified time period. Although regulations to reduce emissions can be effective, critics of these regulations assert that policies that reward lead to better results than those that punish.

Another approach for reducing emissions is regulation of the use and disposal of specific GHGs, such as methane, nitrous oxide, and fluorinated gases. While fluorinated gases comprise only a small percentage of global GHG emissions, these compounds are much more potent than carbon dioxide in contributing to global warming—and they also deplete the stratospheric ozone level. Since their usage is limited to a few industrial processes and consumer products, implementation of policies to reduce fluorinated gas emissions have been both realistic and cost-effective. For example, the European Union bans sulfur hexafluoride (SF_6) in soundproof windows and sporting equipment and it implements strict policies to control leakage during industrial processes. The United States and many Western European countries have developed and implemented regulations to remove biodegradable material from landfills and compost it. They also require landfills to use gas capture systems to reduce methane emissions.

Standardizing processes and technical requirements based on specific regulations from state public service commissions or the Federal Energy Regulatory Commission is an approach that can help electric utilities connect alternative energy sources to the grid and help reduce uncertainties and delays in the development of alternative energy sources. However, this standardization should be balanced with attention to the local context of alternative energy development. Compensating customers who generate alternative energy can encourage public participation while reducing GHG emissions.

Market-Based Mechanisms: Carbon Tax, Cap and Trade, and Carbon Offsets

A *carbon tax*, which targets carbon dioxide emissions from combustion of fossil fuels at any step in the production cycle, is a market-based mechanism in which governments set a price that emitters have to pay for each ton of carbon dioxide they emit. This can be an effective mechanism because prices are, by far, the most efficient way to influence the decisions of both producers and consumers of energy. Using this mechanism is intended to force businesses to avoid paying the tax by reducing their emissions. Revenues from a carbon tax can be used for investments in renewable energy technologies and infrastructure or given to communities that have already been affected by emissions. Several countries have implemented a carbon tax, such as South Africa, which applies a carbon tax on new motor vehicles at the time of sale based on anticipated fuel usage and emissions. China imposes a carbon tax on coal, natural gas, and oil companies. India levies a carbon tax on coal production and importation. Sweden levies carbon taxes in its transport sector.

Although carbon taxes have been popular with economists, many people have criticized carbon taxes because they may be ineffective at reducing emissions if they are not high enough. In addition, the burden of carbon taxes may be placed on consumers, many of whom depend on inexpensive electricity. Some analysts claim that carbon taxes are not effective because they compete with fossil fuel subsidies that encourage carbon dioxide emission.[29]

A *cap-and-trade (emissions trading) program* is another market-based approach to environmental regulation that sets a limit on the amount of pollutant a company can emit. Emission permits can be traded between companies. Therefore, companies can choose to emit carbon dioxide, using their allotted emission permits, or reduce emissions and sell their permits. In effect, the buyer pays for increasing emissions, while the seller is rewarded for reducing emissions. For trading to be consistent among sectors and locations, regulators must carefully measure, report, and verify emissions and issue emission permits. Enforcement is a crucial element of a cap-and-trade policy, with fines and penalties imposed on companies that exceed permitted emissions.

In 2013, California launched its cap-and-trade program encompassing many sectors, including large electric power plants. Since then, statewide GHG emissions have decreased. Revenues generated have been used for programs that further reduce GHG emissions, with some revenues mandated to be given directly to environmentally and disadvantaged communities.[30] China, the Republic of Korea, and the European Union have also implemented cap-and-trade programs to help reduce GHG emissions.

Carbon offsets are buyable and tradable certificates linked to specific projects that may lower carbon dioxide and other GHG emissions. An individual or business can purchase carbon offsets to fund projects—such as for solar energy in low-income communities, clean cooking with biogas, and planting of trees—that reduce GHG emissions instead of or in tandem with lowering their own GHG emissions.

Critics of cap-and-trade and carbon offset programs have argued that these programs trade pollution and enable fossil fuel industries to profit from carbon credits instead of actually reducing GHG emissions. They assert that carbon offsets allow companies to pollute above their mandated limits, buy offsets to finance projects, and continue to pollute instead of actively reducing their GHG emissions. In addition, carbon offsets lack strict regulation, with the only requirements being that a project must be long-term, not lead to pollution elsewhere, and, without the credit, would not have been done.

CHALLENGES TO NET-ZERO TARGETS

One challenge to achieving net-zero energy systems is the intermittency of wind and solar power, which requires other resources and technologies to ensure a steady stream of energy to electric grids. Hydropower can be used to address this problem. Some analysts point to natural gas as a "bridge fuel" because it has more reliability throughout the day and is associated with lower carbon dioxide emissions than coal. However, natural gas (a) is not aligned with a zero-emissions goal; (b) is comprised primarily of methane, a potent GHG; and (c) when used, is associated with methane releases, which are difficult to quantify. Nuclear power could solve the intermittency problem, but it has mixed public acceptance, especially since the Fukushima Daiichi nuclear disaster in Japan in 2011.[31] Battery storage, which enables the energy produced from wind and solar power to be inexpensively stored for later use, has been utilized in many places in the grid. However, it currently provides only 4 hours of charge, a potential problem when the grid needs energy for long periods when it is not windy or sunny.

Another challenge is that energy systems are typically slow to change. Much of the capital stock in an energy system often lasts 50 to 100 years. Consider the aged fuel pipelines, power plants, and electricity transmission lines that are key parts of the U.S. energy system. For most system components, operating costs are low relative to capital costs of new installations—often making it less expensive to use and maintain old equipment for decades despite the availability of new and improved technologies.

One consequence of long-lived capital stock is that transitions from one dominant energy source to another—such as from wood to coal, or from coal to oil—have historically taken decades.[32] This inertia is exacerbated when addressing climate change due to the long atmospheric half-lives of GHGs, most of which remain in the atmosphere and continue to trap heat for almost a century. Even if the periods of energy system transition shorten with new technology, these inertial forces cannot be ignored. Transforming the energy system to address climate change will require persistence, patience, and strong popular and political will that is sustained for decades.

SUMMARY POINTS

- The use of low-carbon energy sources, such as wind and solar power, for electricity generation has been growing rapidly, even as fossil fuels remain the major source of energy globally.
- The adoption of wind and solar power is growing quickly and the costs of these energy resources are decreasing dramatically.
- Maximizing the potential for wind and solar power depends on developing electricity systems with more storage, transmission, and demand-response measures, which respond to electricity demand to curb peak electricity loads.
- National policies, such as the Inflation Reduction Act, put the United States on track to meet the goals of the 2015 Paris Agreement, but much more stringent policies will be needed to get to net-zero emissions.
- Net-zero carbon emissions by 2050 has become an orienting goal for almost all large economies and, if reached, would make the goals of the Paris Agreement achieveable.

ACKNOWLEDGMENT

We acknowledge Haein Kim for her valuable contributions to this chapter.

References

1. Arias PA, Bellouin N, Coppola E, et al. Technical Summary. In V Masson-Delmotte, P Zhai A Pirani, et al. (eds.). Climate Change 2021: The Physical Science Basis. Contribution of Working Group I to the Sixth Assessment Report of the Intergovernmental Panel on Climate Change. Cambridge: Cambridge University Press, 2021, pp. 33–144.
2. Intergovernmental Panel on Climate Change. Summary for Policymakers. In V Masson-Delmotte, P Zhai A Pirani, et al. (eds.). Climate Change 2021: The Physical Science Basis. Contribution of Working Group I to the Sixth Assessment Report of the Intergovernmental Panel on Climate Change. Cambridge: Cambridge University Press, 2021, pp. 3–32.
3. Pathak M, Slade R, Shukla PR, et al. Technical Summary. In PR Shukla, J Skea, R Slade, et al. (eds.). Climate Change 2022: Mitigation of Climate Change. Contribution of Working Group III to the Sixth Assessment Report of the Intergovernmental Panel on Climate Change. Cambridge: Cambridge University Press, 2022, pp. 49–147.
4. Intergovernmental Panel on Climate Change. Summary for Policymakers. In PR Shukla, J Skea, R Slade, et al. (eds.). Climate Change 2022: Mitigation of Climate Change. Contribution of Working Group III to the Sixth Assessment Report of the Intergovernmental Panel on Climate Change. Cambridge: Cambridge University Press, 2022, pp. 1–48.
5. Shukla PR, Skea J, Slade R, et al. (eds.). Climate Change 2022: Mitigation of Climate Change. Contribution of Working Group III to the Sixth Assessment Report of the Intergovernmental Panel on Climate Change. Cambridge: Cambridge University Press, 2022.
6 U.S. Energy Information Administration. International Energy Outlook 2021 with Projections to 2050. Available at: https://www.eia.gov/outlooks/ieo/. Accessed November 1, 2022.
7. U.S. Energy Information Administration. Monthly Energy Review: October 2022. Available at: https://www.eia.gov/totalenergy/data/monthly/. Accessed October 31, 2022.
8. BP. Statistical Review of World Energy. 2022. Available at: https://www.bp.com/en/global/corporate/energy-economics/statistical-review-of-world-energy.html. Accessed October 31, 2022.
9. Warner ES, Heath GA. Life cycle greenhouse gas emissions of nuclear electricity generation. Journal of Industrial Ecology 2012; 16: S73–S92.
10. U.S. Environmental Protection Agency. Inventory of U.S. Greenhouse Gas Emissions and Sinks: 1990–2020. US Environmental Protection Agency, EPA 430-R-22-003. Last updated April 13, 2023. Available at: https://www.epa.gov/ghgemissions/inventory-us-greenhouse-gas-emissions-and-sinks-1990-2020. Accessed October 31, 2022.
11. Mahajan M, Ashmoore O, Rissman J, et al. Modeling the Inflation Reduction Act Using the Energy Policy Simulator. Energy Innovation Policy & Technology LLC, 2022. Available at: https://energyinnovation.org/publication/modeling-the-inflat

ion-reduction-act-using-the-energy-policy-simulator/.   Accessed   November
21, 2022.

12. King B, Larsen J, Kolus H. A Congressional Climate Breakthrough. Rhodium Group,
   July 28, 2022. Available at: https://rhg.com/research/inflation-reduction-act/.
   Accessed November 21, 2022.

13. Jenkins J, Mayfield EN, Farbes J, et al. Preliminary Report: The Climate and Energy
   Impacts of the Inflation Reduction Act of 2022. Princeton, NJ: REPEAT Project,
   2022. Available at: https://repeatproject.org/docs/REPEAT_IRA_Prelminary_Repo
   rt_2022-08-04.pdf. Accessed November 21, 2022.

14. Sarabhai K, Vyas P. The Leapfrogging Opportunity: Role of Education in Sustainable
   Development and Climate Change Mitigation. Centre for Environment Education
   Australia, Incorporated, 2016. Available at: https://unesdoc.unesco.org/ark:/48223/
   pf0000245615. Accessed November 4, 2022.

15. BBC News. Renewables met 97% of Scotland's electricity demand in 2020. March
   25, 2021. Available at: https://www.bbc.com/news/uk-scotland-56530424. Accessed
   November 21, 2022.

16. Ministry of Petroleum and Energy. Renewable Energy Production in Norway. May
   11, 2016. Available at: https://www.regjeringen.no/en/topics/energy/renewable-ene
   rgy/renewable-energy-production-in-norway/id2343462/.   Accessed   November
   21, 2022.

17. International Trade Administration. Costa Rica's Renewable Energy. January 11,
   2022. Available at: https://www.trade.gov/market-intelligence/costa-ricas-renewa
   ble-energy. Accessed November 21, 2022.

18. International Energy Agency. Myanmar Country Profile. Available at: https://www.
   iea.org/countries/myanmar. Accessed November 4, 2022.

19. The ASEAN Post. Electrifying Myanmar. January 4, 2019. Available at: https://theas
   eanpost.com/article/electrifying-myanmar. Accessed November 21, 2022.

20. National Academies of Science, Engineering, and Medicine. Accelerating
   Decarbonization of the U.S. Energy System. Washington, DC: The National
   Academies Press, 2021. Available at: https://nap.nationalacademies.org/catalog/
   25932/accelerating-decarbonization-of-the-us-energy-system. Accessed February
   28, 2023.

21. Redman J. Dirty Energy Dominance: Dependent on Denial – How the U.S. Fossil
   Fuel Industry Depends on Subsidies and Climate Denial. Oil Change International,
   October 3, 2017. Available at: https://priceofoil.org/2017/10/03/dirty-energy-
   dominance-us-subsidies/. Accessed October 31, 2022.

22. Gallagher KS, Grubler A, Kuhl L, et al. The energy technology innovation system.
   Annual Review of Environment and Resources 2012; 37: 137–162.

23. U.S. Environmental Protection Agency. Environmental Justice. September 30,
   2022. Available at: https://www.epa.gov/environmentaljustice. Accessed November
   21, 2022.

24. George Washington University. Equity vs. Equality: What's the Difference? November
   5, 2020. Available at: https://onlinepublichealth.gwu.edu/resources/equity-vs-equal
   ity/. Accessed November 21, 2022.

25. Larson EC, Krannich RS. A great idea, just not near me! Understanding public
   attitudes about renewable energy facilities. Society & Natural Resources 2016;
   29: 1436–1451.

26. Motavalli J. The NIMBY Threat to Renewable Energy. Sierra. September 20, 2021.
   Available   at: https://www.sierraclub.org/sierra/2021-4-fall/feature/nimby-threat-
   renewable-energy. Accessed March 16, 2023.

27. Cho CH, Martens ML, Kim H, Rodrigue M. Astroturfing global warming: It isn't always greener on the other side of the fence. Journal of Business Ethics 2011; 104: 571–587.

28. Stokes LC. Short Circuiting Policy: Interest Groups and the Battle Over Clean Energy and Climate Policy in the American States. New York: Oxford University Press, 2020.

29. Coady D, Parry IWH, Le N-P, Shang B. Global Fossil Fuel Subsidies Remain Large: An Update Based on Country-level Estimates. International Monetary Fund, May 2, 2019. Available at: https://www.imf.org/en/Publications/WP/Issues/2019/05/02/Global-Fossil-Fuel-Subsidies-Remain-Large-An-Update-Based-on-Country-Level-Estimates-46509. Accessed November 21, 2022.

30. Center for Climate and Energy Solutions. California Cap and Trade. Available at: https://www.c2es.org/content/california-cap-and-trade/. Accessed November 4, 2022.

31. Cooper M, Sussman D. Nuclear power loses support in new poll. New York Times, March 22, 2011.

32. Grubler A. Diffusion: Long-term patterns and discontinuities. Technological Forecasting and Social Change 1991; 39: 159–180.

Further Reading

Grubler A, Nakicenovic N, Pachauri S, et al. Energy Primer. Laxenburg, Austria: International Institute for Applied Systems Analysis, 2014. Available at: https://iiasa.ac.at/projects/energy-primer. Accessed February 28, 2023.

This publication provides a basic introduction to fundamental concepts, frameworks, and impacts of energy systems.

Levenda AM, Behrsin I, Disano F. Renewable energy for whom? A global systematic review of the environmental justice implications of renewable energy technologies. Energy Research & Social Science 2021; 71: 101837.

This systematic review of the literature assesses renewable energy technologies from the perspective of distributive, procedural, recognition, and capability interpretations of environmental justice.

National Academies of Science, Engineering, and Medicine. Accelerating Decarbonization of the U.S. Energy System. Washington, DC: The National Academies Press, 2021. Available at: https://nap.nationalacademies.org/catalog/25932/accelerating-decarbonization-of-the-us-energy-system. Accessed February 28, 2023.

This report identifies key technological and socioeconomic goals that must be achieved to put the United States on the path to reach net-zero carbon emissions by 2050.

Nemet GF. How Solar Energy Became Cheap: A Model for Low-Carbon Innovation. Abingdon, Oxon: Routledge, 2019.

This book provides comprehensive and international explanations for how solar energy has become inexpensive.

U.S. Energy Information Administration. Available at: https://www.eia.gov/. Accessed February 20, 2023.

This is a federal agency that collects, analyzes, and disseminates independent and impartial energy information to promote sound policymaking, efficient markets, and public understanding of energy and its interaction with the economy and the environment.

13

Transportation Policy

Kathryn A. Zyla

The transportation sector generates the largest share of carbon dioxide emissions in the United States—approximately one-third—and an even higher fraction in some states, such as California, and some regions, such as the Northeast. Emission reductions from this sector must be a key component of any comprehensive program to reduce greenhouse gases (GHGs). Achieving these reductions requires the involvement of each level of government (federal, state, and local), given their differing legal authorities and the breadth of actions needed.

Local governments oversee land-use decisions and build and maintain roads, transit systems, bicycle paths, and sidewalks. States oversee substantial transportation planning activities and expenditures. States have played leading roles in policies and programs to reduce emissions from the transportation sector. They have set vehicle emissions standards, mandated cleaner fuels, and changed incentives to foster more sustainable land use and reduce miles driven. In some cases, state policies have served as models for federal policies. For example, in 2010, the U.S. Environmental Protection Agency (EPA) and U.S. Department of Transportation (DOT) finalized the first harmonized federal vehicle standards, which were consistent with policies first proposed by California and adopted by several other states. The U.S. federal government sets corporate average fuel economy (CAFE) standards through the DOT, under energy laws, and sets air quality standards for both conventional pollution and GHGs through the EPA. The federal government also sets budget and funding priorities through transportation authorization and appropriations legislation. All levels of government have opportunities to foster changes through the fleets of vehicles that they purchase, operate, and maintain through financial and other incentives that they implement.

This chapter analyzes the role of each level of government in shaping transportation policy related to climate change and identifies opportunities to reduce emissions from each of the following three components of transportation: vehicle technology, fuel content, and vehicle miles traveled (VMT). Since transportation-related emissions are determined by all three components, reductions from each are needed to achieve significant overall emissions reductions and provide health and quality-of-life co-benefits. While most of the

Kathryn A. Zyla, *Transportation Policy* In: *Climate Change and Public Health*. Edited by: Barry S. Levy and Jonathan A. Patz, Oxford University Press. © Oxford University Press 2024. DOI: 10.1093/oso/9780197683293.003.0013 strategies discussed here reduce emissions from personal transportation, medium- and heavy-duty vehicles account for more than one-fourth of transportation emissions of GHGs in the United States[1]; therefore, this chapter also briefly addresses strategies for reducing emissions from freight vehicles.

All of the strategies described in this chapter to reduce GHG emissions from transportation also provide benefits to public health because the control of transportation-related emissions reduces both GHGs and other air pollutants. Strategies that increase active transportation, such as bicycling or walking, or that reduce disparities in access to transportation choices can also provide a broad range of other public health benefits, such as increased physical activity or access to healthcare.[2] (See Box 13-1.)

While not the focus of this chapter, impacts of climate change, such as increased temperature, flooding, and sea-level rise, are already affecting transportation infrastructure. Many of the policies discussed in this chapter are aimed at reducing GHG emissions that contribute to climate change; they can also increase resilience to these impacts. For example, the availability of transportation options from a variety of diverse fuels, such as electric batteries or compressed natural gas, and options such as transit and bicycle trails, provide alternatives when extreme weather limits gasoline supplies. This was the case in the aftermath of hurricanes Katrina and Rita on the U.S. Gulf Coast in 2005, Superstorm Sandy in the northeastern United States in 2013, and Hurricane Ida in multiple states in 2021.

VEHICLE TECHNOLOGY AND EFFICIENCY

The design, assembly, and maintenance of motor vehicles determine how efficiently they consume fuel and use refrigerants in air conditioning, which, in turn, affect the amount of GHGs emitted from combustion and leaking refrigerants. The U.S. federal government and some state governments can affect vehicle performance either by setting mandatory standards, which are authorized by the Clean Air Act, or by encouraging the production and purchase of low-emitting vehicles through financial or other incentives, such as allowing access to high-occupancy vehicle lanes.

California GHG Standards Complement Federal Emissions and Fuel-Economy Regulations

GHG vehicle standards in California are an outgrowth of its unique authority under the Clean Air Act. Under the Act, states are prohibited from adopting or

Box 13-1 Epidemiological Evidence for the Health Co-Benefits of Active Transportation

Natalie Levine and Maggie L. Grabow

Walking and bicycling for a purpose rather than recreation are forms of *active transportation (active travel)* that both increase physical activity (thereby improving mental and physical health) and decrease emission of GHGs and other air pollutants. Some of the epidemiologic evidence supporting these health co-benefits is derived from the following studies.

A systematic review and meta-analysis of 23 prospective cohort studies performed in the United States, Europe, and Asia with 531,333 participants showed that people engaged in active transportation had 8% significantly lower all-cause mortality and 9% lower incidence of cardiovascular disease compared to those who did not. It also found that bicycle commuters had a 24% further reduction in all-cause mortality and a 25% further reduction in cancer mortality than commuters who walked.[1]

A prospective cohort study of adults in the United Kingdom found that bicycling was significantly associated with lower all-cause mortality and lower cancer mortality and cancer incidence. There was a dose-response relationship between weekly commuting distances and these reported reductions in mortality and cancer incidence. This study also found that walking was associated with lower cardiovascular disease mortality and incidence, although to a lower extent than bicycling.[2]

The benefits of active transportation also include improved mental health. The British Household Panel Survey of 17,985 adults found that users of active transportation and/or public transport reported significantly higher psychological well-being than those who traveled by car.[3] A combined cohort and cross-sectional study of European adults found that active transportation was associated with higher energy and lower fatigue, more frequent contact with friends and family members, and less frequent feelings of loneliness. The study also found that people who bicycled reported less stress and higher mental health benefits overall.[4] A cross-sectional study in Austria concluded that children who participated in active transportation perceived improved psychological well-being.[5]

While cross-sectional studies can generally not be used to prove causation, they provide additional valuable information. For example, a cross-sectional study in Nigeria determined that time spent walking or bicycling to work was significantly associated with lower body mass index, lower blood pressure, and lower serum cholesterol.[6] A cross-sectional study based on data from all U.S. states as well as 47 of the 50 largest U.S. cities found that increased physical activity and higher rates of commuting by active transportation were significantly associated with a lower prevalence of both obesity and diabetes; cities with the highest rates of active-transportation commuting had a 20% lower prevalence of obesity and a 23% lower prevalence of diabetes than cities with the lowest rates of active transportation.[7]

Modeling studies have also demonstrated the health co-benefits of active transportation. An analysis of the 11 largest cities in the Upper Midwest region of the United States found that replacing short urban car trips of less than 5 kilometers (3.1 miles) with bicycle trips could improve air quality and increase exercise, which would prevent approximately 1,100 deaths and more than $7 billion in associated costs by subsequent air-quality improvements and increasing exercise.[8] A modeling study of the San Francisco Bay Area found that increasing daily walking and bicycling from 4 to 22 minutes could reduce disability-adjusted life years due to cardiovascular disease and diabetes by 14%, but would increase traffic-related injuries by 39%.[9] However, after the number of bicyclists and pedestrians reaches a threshold, drivers are more aware of the need to share the road with bicyclists and improve their own driving.[10]

Many gaps in research on active transportation remain, including disparities in use of active transportation by race, ethnicity, and socioeconomic status as well as assessment of active transportation interventions to improve health equity.[11]

Box References

1. Dinu M, Pagliai G, Macchi C, Sofi F. Active commuting and multiple health outcomes: A systematic review and meta-analysis. Sports Medicine 2019; 49: 437–452.

2. Celis-Morales CA, Lyall DM, Welsh P, et al. Association between active commuting and incident cardiovascular disease, cancer, and mortality: Prospective cohort study. British Medical Journal 2017: j1456. doi: 10.1136/bmj.j1456.

3. Martin A, Goryakin Y, Suhrcke M. Does active commuting improve psychological wellbeing? Longitudinal evidence from eighteen waves of the British Household Panel Survey. Preventive Medicine 2014; 69: 296–303. doi: 10.1016/j.ypmed.2014.08.023.

4. Avila-Palencia I, Int Panis L, Dons E, et al. The effects of transport mode use on self-perceived health, mental health, and social contact measures: A cross-sectional and longitudinal study. Environment International 2018; 120: 199–206. doi: 10.1016/j.envint.2018.08.002.

5. Stark J, Meschik M, Singleton PA, Schützhofer B. Active school travel, attitudes and psychological well-being of children. Transportation Research Part F: Traffic Psychology and Behaviour 2018; 56: 453–465.

6. Forrest KY, Bunker CH, Kriska AM, et al. Physical activity and cardiovascular risk factors in a developing population. Medicine and Science in Sports and Exercise 2001; 33: 1598–1604.

7. Pucher J, Buehler R, Bassett DR, Dannenberg AL. Walking and cycling to health: A comparative analysis of city, state, and international data. American Journal of Public Health 2010; 100: 1986–1992. doi: 10.2105/AJPH.2009.189324.

8. Grabow ML, Spak SN, Holloway T, et al. Air quality and exercise-related health benefits from reduced car travel in the Midwestern United States. Environmental Health Perspectives 2012; 120: 68–76. doi: 10.1289/ehp.1103440.

9. Maizlish N, Woodcock J, Co S, et al. Health cobenefits and transportation-related reductions in greenhouse gas emissions in the San Francisco Bay Area. American Journal of Public Health 2013; 103: 703–709.

10. Jacobsen PL. Safety in numbers: More walkers and bicyclists, safer walking and bicycling. Injury Prevention 2015; 21: 271–275.

11. Hansmann KJ, Grabow M, McAndrews C. Health equity and active transportation: A scoping review of active transportation interventions and their impacts on health equity. Journal of Transport & Health 2022; 25: 101346. doi: 10.1016/j.jth.2022.101346.

attempting to enforce any standard relating to the control of emissions from new motor vehicles or new motor vehicle engines. However, because of its historic leadership role and its unique air pollution problems, California may apply for a waiver from this provision and adopt its own standard, and other states may choose to adopt the California standard if a waiver is granted. This process has played an important role as federal GHG standards have evolved during recent changes in administrations and policy approaches.

California has received more than 100 waivers for mobile sources through this provision since 1968.[3] The EPA has only denied the waiver once, in 2008, during the George W. Bush administration, by EPA Administrator Stephen Johnson.[4] Soon afterward, President Barack Obama ordered the EPA to reassess its denial of California's waiver request.[5] Several parties had previously challenged the waiver, including auto manufacturers, which insisted that the alternate standards would create an unmanageable patchwork of regulations.[6,7] In 2009, the challenging parties, the State of California, the EPA, and the DOT reached an agreement to resolve current and potential disputes over standards through Model Year (MY) 2016.

In accordance with the agreement, the EPA and the National Highway Traffic Safety Administration (NHTSA) of DOT created a joint program to establish GHG and fuel-economy standards for passenger vehicles, which would achieve GHG reductions equivalent to or greater than the California regulations for MYs 2012 to 2016.[8] Manufacturers agreed to drop current challenges to the waiver and forgo any future challenges. The California waiver was granted in June 2009.[9] The EPA then issued a rule aimed at achieving a standard of 250 grams of carbon dioxide equivalent per mile in MY 2016, coupled with harmonized fuel efficiency standards from the DOT that set CAFE standards at 34.1 miles per gallon (mpg) by MY 2016.[9]

In 2012, EPA and NHTSA, in close consultation with California's Air Resources Board (ARB), issued another rule to further reduce GHGs and improve fuel economy for light-duty vehicles for MYs 2017–2025.[10] However, in 2019, under the Trump Administration, the EPA and NHTSA issued a joint regulation (the Safer Affordable Fuel-Efficient ["SAFE"] Vehicles Rule), which both revised federal vehicle fuel economy and emission standards and revoked the waiver of California's "Advanced Clean Cars" GHG and zero-emission vehicle program issued in 2013.[11] (No waiver had ever previously been revoked. Waivers do not expire.) In December 2021, the EPA under the Biden Administration finalized the revised national GHG standards for passenger cars and light trucks for Model Years 2023–2026, aimed at achieving a target of 161 grams of carbon dioxide per mile in MY 2026—47 grams of carbon dioxide per mile lower than the standards that they replaced.[12] The EPA estimated that between $8 and $19 billion of the total benefits it projected through 2050 would be from improved public health as a result of reduced emissions of non-GHG pollutants, including oxides of nitrogen and fine particulate matter.[13]

In March 2022, the EPA reinstated California's waiver so that it could implement its own GHG standards for cars and light trucks.[14] As a result, other states may once again choose to adopt and enforce California's GHG standard instead of the federal standard. In August 2022, California's ARB approved the "Advanced Clean Cars II" rule, which establishes a roadmap for all new cars and trucks sold in California to be zero-emission vehicles.[15] Several states, including Washington, Oregon, Vermont, and New York, have already indicated their intention to follow California's new standards; others, such as Massachusetts and Virginia, are required to do so by their state laws.[16]

Promoting Electric Vehicles and Other Alternatives

The federal government and many state and local governments are encouraging the use of lower-emitting vehicles and electric vehicles (EVs), which are gaining in market share. In the first quarter of 2022, EV registrations in the United States increased 60% even as new car registrations overall decreased 18%.[17] The lifecycle costs of some of these vehicles can be less than conventional cars, even when up-front costs are higher.[18] Policies designed to encourage consumer acceptance include state and federal tax credits and rebates, access to high-occupancy lanes on highways, and preferential parking. Cities and states are also developing EV charging and fueling networks and removing regulatory barriers, such as complicated permitting processes for EV charging. Public charging infrastructure can be helpful, as can fleet procurement requirements. Collaborations among states in several regions have helped to enable long-distance EV travel and improve users' experience by providing accessible charging stations marked with consistent signage.[19-21]

In addition, California and several other states are moving forward with zero- or low-emission vehicle standards and agreements. For example, California has created programs to encourage EVs and fuel-cell vehicles, which convert hydrogen gas to electricity to drive an electric motor.[22] It is collaborating with other states through the Clean Air Act and memoranda of understanding. For example, in 2013, the governors of California and seven other states agreed to work together to put 3.3 million zero-emission light-duty vehicles on the road by 2025; two other states have since joined this initiative.[23]

FUEL STANDARDS AND EMISSIONS CAPS

In addition to vehicle design, the choice of transportation fuel has a significant effect on GHGs emitted per mile driven. Federal energy laws have set timetables for specific volumes of renewable fuels, some with GHG thresholds, through a renewable fuels standard (RFS). The RFS allows a significant amount of fuel to come from corn-based ethanol, with a desired shift to other sources (such as cellulosic) over time. (See Chapter 14.)

California aims to reduce the carbon content of fuels through its low-carbon fuel standard (LCFS). The state legislature provided authority for its ARB to adopt an LCFS, pursuant to California Assembly Bill (AB) 32 in 2006 and a Governor's Executive Order in 2007. In 2009, the California ARB approved, and 2 years later began implementing, its first LCFS regulation to reduce the carbon intensity of transportation fuels used in California by an average of 10% by 2020. The ARB readopted and adjusted the program in 2015 (effective in 2016) to align the LCFS with California's statutory 2030 GHG reduction target.[24] In 2016, Oregon launched its own clean fuels program, and, in 2021, Washington's state legislature approved a clean fuel standard, which is linked to the programs in California and Oregon and to a similar program in British Columbia.[25,26]

In addition to the LCFS, AB 32's comprehensive cap-and-trade program has covered transportation fuels since 2015. Allowance auction proceeds also support clean transportation projects and programs that meet other objectives under AB 32.

REDUCING VEHICLE MILES TRAVELED

GHGs can be reduced by developing policies and personal behavioral changes, such as driving less often, driving shorter distances, and spending less time idling. Land-use patterns, policies, and incentives can also play important roles, as can the availability of attractive alternatives to driving single-occupancy vehicles. A 2009 study found that aggressively implementing a full range of strategies aimed at reducing vehicle miles traveled (VMT) could reduce on-road GHG emissions by 18–24% by 2050.[27] Effectiveness of actions depends on multiple factors, including who is driving, where they are going, and what alternative modes and destinations are available.[28]

VMT reduction strategies depend largely on local and regional government entities, such as metropolitan planning organizations. Local governments have important roles in zoning and providing alternatives to single-occupancy vehicles, such as mass transit systems, bike lanes, and assistance with carpools. States can provide important financial incentives and enact enabling laws to give local governments the authority and the financial resources to make positive changes. Often these alternative modes of transportation are accompanied by health co-benefits and improve social capital through connecting individuals with each other, their neighborhoods, and their communities.

VMT strategies can be grouped into three categories:

- Land-use planning that promotes "smart growth"
- Improving and expanding transit options
- Driving and parking pricing strategies.

Each of these is described below.

Land-Use Planning That Promotes "Smart Growth"

While land use is often a local government concern, states are working to promote *smart growth* for the variety of benefits it brings. Smart-growth planning promotes efficient and sustainable land development, reuse of existing infrastructure, and smaller development footprints.[29] It is often associated with compact, walkable communities that minimize sprawl. California's SB 375 lays out a planning framework that encourages metropolitan areas to reduce GHG emissions through planning requirements and incentives.[30] On the East Coast, some states are implementing "smart-growth" policies to encourage development in dense areas, with maintained sidewalks and bicycle paths and near transit hubs to reduce the need to drive long distances.[31] (See Chapter 16.) The Colorado Transportation Commission approved a new standard in 2022 to reduce GHG emissions from the transportation sector, improve air quality, reduce smog, and provide more travel options. The standard requires the state department of transportation and Colorado's five metropolitan planning organizations to determine the total GHG emissions expected from future transportation projects and reduce emissions by set amounts.[32]

Within these frameworks, many land-use strategies are available. Zoning can encourage *transit-oriented development* (TOD), the construction of higher-density housing and commercial development convenient to transit stations. For example, the City of Seattle created an overlay zoning district, which prohibits low-density uses around new light-rail and monorail corridors.[33]

Removing minimum parking space requirements or converting to maximum parking allotments can discourage driving when alternatives are available and create additional space for transit-oriented development. Reducing the number of parking spaces can also save on construction costs and make TOD more affordable. A case study of six San Francisco neighborhoods found that the standard requirement for off-street parking increased costs for single-family homes and condominiums by more than 10%.[34] Smart-growth planning, including requirements that streets safely accommodate pedestrians and bicyclists, can reduce trips made by car (Figure 13-1).

Several states have enacted legislation requiring the incorporation of pedestrian- and bicycle-friendly designs into all new construction projects. For example, a Florida statute requires pedestrian- and bicycle-oriented designs for new road construction projects.[35] More than 1,600 "complete streets" policies have been passed in the United States, aimed at enabling safe access to streets for all users, including pedestrians, bicyclists, motorists, and transit riders of all ages and abilities[36]; extensive networks of bicycle lanes[37,38]; and popular bicycle-sharing programs. While development changes take time, these solutions have benefits beyond GHG reduction. They reduce local air pollution, ozone exceedances, and the urban heat island effect. There are direct public health benefits from active transportation (Figure 13-2) and societal benefits from neighborhood cohesion. In addition, investment in additional transportation options provides for redundancies and diversity that can promote resilience to the impacts of climate change.

Figure 13-1 Bicyclists in Copenhagen. (Photograph by Barry S. Levy.)

Improving and Expanding Transit Options

Improving and expanding the use of transit systems can be a cost-effective way of reducing GHGs. Transportation services could make more routes available, operate routes with greater frequency, increase service hours, lower fares, and start park-and-ride programs.[39] Provision of real-time arrival information can also enhance transit service and increase ridership. Light rail and bus rapid transit can increase ridership in transit systems and promote economic development. In rural areas, vanpools offer alternatives to single-passenger vehicles.

Driving and Parking Pricing Strategies

Pricing strategies have some of the greatest potential to reduce GHG emissions. A 2009 study found that the largest reductions achievable through state policies would be achieved by congestion pricing or pay-as-you drive insurance.[27] Pricing programs for road travel and parking can reduce GHG emissions by discouraging people from driving, and, for those who still drive, by improving traffic flow, which produces lower GHG emissions than highly congested, stop-and-go driving. Several types of pricing strategies can be used, including congestion

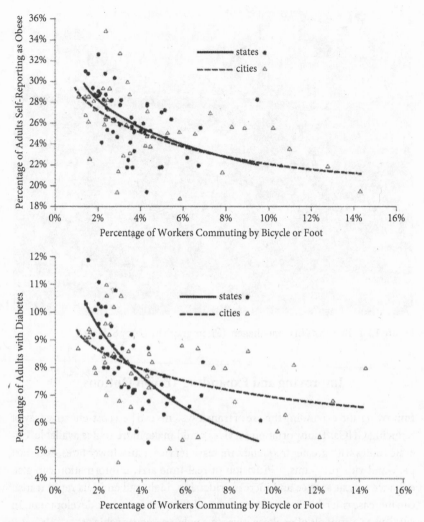

Figure 13-2 Relationships, in U.S. states and cities, of percentage of workers commuting by walking or bicycling and (upper graph) percentage of adults self-reporting as obese, and (lower graph) percentage of adults with diabetes. (Sources: Xu F, Town M, Balluz LS, et al. Surveillance for certain health behaviors among states and selected local areas—United States, 2010. Morbidity and Mortality Weekly Report 2013; 62: 1–247; U.S. Census Bureau. American Community Survey, 2009–2011; and Centers for Disease Control and Prevention. Behavioral Risk Factor Surveillance Systems Survey [BRFSS]. Atlanta: CDC, 2011.)

pricing (such as time-variable tolls), cordon pricing (charging all vehicles entering high-use areas), and increases in parking fees.

A mileage-based user fee (VMT fee) imposes a cost based on mileage (rather than on gasoline consumption); early pilot programs indicate that it can reduce the number of miles driven.[40] This approach is often considered an alternative to the gasoline tax for raising revenue to support state transportation programs and infrastructure investment, but, depending on how it is administered, it may raise concerns about privacy and implementation.[41] Unlike gasoline taxes, fees based solely on VMT do not promote vehicle efficiency.

ENCOURAGING TELECOMMUTING AND ECODRIVING

Working with employers to promote telecommuting can offer some emissions reductions, although most vehicle trips are not related to work. (One study found that commuting accounted for only 16% of "household person trips."[1]) Driver education can promote *ecodriving*[42] skills, such as maintaining steady speeds and anticipating changes in traffic flow. These steps can reduce GHG emissions while saving gasoline and money.[42] Ecodriving programs can reduce carbon dioxide emissions by 340 pounds per driver per year.[27,42] Driver education can also stress the importance of proper tire inflation and other measures to enhance vehicle efficiency and reduce GHG emissions.

LOWERING MAXIMUM SPEED LIMITS

GHG emissions decrease as a car's speed increases up to approximately 55 miles per hour (mph), above which GHG emissions increase.[43] Generally, 55 mph is the most GHG-efficient maximum speed limit; however, most states allow significantly higher speed limits. Not including economy-wide strategies, speed limit reductions to 55 mph can create the greatest short-run emission reductions.[27] And lower speed limits save lives.

IMPLEMENTING FEDERAL INCENTIVES

While most of these strategies are implemented at the local or state level, the federal government can also play an important role. During the Obama administration, the Partnership for Sustainable Communities, formed by the DOT, EPA, and the U.S. Department of Housing and Urban Development (HUD),[44] helped communities to improve access to affordable housing and transportation while protecting the environment. The three agencies developed livability principles that are incorporated into federal funding programs, policies, and future legislative proposals. Federal legislation could create incentives that tie financial support to reductions in GHG emissions and public health benefits of a cleaner transportation system. This could be done by setting performance goals and linking funding to meeting these goals.

REDUCING EMISSIONS FROM TRUCKS

Heavy-duty truck travel is largely determined by demand for freight, with both payload size and length of haul affecting fuel consumption and emissions. It is estimated that freight truck VMT will increase by 55% between 2019 and 2050.[45] Therefore, increasing fuel efficiency is critical to reducing GHG emissions from freight trucks. In 2022, a study by the National Renewable Energy Laboratory found that, with continued improvements in vehicle and fuel technologies, zero-emission vehicles could reach total-cost-of-driving parity with conventional diesel vehicles by 2035 for all medium- and heavy-duty vehicle classes, without incentives.[45] In April 2023, the EPA proposed standards to reduce GHG emissions from heavy-duty vehicles starting in FY2027. These will accompany standards finalized in December 2022 for emissions of smog- and soot-forming nitrogen oxides and other air pollutants from heavy-duty vehicles and engines.[46]

These standards are expected to improve public health, given that (a) heavy-duty on-road diesel vehicles are the leading contributor to transportation-sector air pollution and related adverse health effects, and (b) the estimated 72 million people who live in communities near heavy-duty truck traffic suffer a disproportionate share of these adverse health effects. These people are more likely to be people of color and those with lower incomes.[47-49]

In addition to fuel taxes and vehicle standards, fuel consumption by freight trucks can be reduced by changes to traffic policy and transportation infrastructure. For example, to maximize efficiency, many countries set a lower speed limit for freight trucks than for personal vehicles.[1] Infrastructure changes, either physical improvements, such as dedicated lanes for carrying long-haul trucks through congested areas, or organizational improvements, such as those that increase traffic flow, can also reduce fuel consumption and GHG emissions.[1] Electrification of rest stops and other strategies for reducing idling can also substantially reduce the fuel consumed and emissions resulting from truck operations.[50]

Emissions from freight can also be reduced by shifting to alternative modes of travel. Trucks transport more than two-thirds of all goods (by weight) in the United States and account for more than three-fourths of GHG emissions from freight transportation.[51] Trucks also consume significantly more energy per ton-mile shipped than other modes (approximately eight times more in one study);[52] therefore, shifting freight transport to rail may significantly reduce emissions from goods movement.

Use of alternative fuels, such as electricity, natural gas, propane, and biodiesel, can decrease GHG emissions from the freight sector,[53] with recent efforts focused on zero-emission vehicles that produce no GHG emissions from the tailpipe. Since 2020, the governors of California and 16 other states, and the mayor of the District of Columbia have announced a collaboration, based on their light-duty vehicle partnership, to advance medium- and heavy-duty zero-emission vehicles.[54]

These improvements can not only reduce GHG emissions, but can also yield health co-benefits. For example, engine emissions are believed to be responsible for about 70% of California's estimated known cancer risk attributable to toxic air contaminants. And the California ARB estimates that diesel particulate matter contributes to approximately 1,400 premature deaths from cardiovascular disease annually in California.[55]

TRANSPORTATION'S ROLE IN IMPROVING RESILIENCE

Diversifying transportation options not only helps to reduce GHG emissions, but also offers opportunities to increase communities' resilience to the impacts of climate change. During Superstorm Sandy in 2012, when petroleum supplies were low, compressed natural gas (CNG) buses were used in Atlantic City, New Jersey, to evacuate elderly and disabled residents from vulnerable areas.[56] Bicycles were used to bring supplies to affected areas as alternatives to cars and mass transit, which was also adversely affected by the storm.[57] Some electric vehicle drivers who were able to charge their vehicles before the storm (or to find locations that had not lost power) were able to use the energy stored in the vehicles to drive and to operate or charge small appliances when their homes lost power.[58]

SUMMARY POINTS

- To significantly reduce energy use and emissions from the transportation sector, all levels of government have important roles to play.
- There are many and varied opportunities to make these reductions while achieving other societal and health goals.

- There are also many benefits to reducing energy use and emissions, such as improved air quality, fewer temperature and weather extremes, and health co-benefits from alternative transportation modes, such as bicycling and walking.
- Measures to reduce GHG emissions and other air pollutants from heavy-duty on-road traffic can reduce disparities concerning exposure to air pollutants and related adverse health effects.
- Offering transportation alternatives can promote stronger, more sustainable, and more resilient communities.

ACKNOWLEDGMENT

This chapter is based on the "Transportation Policy" chapter by Vicki Arroyo and Kathryn A. Zyla from the first edition of this book.

References

1. U.S. Environmental Protection Agency. Fast Facts on Transportation Greenhouse Gas Emissions. 2020. Available at: https://www.epa.gov/greenvehicles/fast-facts-transportation-greenhouse-gas-emissions. Accessed August 11, 2022.
2. U.S. Department of Transportation Federal Highway Administration. How Does Transportation Affect Public Health? May/June 2013. Available at: https://highways.dot.gov/public-roads/mayjune-2013/how-does-transportation-affect-public-health. Accessed December 10, 2022.
3. U.S. Environmental Protection Agency. Vehicle Emissions California Waivers and Authorizations. Available at: https://www.epa.gov/state-and-local-transportation/vehicle-emissions-california-waivers-and-authorizations. Accessed August 11, 2022.
4. U.S. Environmental Protection Agency. California State Motor Vehicle Pollution Control Standards; Notice of Decision Denying a Waiver of Clean Air Act Preemption for California's 2009 and Subsequent Model Year Greenhouse Gas Emission Standards for New Motor Vehicles. Federal Register, March 6, 2008; 73: 12156–12169. Available at: https://www.govinfo.gov/content/pkg/FR-2008-03-06/pdf/E8-4350.pdf. Accessed August 15, 2022.
5. The White House. Memorandum for the Administrator of the Environmental Protection Agency: State of California Request for Waiver Under 42 U.S.C. 7543(b) the Clean Air Act. Jan. 26, 2009. Available at: https://www.govinfo.gov/content/pkg/CFR-2010-title3-vol1/pdf/CFR-2010-title3-vol1-other-id198.pdf. Accessed August 15, 2022.
6. Central Valley Chrysler-Jeep v. Goldstene, 529 F.Supp.2d 1151 (E.D. Ca. 2007). Available at: https://case-law.vlex.com/vid/central-valley-chrysler-jeep-884585836. Accessed August 15, 2022.
7. Lincoln-Dodge v. Sullivan, 588 F.Supp.2d 224 (Dist. R.I. 2008). Available at: http://www.leagle.com/decision/2008812588agfsupp2d224_1780. Accessed August 15, 2022.

8. U.S. Environmental Protection Agency, Department of Transportation. Notice of Upcoming Joint Rulemaking to Establish Vehicle GHG Emissions and CAFE Standards. Federal Register May 22, 2009; 47: 24007–24012. Available at: http://www.gpo.gov/fdsys/pkg/FR-2009-05-22/html/E9-12009.htm. Accessed August 15, 2022.

9. U.S. Environmental Protection Agency. Notice of Decision Granting a Waiver of Clean Air Act Preemption for California's 2009 and Subsequent Model Year Greenhouse Gas Emission Standards for New Motor Vehicles. Federal Register July 8, 2009; 74: 32744–32784. Available at: http://www.gpo.gov/fdsys/pkg/FR-2009-07-08/pdf/E9-15943.pdf. Accessed August 15, 2022.

10. U.S. Environmental Protection Agency. 2017 and Later Model Year Light-duty Vehicle Greenhouse Gas Emissions and Corporate Average Fuel Economy Standards. Federal Register October 15, 2012; 77: 62624–63200. Available at: http://www.gpo.gov/fdsys/pkg/FR-2012-10-15/pdf/2012-21972.pdf. Accessed August 15, 2022.

11. U.S. Environmental Protection Agency. The Safer Affordable Fuel Efficient (SAFE) Vehicles Final Rule for Model Years 2021–2026. 2020. Available at: https://www.epa.gov/regulations-emissions-vehicles-and-engines/safer-affordable-fuel-efficient-safe-vehicles-final-rule. Accessed August 12, 2022.

12. U.S. Environmental Protection Agency. Regulations for Greenhouse Gas Emissions From Passenger Cars and Trucks. 2021. Available at: https://www.epa.gov/regulations-emissions-vehicles-and-engines/regulations-greenhouse-gas-emissions-passenger-cars-and. Accessed August 12, 2022.

13. U.S. Environmental Protection Agency. Revised 2023 and Later Model Year Light-duty Vehicle Greenhouse Gas Emissions Standards by the Numbers. 2021. Available at: ttps://nepis.epa.gov/Exe/ZyPDF.cgi?Dockey=P1013NRF.pdf. Accessed August 12, 2022.

14. U.S. Environmental Protection Agency. EPA Restores California's Authority to Enforce Greenhouse Gas Emission Standards for Cars and Light Trucks. March 2022. Available at: https://www.epa.gov/newsreleases/epa-restores-californias-authority-enforce-greenhouse-gas-emission-standards-cars-and. Accessed August 12, 2022.

15. California Air Resources Board. California moves to accelerate to 100% new zero-emission vehicle sales by 2035. August 25, 2022. Available at: https://ww2.arb.ca.gov/news/california-moves-accelerate-100-new-zero-emission-vehicle-sales-2035. Accessed December 10, 2022.

16. Bright Z. States ride shotgun with California to rev up clean cars rules. September 2, 2022. Available at: https://news.bloomberglaw.com/environment-and-energy/states-ride-shotgun-with-california-to-rev-up-clean-cars-rules. Accessed December 10, 2022.

17. Blanco S. Electric cars' turning point may be happening as U.S. sales numbers start to climb. Car and Driver. August 8, 2022. Available at: https://www.caranddriver.com/news/a39998609/electric-car-sales-usa/. Accessed August 12, 2022.

18. Raustad R. Electric Vehicle Lifecycle Cost Analysis. Electric Vehicle Transportation Center. University of Central Florida February 2017. Available at: https://rosap.ntl.bts.gov/view/dot/31875/dot_31875_DS1.pdf. Accessed August 12, 2022.

19. Northeast Electric Vehicle Network, Transportation and Climate Initiative. Available at: https://www.transportationandclimate.org/content/northeast-electric-vehicle-network. Accessed August 12, 2022.

20. Michigan.gov. Governor Whitmer announces partnership with Midwest governors to coordinate electric vehicle charging infrastructure, grow jobs, and futureproof

regional commerce. September 30, 2021. Available at: https://www.michigan.gov/whitmer/news/press-releases/2021/09/30/governor-whitmer-announces-part nership-with-midwest-governors-to-coordinate-electric-vehicle-chargi. Accessed August 12, 2022.

21. National Association of State Energy Officials. REV West. Available at: https://www.naseo.org/issues/transportation/rev-west. Accessed August 12, 2022.

22. U.S. Department of Energy. Fuel Cell Vehicles. Available at: http://www.fueleconomy.gov/feg/fuelcell.shtml. Accessed August 15, 2022.

23. Northeast States for Coordinated Air Use Management (NESCAUM). State Zero-emission Vehicle Programs Memorandum of Understanding. October 24, 2013. Available at: https://www.nescaum.org/documents/zev-mou-10-governors-signed-20191120.pdf/. Accessed August 12, 2022.

24. California Air Resources Board. Low Carbon Fuel Standard: About. Available at: https://ww2.arb.ca.gov/our-work/programs/low-carbon-fuel-standard/about. Accessed August 12, 2022.

25. Oregon Department of Environmental Quality. Oregon Clean Fuels Program. Available at: https://www.oregon.gov/deq/ghgp/cfp/Pages/default.aspx. Accessed August 12, 2022.

26. Washington Department of Ecology. Clean Fuel Standard. Available at: https://ecol ogy.wa.gov/Air-Climate/Climate-change/Reducing-greenhouse-gases/Clean-Fuel-Standard#:~:text=In%20Washington%2C%20the%20Clean%20Fuel,below%202 017%20levels%20by%202038. Accessed August 12, 2022.

27. Cambridge Systematics, Inc. Moving Cooler: An Analysis of Transportation Strategies for Reducing Greenhouse Gas Emissions. July 2009. Available at: http://www.reconnectingamerica.org/assets/Uploads/2009movingcoolerexecsumandapp end.pdf. Accessed August 15, 2022.

28. Salon D. The Effect of Land Use Policies and Infrastructure Investments on How Much We Drive: A Practitioner's Guide to the Literature. UC Davis: National Center for Sustainable Transportation, 2015. Available at: https://escholarship.org/uc/item/54d4567m. Accessed August 12, 2022.

29. American Planning Association. APA Policy Guide on Smart Growth. April 15, 2002. Available at: https://www.planning.org/policy/guides/adopted/smartgrowth.htm. Accessed August 15, 2022.

30. California Air Resources Board. Sustainable Communities. Available at: https://ww2.arb.ca.gov/our-work/topics/sustainable-communities. Accessed August 15, 2022.

31. Commonwealth of Massachusetts. GHG Emissions and Mitigation Policies. Available at: https://www.mass.gov/info-details/ghg-emissions-and-mitigation-policies. Accessed August 12, 2022.

32. Colorado Department of Transportation. Colorado's New Greenhouse Gas Standard for Transportation Planning. December 2021. Available at: https://www.codot.gov/programs/environmental/greenhousegas/assets/ghg-standard-fact-sheet.pdf. Accessed August 12, 2022.

33. Seattle Municipal Code. Chapter 23.61 - Station Area Overly District. Available at: https://library.municode.com/wa/seattle/codes/municipal_code?nodeId= TIT23LAUSCO_SUBTITLE_IIILAUSRE_CH23.61STAROVDI. Accessed August 25, 2023.

34. California Department of Transportation. Statewide Transit-Oriented Development (TOD) Study: Factors for Success in California. Parking and TOD: Challenges and

Opportunities (Special Report). Available at: https://www3.drcog.org/documents/archive/Parking%20and%20TOD.pdf. Accessed August 15, 2022.

35. Online Sunshine. The 2022 Florida Statutes. 335.065. Bicycle and Pedestrian Ways Along State Roads and Transportation Facilities. Available at: http://www.leg.state.fl.us/statutes/index.cfm?App_mode=Display_Statute&Search_String=&URL=0300-0399/0335/Sections/0335.065.html. Accessed August 15, 2022.

36. Smart Growth America. National Complete Streets Coalition. Available at: https://smartgrowthamerica.org/what-are-complete-streets/. Accessed August 12, 2022.

37. City of Boston. Boston Bike Network Plan. Available at: https://www.cityofboston.gov/images_documents/Boston%20Bike%20Network%20Plan%2C%20Fall%202013_FINAL_tcm3-40525.pdf. Accessed August 15, 2022.

38. Seattle Department of Transportation. Protected Bike Lanes. Available at: https://www.seattle.gov/transportation/projects-and-programs/programs/bike-program/protected-bike-lanes. Accessed August 15, 2022.

39. Pratt RH. Traveler Response to Transportation System Changes: An Interim Handbook (TCRP Web Document 12 [Project B-12]: Contractor's Interim Handbook). March 2000. Available at: http://onlinepubs.trb.org/onlinepubs/tcrp/tcrp_webdoc_12.pdf. Accessed August 15, 2022.

40. Oregon Department of Transportation. Oregon's Mileage Fee Concept and Road User Fee Pilot Program: Final Report. November 2007. Available at: https://www.myorego.org/wp-content/uploads/2017/07/RUFPP_finalreport.pdf. Accessed August 15, 2022.

41. I-95 Corridor Coalition. A 2040 Vision for the I-95 Coalition Region: Supporting Economic Growth in a Carbon-constrained Environment. December 2008. Available at: https://tetcoalition.org/wp-content/uploads/2015/03/2040_Vision_for_I-95_Region_Executive_Summary.pdf?x70560. Accessed August 15, 2022.

42. Ecodrive.org. What is Ecodriving? Available at: https://www.ecodrive.org/en/what_is_ecodriving/. Accessed August 15, 2022.

43. U.S. Environmental Protection Agency. HUD, DOT, and EPA Partnership for Sustainable Communities. June 16, 2009. Available at: https://archive.epa.gov/epa/smartgrowth/hud-dot-epa-partnership-sustainable-communities.html#:~:text=On%20June%2016%2C%202009%2C%20EPA,the%20environment%20in%20communities%20nationwide. Accessed August 15, 2022.

44. Partnership for Sustainable Communities. Available at: https://obamawhitehouse.archives.gov/sites/default/files/uploads/SCP-Fact-Sheet.pdf. Accessed August 15, 2022.

45. Ledna C, Muratori M, Yip A, et. al. Decarbonizing Medium- & Heavy-duty On-road Vehicles: Zero-emission Vehicles Cost Analysis. National Renewable Energy Laboratory. March 2022. Available at: https://www.nrel.gov/docs/fy22osti/82081.pdf. Accessed August 12, 2022.

46. U.S. Environmental Protection Agency. Regulations for Greenhouse Gas Emissions from Commercial Trucks & Buses. Available at: https://www.epa.gov/regulations-emissions-vehicles-and-engines/regulations-greenhouse-gas-emissions-commercial-trucks. Accessed August 25, 2023.

47. American Lung Association. Fact Sheet: Medium and Heavy Duty Vehicles. Available at: https://www.lung.org/getmedia/bb0d60ba-eff2-4084-907b-916839ae985d/Medium-and-Heavy-Duty-Vehicles-Fact-Sheet. Accessed December 10, 2022.

48. The International Council on Clean Transportation. Air Quality and Health Impacts of Heavy-duty Vehicles in G20 Economies. July 22, 2021. Available at: https://thei

cct.org/publication/air-quality-and-health-impacts-of-heavy-duty-vehicles-in-g20-economies/. Accessed December 10, 2022.

49. U.S. Environmental Protection Agency. EPA Proposes Stronger Standards for Heavy-duty Vehicles to Promote Clean Air, Protect Communities, and Support Transition to Zero-emissions Future. March 7, 2022. Available at: https://www.epa.gov/newsrelea ses/epa-proposes-stronger-standards-heavy-duty-vehicles-promote-clean-air-prot ect. Accessed December 10, 2022.

50. U.S. Department of Transportation, Federal Highway Administration. Assessing the Effects of Freight Movement on Air Quality at the National and Regional Level. July 6, 2011. Available at: http://www.fhwa.dot.gov/environment/air_quality/publications/ effects_of_freight_movement/chapter04.cfm. Accessed August 15, 2022.

51. U.S. Department of Transportation, Federal Highway Administration. 2007 Commodity Flow Survey. April 2010. Available at: https://rosap.ntl.bts.gov/view/dot/ 6345. Accessed December 10, 2022.

52. Winebrake JJ. Achieving Emissions Reductions in the Freight Sector: Understanding Freight Flows and Exploring Reduction Options. March 21, 2012. Available at: http:// www.fhwa.dot.gov/planning/freight_planning/talking_freight/talkingfreight03_21_ 12jw.pdf. Accessed August 15, 2022.

53. U.S. Department of Transportation, Federal Highway Administration. Freight and Air Quality Handbook. Available at: http://www.ops.fhwa.dot.gov/publications/ fhwahop10024/sect3.htm. Accessed August 15, 2022.

54. Northeast States for Coordinated Air Use Management (NESCAUM). Multi-state Medium- and Heavy-duty Zero-emission Vehicle Memorandum of Understanding. July 10, 2020. Available at: https://www.nescaum.org/documents/mhdv-zev-mou-20220329.pdf/. Accessed August 12, 2022.

55. California Air Resources Board. Summary: Diesel Particulate Matter Health Impacts. Available at: https://ww2.arb.ca.gov/resources/summary-diesel-particulate-matter-health-impacts#footnote3_smsqyp4. Accessed August 12, 2022.

56. Motorweek. Sandy recovery. March 21, 2013. Available at: http://www.motorweek. org/features/auto_world/sandy_recovery. Accessed August 15, 2022.

57. Keyes M. Adaptive transportation: Bicycling through Sandy's aftermath. Project for Public Spaces, November 27, 2012. Available at: https://www.pps.org/article/adapt ive-transportation-bicycling-through-sandys-aftermath. Accessed June 24, 2014.

58. Lavrinc D. EV hack keeps homes humming after Hurricane Sandy. Wired. November 8, 2012. Available at: https://www.wired.com/2012/11/sandy-ev-powered-home/. Accessed August 15, 2022.

Further Reading

Transportation Research Board. Policy Options for Reducing Energy Use and Greenhouse Gas Emissions from U.S. Transportation. 2011. Available at: http://onlinepubs.trb.org/onl inepubs/sr/sr307.pdf. Accessed August 15, 2022.

This report examines consumption of petroleum fuels in the U.S. transportation sector and reviews policy options to reduce GHG emissions and fuel consumption in the next 50 years.

Cambridge Systematics, Inc. Moving Cooler: An Analysis of Transportation Strategies for Reducing Greenhouse Gas Emissions. July 2009. Available at: http://www.reconnec

tingamerica.org/assets/Uploads/2009movingcoolerexecsumandappend.pdf.   Accessed
August 15, 2022.

This report provides information on the effectiveness and costs of almost 50 strategies
to reduce transportation GHG emissions by reducing vehicle miles traveled and
making changes to the transportation system broadly.

Transportation and Climate Initiative. Indicators to Measure Progress in Promoting
Sustainable Communities. Available at: http://www.transportationandclimate.org/indicat
ors-measure-progress-promoting-sustainable-communities. Accessed August 15, 2022.

This set of research papers examines 11 potential indicators that could help measure
progress in promoting sustainable communities and demonstrate the benefits of such
policies.

14

Agriculture Policy

Valerie J. Stull and Jonathan A. Patz

Agriculture policy overlaps with a wide range of environmental and health issues linked to climate change, including farm management, food prices, diets, land use, and energy production. It plays an important role in promoting food security, human health, and environmental sustainability.

Agriculture is both being impacted by and influencing climate change. It is extremely sensitive to climate shifts—shaped by environmental factors including changes in temperature, precipitation, and pest pressure. Adverse climate impacts are expected to place more people at risk of food insecurity and poor health. But to produce food and goods for a growing population, agriculture will likely be expanded and intensified, which may exacerbate climate change. Agricultural activities—including land clearing, production and use of synthetic fertilizers, enteric fermentation from livestock, soil and manure management, and fossil fuel combustion—can contribute to climate change by emitting substantial amounts of greenhouse gases (GHGs). But agricultural activities can also mitigate climate change by creating natural reservoirs that act as *carbon sinks*, which accumulate and sequester carbon-containing chemical compounds in the soil or vegetation, thereby removing them from the atmosphere.

In the United States, agriculture contributed almost 11% of all human-induced GHGs in 2020, primarily from livestock farming, soil management, and rice production.[1] Since 1990, total GHG emissions from agriculture have increased about 6%, including a 62% increase in methane and nitrous oxide from systems that manage livestock manure.[1] Manure storage methods, which determine exposure of manure to moisture and oxygen, influence the pathways and magnitude of GHG production and release.

Agriculture is an especially important source of non-carbon dioxide GHG emissions, such as methane and nitrous oxide. Excluding carbon dioxide, agriculture was the leading industrial sector in GHG emissions globally in 2015, responsible for about 48% of methane and nitrous oxide emissions.[2] By weight, methane, nitrous oxide, and fluorinated gases have greater impacts on climate change than carbon dioxide due to their increased *global warming potential*—the amount of energy a gas will absorb over a period of time (Table 1-1 in Chapter 1). These gases stay in the atmosphere for less time than carbon

Valerie J. Stull and Jonathan A. Patz, *Agriculture Policy* In: *Climate Change and Public Health*.
Edited by: Barry S. Levy and Jonathan A. Patz, Oxford University Press. © Oxford University Press 2024.
DOI: 10.1093/oso/9780197683293.003.0014

dioxide, but they are more potent. Methane is approximately 28 times as potent as carbon dioxide in trapping heat in the atmosphere; nitrous oxide is about 300 times more potent. Fluorinated gases, almost all of which are emitted from human activities, are the most potent and longest lasting GHGs.[3]

Food systems—which consist of food production, land-use change, and food distribution, preparation, consumption, and disposal—are a huge source of GHG emissions. They contributed about 34% of global anthropogenic GHG emissions in 2015—18 gigatons of carbon dioxide equivalents ($GtCO_{2e}$).[4]

Carbon sequestration in the soils of grazing land, rangeland, and cropland represents the greatest potential for climate change mitigation from agriculture. If best management practices are implemented globally, soils could store up to 4 $GtCO_{2e}$ annually.[5] Carbon-intensive farm activities, such as tillage, irrigation, and use of fertilizers and pesticides are activities that generate carbon dioxide because (a) they require burning fossil fuels for energy or, (b) like tillage, lead to the release of carbon dioxide from the soil. These activities can be limited and managed to preserve soil carbon and reduce emissions. Especially effective are no-till systems, integrated pest management, cover cropping, and adoption of drip and sub-irrigation techniques. Trees and perennial crops also remove carbon dioxide from the atmosphere and can sequester it in the soil for long periods (Chapter 17).

Climate change exerts both positive and negative impacts on agriculture. But its net impact is expected to be negative in the long term. Climate change has already adversely affected food production in some regions due to increased pest pressure, higher temperatures, and more extreme weather.[6] Variable impacts are likely to be seen across different crops (Figure 8-2 in Chapter 8). Future maize, soybean, and rice yields, for example, are projected to decline in many countries, placing vulnerable populations at greater risk of hunger.[6] On the other hand, wheat production is likely to increase because of the crop's positive response to higher carbon dioxide concentrations and expanded regions for farming in higher latitudes due to rising temperatures.[7] Some countries, mainly high-income countries in northern latitudes, may experience increased crop productivity in the near term due to (a) greater carbon dioxide concentrations in the atmosphere, because of the *carbon dioxide fertilization effect* seen in some plants; (b) warming temperatures that extend the growing season; and (c) robust agriculture infrastructure, including irrigation and farmer support. However, low-income countries in latitudes near or below the Equator, with reduced infrastructure and hotter climates, will likely experience adverse impacts on crop production in the near term.

The food system plays such a significant role in climate change that, even if fossil fuel emissions from other industries were immediately halted, current agricultural production (including land clearing) and supply chains alone could preclude the success of the global campaign to limit temperature increases to 2.0°C (3.6°F) above preindustrial levels.[8]

HUMAN HEALTH AND NUTRITION

Decreased agricultural productivity due to climate change increases the risk of malnutrition, weakens the immune system, and is associated with illness and death (Chapter 8). Nutrition is perhaps the most direct human health link between agriculture, agriculture policy, and climate change. The food people eat has a tangible impact on health and farming practices, which both mitigate and drive climate change. Despite a recent increase in crop yields and total crop biomass globally, the nutrient composition of some crops is declining, partly due to changes in plant chemistry;[9] specifically, experiments have found that the ratio of carbon to nitrogen (C:N) may increase with more carbon dioxide in the atmosphere for some important staple crops.

The result of increasing atmospheric GHG levels and declining nutritional quality of staple foods will likely be adverse health outcomes — especially for the most vulnerable populations. Elevated carbon dioxide levels enhance sugar production in plants, which dilutes nutrients incorporated into tissues and fruits. Ultimately, high carbon dioxide levels could reduce crop nitrogen (a proxy for protein content) and potentially contribute to protein energy malnutrition, especially among populations that are already at risk of inadequate protein intake. Cereal grains and some tubers grown while exposed to excess carbon dioxide in experiments contain less protein, suggesting that future increases in atmospheric carbon dioxide may decrease protein content in staple crops.[10,11] Specifically, under experimental conditions with elevated carbon dioxide levels, some staple crops like wheat and rice generate about 10% less protein, 5–10% less iron and zinc, and a 30% decrease in individual B vitamins.[9,12,13] An estimated 175 million more people may become zinc-deficient and 122 million more people may become protein-deficient in the next 30 to 80 years if atmospheric carbon dioxide levels surpass 550 ppm.[14]

Changes in precipitation and temperature affect ecological systems, altering pest and pathogen populations, therefore impacting agricultural management practices. Farmers responding to increased populations of pests, weeds, invasive plant species, and insects may apply substantially more pesticides, possibly threatening human health by contaminating air, water, and soil.[15] In 2022, about 2.7 million tons of active ingredients were used in agricultural pesticides globally, part of a trend of increased pesticide use in the past decade despite a preceding plateau.[16]

Agricultural emissions are a major source of air pollutants, especially ammonia (from fertilizer use and animal husbandry) and fine particulate matter ($PM_{2.5}$). These pollutants can cause respiratory tract irritation, lung damage, and even death.[17] Reducing agricultural GHG emissions could prevent about 250,000 deaths globally per year.[18]

ENERGY POLICY IMPACTS ON AGRICULTURE

Global agricultural policy fluctuates, largely due to political pressures and budgetary constraints along with international trade agreements, treaties, and population growth. But agriculture policy is often not recognizable as such. Modern agriculture produces not only food and fiber, but also energy. About one-third of ice-free land surface globally is used for cultivation or grazing[19]—10.6 billion acres. Growing crops for food, fuel, and fiber accounts for most land use. It is the primary way that people alter the Earth's surface.

The Renewable Fuel Standard

The Renewable Fuel Standard (RFS) is the primary policy dictating agriculture and land use for biofuel production in the United States. Established under the Energy Policy Act of 2005, the RFS originally required that 7.5 billion gallons of renewable fuel be blended with gasoline by 2012.[20] After expansion under the Energy Independence and Security Act in 2007, the RFS required that all transportation fuel sold in the United States contain a minimum blend of renewable fuel, with the proportion increasing annually.

The RFS was intended to help reduce GHG emissions, pollution, and dependence on imported petroleum while expanding the renewable fuels industry.[20] Fuel-grade ethanol, which is derived from plant biomass and produced through a fermentation process, is a common biofuel. Corn (maize) is the primary crop used to generate ethanol for the RFS. About 40% of all corn grown in the United States is used to make fuel,[21] and, in 2021, corn ethanol production reached about 15 billion gallons.[22] In 2021, ethanol, which was included in 98% of gasoline, accounted for about 10% of gasoline used in the United States by volume.[23,24]

When used in "E85 gasoline" (85% ethanol and 15% gasoline), the production potential of ethanol is about 32% of total gasoline consumption globally.[25] The U.S. Department of Agriculture (USDA) and proponents of the RFS argue that the GHG profile of corn-based ethanol is much lower than that of gasoline, suggesting that there are substantial pollution benefits to burning it over fossil fuels.[26] However, these estimates are contentious, and there remains controversy regarding the overall impact of corn-ethanol production on the environment and human health.

The GHG profile of corn ethanol is tightly connected with land-use change, which is inherently difficult to quantify. If growing corn for fuel requires bringing significant land into agricultural production that was previously not cultivated, via deforestation or converting wetlands or marginalized land to cropland, then

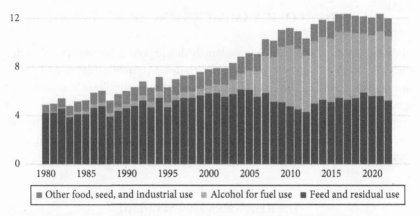

Figure 14-1   U.S. domestic corn use (in billions of bushels), updated September 2022. (*Source*: National Agricultural Statistics Service, U.S. Department of Agriculture.)

emissions from these processes and the subsequent burning of corn ethanol in cars could outweigh the emissions savings of burning ethanol instead of gasoline.[27] In addition, using corn to generate ethanol could reduce global supplies of corn for food and thereby increase food prices, incentivizing change of land use to crop production in other parts of the world.[27]

The United States leads all other countries in growing, consuming, and exporting corn, sending 10–20% of its production volume to other countries.[21] Corn is also the most prolific crop grown in the United States, covering about 23% of cropland—95 million acres, an area about the size of California.[28] Its production expanded substantially after the RFS was established. Corn remains the primary biomass used for ethanol. In 2022, almost half of all the corn produced in the United States was used in biofuel production (Figure 14-1).[29]

Producing corn consistently on such a wide scale requires extensive fertilizer, equipment, and land while also generating air pollution. Farmers apply almost half of all fertilizer used in the United States just to corn, making it a major source of emissions of ammonia, which contributes to the formation of fine particulate matter.[30] Scientists have estimated that reduced air quality from corn production in the United States is annually associated with 4,300 premature deaths, with estimated damages ranging from $14 to $64 billion.[31]

Corn Production and Nitrates

Corn requires more nitrogen per acre than any other primary crop in the United States.[32] Nitrogen from synthetic fertilizers or cow manure is incredibly effective at boosting the productivity of corn and other staple crops, contributing to high yields that are important for nutrition and health. However, because it is often difficult to determine the exact amount of nitrogen needed during one planting season, farmers may sometimes overapply fertilizers. As environmental nitrogen levels continue to rise due to human activity, the net health effects are expected to be increasingly negative (Figure 14-2).[33]

Excess fertilizer use can lead to nutrient runoff and contamination,[32] contributing to ecosystem damage, such as harmful algal blooms and poor health outcomes. In 2021, approximately 2.4% of community water systems in the United States were "significant violators" of the safe 10 mg/L threshold for nitrate in drinking water,[34] and between 10% and 20% of public water supplies exceeded this threshold.[35] People in about 2 million households in the United States, often in agricultural areas of lower socioeconomic status,[34] drink from water supplies

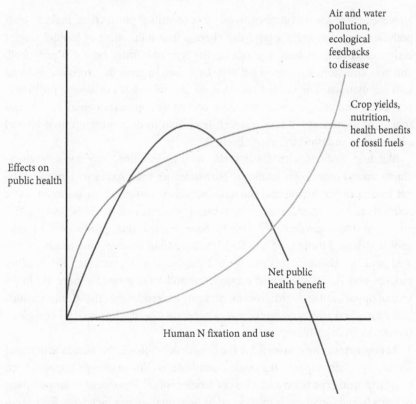

Figure 14-2 Conceptual model depicting the public health impact of increasing anthropogenic fixation and use of atmospheric nitrogen (N_2). (*Source*: Townsend AR, Howarth RW, Bazzaz FA, et al. Human health effects of a changing global nitrogen cycle. Frontiers in Ecology and the Environment 2003; 1: 240–246.)

that fail to meet the federal standard for nitrate.[36] Although nitrates and nitrites are naturally occurring, excessive exposure via consumption has been linked to some cancers, reproductive health impacts, hypothyroidism, and methemoglobinemia ("blue baby syndrome").[37]

More on Biofuel Production

The impetus to generate biofuel stems from two purported benefits:

- They are made from relatively renewable energy sources derived from biodegradable living matter, can be produced and refined in the United States, and promote energy independence.
- They can burn cleaner and may emit fewer GHGs than gasoline.

However, there are potential consequences of biofuel production, including air pollution and detrimental land-use change. The total effect of biofuel on net carbon dioxide emissions depends on the type of biofuel, how it is produced, and how emissions associated with the land used to grow the crops are factored into calculations. The GHG balance from biofuel use is not always positive—producing and using biofuels may also have adverse environmental, health, and social consequences.[38] More research is needed to determine optimal biofuel crop sources and production methods.

Blending biofuels with fossil fuels, such as gasoline, may reduce vehicle emissions.[24] Ethanol, for example, burns cleaner than gasoline. However, the net impact of producing and burning biofuels is complex. For example, some researchers have found that corn ethanol generates up to 52% lower GHG emissions than gasoline.[39,40] Others have argued that production of ethanol is at least 24% more fossil fuel-intensive than gasoline due to (a) associated land-use changes, processing, and combustion unaccounted for in other models, and (b) its use could increase ground-level ozone.[41] There are likely several opportunities for improvement; some experts believe that future second-generation and advanced biofuels may generate 47–70% lower GHG emissions than those from gasoline.[42]

As government incentives drive the growth of biodiesel, the increased demand for oil, especially soybean oil, could contribute to climate change impacts, such as agricultural expansion and a loss of biodiversity.[43] The extent of air pollution benefits from biofuel use is influenced by how they are produced, the feedstocks used (such as corn or soybeans), where and how feedstocks are grown, and if land conversion takes place. There are major differences in these factors among fuel types, engine types, operating conditions, blend ratio, and biofuel feedstock, making it difficult to determine associations between biofuel use and reduced air pollution and improved health.[44]

The extent of air pollution from biofuel combustion is still controversial. Lower-proportion ethanol blends tend to decrease emissions of respirable particulate matter (PM_{10}), hydrocarbons, and carbon monoxide, but they also often increase oxides of nitrogen. Ethanol blends with high volatility lead to evaporative emissions, which may increase ozone formation in ambient air.[45,46] Some argue that biofuel blends yield higher emissions than fossil fuels, partly because of the increased fuel required to travel the same distance. Ethanol provides about 33% less energy per volume compared to gasoline.[47] The ultimate effect of ethanol on fuel economy is a result of the content mixed into fuel and whether the engine is optimized to run on ethanol. Others assert that (a) because ethanol has a high octane level, blending it with gasoline can improve the performance of gasoline, and (b) ethanol improves fuel combustion in vehicles and reduces emissions of carbon monoxide, unburned hydrocarbons, and carcinogens. Likewise, adding ethanol to gasoline alters its formulation, lowering aromatic compounds and tailpipe emissions, which could reduce rates of cancer and other diseases and lower healthcare costs.[48] In addition, compared with gasoline, pure ethanol contains negligible amounts of sulfur, so supplementing gasoline with ethanol generally results in lower sulfur dioxide emissions.

First-Generation and Second-Generation Biofuels

First-generation biofuels come primarily from agricultural crop biomass feedstocks, such as sugar and corn, using fermentation to extract the sugars during ethanol production. The United States (using corn) and Brazil (using sugar cane) have been the leading producers of first-generation biofuels (such as ethanol). Because of potential adverse environmental and health impacts of increasing corn production, research has focused on whether second-generation biofuels can be produced from other (non-food-based) feedstocks. These potential feedstocks include dedicated cellulosic biomass/energy crops (such as switchgrass or miscanthus), crop waste or residues (such as corn stalks and husks), and woody biomass (such as hybrid poplar and willow). Since 2015, second-generation biofuels have been commercially produced, and market demand is projected to increase during the next decade.

Since biofuel refineries for second-generation biofuels are expensive, continued government support to finance these projects is critical. Policy is required to help develop the market for second-generation biofuels. The federal RFS should be revised to mandate required levels of advanced biofuels mixed with gasoline. Note that third-generation biofuels, which are derived from algae, exist and also require more research and support in order to be a viable alternative to first-generation options.

Land-Use Change from Agricultural Policies

Agricultural policy leads to changes in land use. Significant carbon dioxide is released when rainforests are cleared, for example for growing biofuel crops (such as sugarcane in Brazil), and when land formerly part of the Conservation Reserve Program (CRP) is converted to corn or soy production.[27] CRP, the largest agricultural land retirement and conservation program in the United States, financially supports farmers in removing environmentally sensitive land from agricultural production by planting species that will improve environmental health and quality. Moving marginal or sensitive land out of production is a healing and restorative process, allowing soil carbon to rebuild and reducing overapplication of fertilizers, which can run off and pollute waterways. CRP contracts, which typically last 10–15 years, allow land to heal, improve water quality, prevent soil loss, and create wildlife habitat.

Between 2010 and 2013, an estimated 1.3 million acres of land from CRP was converted back to crop production.[49] About two-thirds of the land that was converted had been grasslands and about one-third had been wildlife habitat and wetlands.[49] This conversion threatened waterways, wetlands, and the animals that live there. Land-clearing releases carbon dioxide by disrupting carbon sinks, such as rainforests, peatlands, savannas, and grasslands—releasing 17-420 times more carbon dioxide than the GHG emissions that would be prevented by biofuels displacing fossil fuels, a *carbon debt* that could take decades to repay.[50]

When monoculture row crops replace diverse ecosystems, land-use change is associated with substantial losses in ecological biodiversity. Soil and water quality often also decline under industrial production.

In summary, the intersection of energy policy with agriculture has led to substantial shifts in crop production and land use, with the following associated environmental and health outcomes:

- Biofuels are sourced from renewable living matter that is fully biodegradable and can be produced in the United States.
- Corn production for ethanol in the United States has quadrupled since 2005, which may intensify fertilizer and pesticide use, soil erosion, and GHG emissions.[51]
- While biofuels may burn cleaner than gasoline, the GHG benefits of substituting biofuels for petroleum-derived gasoline depend on a variety of factors involved in growing crops for energy, including the extent of land-use change, land management practices, and displacement of food crops.
- Biofuels may contribute to energy independence, but they are not risk-free. Production and use can adversely affect health via air pollution,

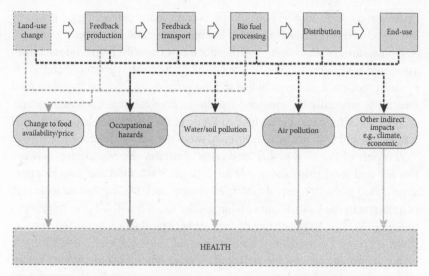

Figure 14-3 Schematic showing the life-cycle and primary pathways between biofuel production and use and human health. (*Source*: Scovronick N, Wilkinson P. Health impacts of liquid biofuel production and use: A review. Global Environmental Change 2014; 24: 155–164.)

occupational hazards (such as pesticides), and exposure to pollutants in soil and water.

- Availability of biofuel is related to the price of food, which can adversely affect nutrition and food security.[44]
- Use of liquid biofuels may improve air quality in some areas while worsening nutrition in others (Figure 14-3).
- Conservation programs, such as CRP, remove environmentally sensitive land from production, improving environmental health. Questions remain regarding the impact of converting these lands back to crop production for biofuels.

FOOD PRICE SHOCKS AND FOOD RIOTS

To promote sustainable energy and energy independence, policies supporting biofuel production and research have been developed. These policies operate in a global marketplace for agricultural goods and have influenced the trade and value of several staple crops (Chapter 8). For example, policies that promote biofuel production contribute to volatility of the prices of grain crops, such as corn (maize) and wheat. Biofuel policy impacts commodity prices of food grains by linking oilseed to biodiesel prices and linking corn to ethanol prices.

Biofuels are associated with health—not only in *how* they are produced (with associated environmental impact), but also in *how much* is produced and *when*. Biofuel production has been linked to food price shocks, which aggravate food insecurity, especially for low- and middle-income countries, whose residents spend most of their income on food. Because of the association between biofuel production and food prices, a "fuel versus food" debate has developed.

A review of the 2008 global food crisis illustrates the relationship between biofuels and food price shocks. From 2007 to 2008, food and energy prices skyrocketed globally, nearly doubling in many places. During that same period, corn grown in the United States accounted for about one-third of corn production globally.[52] In 2008, the rate of corn ethanol production increased dramatically, sparking a diversion of corn to ethanol feedstock and contributing to high corn prices and an unprecedented shock to other food commodity prices. Diverting food and feed to biofuel production substantially increased food prices globally. (In 2011, biofuels accounted for 20–40% of the increases in food prices.[53]) Corn ethanol is, by far, the most important contributor to global corn price and aggregate commodity spikes due to the unprecedented speed with which the ethanol industry expanded in the several years preceding the 2008 crisis.[54]

The average household in the United States spends a lower proportion of income on food than households in any other country—only about 13% in 2021[55] compared to the poorest households in low-income countries, which may spend as much as 80% of income on food.[56] Price shocks have a greater impact on households in low-income countries. Undernutrition and food insecurity also perpetuate a vicious cycle of poverty, poor health, and low social capital (Chapter 8).[57] Volatility in commodity prices and sudden unpredictable price spikes seriously threaten global food security. In many parts of the world, the poorest people are farmers, who may benefit financially from incremental increases in global crop prices but not from unexpected dramatic changes.[57] Poor people in urban areas are especially vulnerable to food price shocks because they purchase most of their food rather than growing it. Spikes in global commodity prices are strongly correlated with food riots because desperation, distrust, and political instability can spark violence (Figure 14-4).

MAJOR AGRICULTURAL POLICY IN THE UNITED STATES

The "Farm Bill," first implemented in 1933, is the generic name for one of the most significant pieces of legislation affecting agriculture, food, and conservation

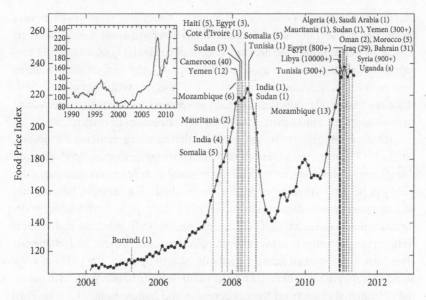

Figure 14-4 World food prices and food riots in the Middle East, Africa, and elsewhere, 2004-2011. The insert graph depicts the world food price index from 1990 to 2011. (*Source*: Lagi M, Bertrand KZ, Bar-Yam Y. The Food Crises and Political Instability in North Africa and the Middle East, August 10, 2011. Available at: https://arxiv.org/abs/1108.2455. Accessed March 14, 2023.)

in the United States—influencing what Americans grow and eat, how it is grown, who will profit, and the nature and extent of environmental impacts.[58] This omnibus bill, which is renewed regularly (about every 5 years), provides governing policy for many aspects of agriculture, including farm income support, food assistance, and trade. The Farm Bill provides an incredibly powerful, but underutilized, mechanism to promote public health via nutrition and to set land-use regulations that reduce disparities and decrease GHG emissions. Currently, the law allocates most of its funds to nutrition and commodity programs. Its domain also includes bioenergy and conservation—critical considerations for health and climate change. In 2018, the budget for the Agriculture Improvement Act of 2018 (the Farm Bill) declined from the 2014 version, as did support allocated to nutrition (76% of the budget).[59] On the other hand, conservation outlays were projected to be slightly higher (7% of the $428 billion total).[59]

Agricultural Subsidies

Agricultural subsidies provide financial and in-kind support to farmers and agribusinesses to supplement income and manage or influence the supply and cost of commodities. Subsidies include direct payments to farmers, price supports, regulations that set minimum prices, and funds for crop insurance and disaster response, as well as import barriers in the form of tariffs and quotas. Subsidies provide incentives to farmers to grow specific crops, thereby profoundly influencing food production, agricultural management, land use, and availability of food. GHG emissions from food production are tightly linked to a few commodities. Beef, dairy, and rice account for more than 80% of total GHG emissions from agriculture.[60]

Globally, government tariffs and subsidies that support agriculture are substantial. In 2017, governments spent about $233 billion on agriculture.[61] Subsidies, although well-intentioned, can incentivize unsustainable behaviors that emphasize yield but pay little attention to production methods. For example, farmers may be motivated by subsidies to clear forests to produce soy or beef, leading to deforestation and extensive GHG emissions. Alternatively, subsidies that support fertilizers or pesticides may lead to overapplication of environmentally damaging chemicals. Conventional farming methods used to optimize the yields of a few selected crops are sometimes the best fit within subsidy frameworks, with potential detrimental impacts on the environment and human health. Conventional mechanized agriculture emits large amounts of GHGs. Overapplication of pesticides damages soil fertility. Aligning subsidies with a clear set of environmental, social, and health goals would be widely advantageous.[61,62]

Consequences of Overproduction and Overconsumption

Some experts argue that subsidies have contributed to overproduction of specific crops, resulting in unintended adverse effects on the environment and public health. These effects are especially evident when farmers cultivate subsidized crops on marginal farmland, where thin top soil that has been depleted of nutrients from poor soil management requires intensive inputs. Subsidies may incentivize overuse of fertilizers and pesticides to boost yields on both marginal and fertile land while helping to bring substandard land into production.[63] Pesticides applied to increase yields contribute to GHG emissions. Their use is expected to increase. Almost all synthetic pesticides are derived from fossil fuels and their overuse puts vulnerable populations at risk, especially farmworkers, rural communities, and people living near to pesticide production facilities or heavy application sites.

Meat Production

The primary recipients of crop insurance payments under the 2018 Farm Bill are producers of corn, soy, wheat, and cotton.[59] Grains (especially corn) and oilseeds (especially soy) serve as low-cost feed for livestock, reducing the cost of raising poultry, cattle, and swine. Subsidies therefore translate into relatively low prices for meat products[43,65] because grain-fed animals gain weight more quickly.[64] Beef is especially damaging to the environment, requiring more land and freshwater than all plant-based proteins or any other animal-based protein on the basis of tons of protein produced.[66] In addition, beef accounts for more GHG emissions than other plant and animal-based proteins. Livestock production accounts for about 15% of human-induced GHGs globally.[67] On average, Americans consume about 273 pounds of meat, including 97 pounds of beef, annually[68]—substantially more than what is required for a balanced diet. Fatty meat consumption is associated with hypercholesterolemia, coronary artery disease, and type 2 diabetes.[64] Globally, increased demand for meat products by the increased number of middle-class people in middle-income countries (and sustained demand from high-income countries) is contributing to the oversupply of commodities required to feed them. Therefore, the harmful effects of prolific grain and oilseed production are widespread but also indirect.[69] This is not to say that all meat production is bad. Animals provide an important source of nitrogen to farmers and landscapes via manure. Cattle, for example, can be integrated more sustainably into cropping systems via techniques such as agroforestry and rotational grazing. Globally, for many people, livestock serve as an informal savings account, supply farm labor, and provide an important source of fertilizer. In addition, meat consumed in moderation is an excellent source of nutrients. How meat is produced is most relevant to the overlap between agriculture policy and human health.

Lack of Support for Fruits and Vegetables

Although a diet rich in fruits and vegetables helps lower the incidence of and mortality from many chronic diseases,[70] U.S. agriculture policy has generally failed to offer incentives or support for fruit and vegetable production at the scale of other crops. Farmers may be "penalized" in a way for growing "specialty crops" (such as fruits and vegetables) if they have received farm payments from the federal government to grow other crops.[71] While farmers might generate higher marketplace revenue from growing fresh produce, they have substantially lower economic security—so producing fruits and vegetables is a risky strategy in an already-risky industry.[69]

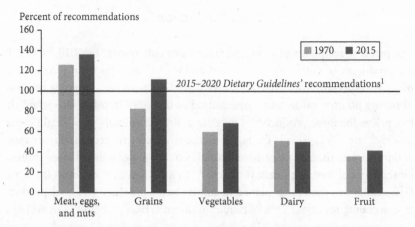

Figure 14-5 Estimated average U.S. consumption of food groups, compared to recommendations, 1970 and 2015, based on a 2,000-calorie-per-day diet. Note: Loss-adjusted food availability data are proxies for consumption. (*Source*: Economic Research Service, U.S. Department of Agriculture.)

Nearly 90% of the U.S. population does not eat the recommended amount of vegetables and 80% do not consume enough fruit to meet dietary guidelines.[72] Support for commodities like corn and soy does not preclude support for fruits and vegetables, but the government's budget is limited. Although the 2014 Farm Bill did increase programs supporting specialty crops from previous budgets, the 2018 Farm Bill did not follow suit and instead maintained most of these provisions instead of expanding them.[73] Although fruit and vegetable consumption has increased since 1970, most Americans, in 2015, did not adhere to dietary recommendations. Meat, egg, and nut consumption has consistently remained above recommended levels (Figure 14-5).

Better policy support for fruit and vegetable production, as well as increased consumption, could have tangible climate change and health benefits. Models of changes to food production show that transitioning to plant-based diets would benefit the environment by reducing diet-related GHG emissions by 49% and land-use by 76%.[74] Plant-based diets may also be associated with a substantially lower risk of developing type 2 diabetes.[75] In a groundbreaking 2019 study, an international research team modeled the impact on global health of adopting a plant-centered *planetary health reference diet* (similar to a Mediterranean diet). It found that reducing consumption of unhealthy foods (red meat and sugar) by 50%, and increasing healthy foods (nuts, fruits, and legumes) by more than 100% could prevent about 11 million deaths per year in 2030 from diet-related conditions such as coronary heart disease, stroke, type 2 diabetes, and some cancers.[76] In addition, such a shift would reduce global GHG emissions from food production. Changing farming practices alone could reduce GHG emissions 10% by 2050, but shifts toward plant-based diets could reduce agricultural emissions by up to 80%.[77]

CONCENTRATED ANIMAL FEEDING OPERATIONS

Concentrated animal feeding operations (CAFOs) are agricultural operations where animals are kept and raised in confined areas, with feed brought to the animals rather than the animals grazing or foraging for food. Animals in CAFOs are confined at least 45 days each year and have no grass or other vegetation present during the normal growing season. CAFOs enable fast production of animal products, including poultry, beef, swine, and dairy cows, ensuring that there is ample supply of meat in the U.S. food system. CAFOs comprise about 15% of all animal feeding operations in the United States.[78] CAFOs are heavy emitters of GHGs, especially methane, and livestock operations account for about 7% of GHGs in the United States.[79] Regulation of emissions from CAFOs are limited, despite GHG emissions—methane, nitrous oxide, and carbon dioxide—released from these operations. Global meat demand is rising, reaching a per-capita consumption rate of 34.7 kilograms (76.5 pounds) in 2018.[80]

Livestock cattle in CAFOs eat huge amounts of soy and corn, another factor that should be considered when calculating impact. In addition, CAFOs must also deal with millions of tons of manure annually, which poses a risk to the environment and public health if not properly managed. Along with valuable nutrients, such as nitrogen and phosphorus, manure contains a variety of potential contaminants, including antibiotics, growth hormones, and pathogens, such as toxigenic *Escherichia coli*. Poorly managed CAFOs can generate water contaminants, air pollutants, and offensive odors while also posing a significant risk to water quality and aquatic ecosystems.

U.S. agriculture policy indirectly encourages CAFOs over more-natural feed systems, such as grassfed livestock-rearing operations. Government crop subsidies support establishment and maintenance of CAFOs by keeping very low the cost of animal feed, especially corn and soy.

A VISION FOR THE FUTURE

As this chapter has documented, agricultural policy strongly influences food production, land use and management, diet, and the contribution of agriculture to climate change. Substantial shifts in land use and agricultural management may be critical strategies to limit climate change in the future.[81] Broadly, future agricultural policies should

- Reevaluate the long-term impact of agriculture on the environment and health
- Support sustainable practices, such as crop rotation, use of cover crops, and no-till and reduced tillage farming, which protects the soil, can reduce dependency on nitrogen chemical fertilizers, and diminishes GHG emissions
- Encourage crop diversity, better nutrient management, and integrated pest management to reduce risk of pesticide and herbicide overuse
- Encourage rotational grazing for livestock
- Discourage environmentally damaging practices, such as CAFOs, and instead encourage more sustainable livestock production, such as via rotational grazing
- Support research funding to improve pasture-based and grassfed animal production methods.

Reducing livestock production and consumption to mitigate cattle-related methane emissions, reducing deforestation from creating more pastureland, and limiting overconsumption of red meat would have a beneficial effect on human health. To mitigate harmful effects of livestock production on climate change and health, some have proposed 90 grams per person per day as a global target, with a maximum of 50 grams per day coming from ruminant red meat.[82] Others have recommended only about 14 grams per day of red meat.[76] In any event, reduced meat production, to any extent, would result in less energy expended for food production. The conversion efficiency of plant matter is about 10%—that is, about 90% of the energy in plant matter is lost when moving up the food chain one step. Therefore, more people could be fed from the same amount of land if they consumed plant-based diets, which are more energy efficient.[83]

Because climate change is affecting the hydrological cycle, ensuring adequate access to freshwater to produce food is a concern. Meat production is water-intensive, requiring two to five times more water than most plant crops. Beef is especially water-intensive, requiring more freshwater than any other animal or plant protein source. Since about 15–23% of freshwater resources globally are used for livestock production, reducing cattle production would save substantial amounts of freshwater.[84] Allocating more freshwater to other uses, such as human drinking water, sanitation, and vegetable crop production could have marked benefits to human health.

Future energy and agriculture policy should promote development of new biofuels that ensure food availability and promote environmental sustainability. Moving away from corn-based ethanol could also reduce pressure on staple food crops and the global grain trade, thereby partially decoupling biofuel from food price shocks. The United States should carefully reevaluate its RFS, given the complex relationship between global food prices, agricultural management, bio-diversity, and long-term sustainability.

One way to quickly reduce GHG emissions would be to encourage biofuel producers to practice *conservation agriculture*, reducing inputs for and emissions from corn-ethanol crop production. Government incentives for conservation agriculture and healing marginalized land should outweigh incentives to cultivate corn on marginal land. Subsidies and other incentives that allow marginalized or damaged land to be restored, such as the CRP, should be greater than incentives to grow corn or produce ethanol.

Further development of second-generation biofuels should be promoted. Some trees and grasses can be used as biofuel and can help diversify agroecosystems, sequester carbon dioxide, and restore soil on degraded land (Chapter 17).[38] The natural processes required to convert perennials and other cellulosic sources, such as stems and leaves, to ethanol rely on bacteria or fungi, which produce helpful enzymes. These organisms are difficult to manage, and the technology needed to ensure that the processes are energy-efficient and reliable is still in its nascent phase. Second-generation cellulosic biofuels may offer a greater energy return than grain-based sources—and may do so without competing with human food resources.

STRATEGIC POLICY PLANNING

Agriculture policymaking should include the voices of stakeholders from multiple sectors. Public discussion among farmers, healthcare professionals, and members of environmental groups and other public interest organizations should be promoted. To achieve sustainable agriculture policies that consider climate change and human health, the public must be educated and encouraged to serve as advocates. Programs, such as agricultural extension services, which connect the public with research institutions and promote education, can facilitate education and advocacy.

Ensuring food security requires adoption of climate-smart and regenerative agricultural practices and policies. *Climate-smart agriculture* sustainably increases resilience and productivity, reduces GHG emissions, and enhances food security. It includes improved management practices that increase soil organic matter, thereby reducing requirements for chemical inputs, such as nitrogen fertilizers, and helping to maintain vital ecosystem functions, such as the hydrological and nutrient cycles—especially important, as climate change places increased pressure on food production.[85] Climate-smart agriculture also promotes more diverse agricultural systems, builds soil, and reduces use of synthetic fertilizers. *Regenerative agriculture*, which is designed to improve soil

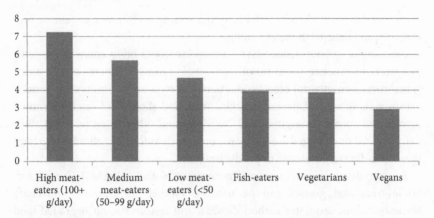

Figure 14-6 Estimated greenhouse gas emissions by categories of food consumers (in kilograms of carbon dioxide equivalents per day). (*Source*: Scarborough P, Appleby P, Mizdrak A, et al. Dietary greenhouse gas emissions of meat-eaters, fish-eaters, vegetarians and vegans in the UK. Climate Change 2014; 125: 179–192.)

health, can enhance biodiversity, water quality, and climate change.[86] The nutrient composition of food is related to farming practices and how these practices change the soil.[87,88] Policy planning, financial support, and research will help farmers adopt climate-smart and regenerative agricultural practices.

New regulations could help make fruits and vegetables more affordable and more widely available—with co-benefits of reducing GHG emissions. For example, altering subsidy provisions under the U.S. Farm Bill could help support production of specialty crops other than corn. Not expanding the Food Insecurity Nutrition Incentive (FINI) program is a missed opportunity. If expanded, FINI could increase consumer demand for specialty crops like fruits and vegetables, whole grains, and beans, providing farmers a broader market for these crops with sales linking back to the program.[73]

Agriculture policy should better support and encourage a plant-based diet. Compared to meat-heavy diets, plant-based diets require fewer natural resources to produce and are less damaging to the environment. Plant-based foods yield fewer detrimental environmental impacts than animal-sourced foods. Even the lowest-resource animal products likely have greater environmental impacts than vegetable alternatives.[89] Policies favoring adoption of a plant-based diet could have multiple benefits, including reduced pressure on agricultural and environmental resources, lower GHGs emissions, improved health, and more-equitable food systems, in which poor people are not excluded by high prices. Figure 14-6 compares GHG emissions of six different types of diets.

SUMMARY POINTS

- Climate change and agriculture share a bidirectional relationship; agricultural activities are both driving climate change and are directly impacted by it.
- Policies and farming activities that lead to deforestation or the conversion of marginal lands to crop production are damaging, often emitting extensive GHGs and increasing environmental pollution, including hazardous nutrient runoff.
- Agriculture policy is often embedded in energy policy; production of crops for energy (biofuels) is a critical consideration for both human health and climate change.
- Subsidies and other agriculture policies in the United States incentivize widespread production of a few crops, especially corn, which leads to homogeneous land use and high fertilizer use, impacting pollution, food availability, energy independency, and human health.
- A healthy food system is diverse and resilient, supported by policies that promote crop rotation, reduce soil disturbance, lower reliance on resource-intensive inputs, and encourage diverse diets.

ACKNOWLEDGMENTS

We thank Aimee Puz, Hannah Stern, Jack Buchanan, and Monica Laurent for their valuable contributions to this chapter.

References

1. U.S. Environmental Protection Agency. Inventory of U.S. Greenhouse Gas Emissions and Sinks: 1990–2020. Available at: https://www.epa.gov/ghgemissions/inventory-us-greenhouse-gas-emissions-and-sinks-1990-2020. Accessed February 23, 2023.
2. U.S. Environmental Protection Agency. Global Non-CO2 Greenhouse Gas Emission Projections & Mitigation Potential: 2015–2050. Available at: https://www.epa.gov/global-mitigation-non-co2-greenhouse-gases/global-non-co2-greenhouse-gas-emission-projections. Accessed February 23, 2023.
3. U.S. Environmental Protection Agency. Overview of Greenhouse Gases. Available at: https://www.epa.gov/ghgemissions/overview-greenhouse-gases. Accessed February 23, 2023.
4. Crippa M, Solazzo E, Guizzardi D, et al. Food systems are responsible for a third of global anthropogenic GHG emissions. Nature Food 2021; 2: 198–209.
5. Paustian K, Larson E, Kent J, et al. Soil C sequestration as a biological negative emission strategy. Frontiers in Climate 2019; 1: 8. doi.org/10.3389/fclim.2019.00008.

6. Field CB, Barros VR, Dokken DJ, et al. (eds.). Climate Change 2014: Impacts, Adaptation, and Vulnerability. Part A: Global and Sectoral Aspects. Contribution of Working Group II to the Fifth Assessment Report of the Intergovernmental Panel on Climate Change. Cambridge: Cambridge University Press, 2014.

7. Jägermeyr J, Müller C, Ruane AC, et al. Climate impacts on global agriculture emerge earlier in new generation of climate and crop models. Nature Food 2021; 2: 873–885.

8. Clark MA, Domingo NGG, Colgan K, et al. Global food system emissions could preclude achieving the 1.5° and 2°C climate change targets. Science 2020; 370: 705–708.

9. Ebi KL, Anderson CL, Hess JJ, et al. Nutritional quality of crops in a high CO_2 world: An agenda for research and technology development. Environmental Research Letters 2021; 16: 064045. https://doi.org/10.1088/1748-9326/abfcfa.

10. Fernando N, Panozzo J, Tausz M, et al. Rising atmospheric CO2 concentration affects mineral nutrient and protein concentration of wheat grain. Food Chemistry 2012; 133: 1307–1311.

11. Ainsworth EA, Long SP. 30 years of free-air carbon dioxide enrichment (FACE): What have we learned about future crop productivity and its potential for adaptation? Global Change Biology 2021; 27: 27–49.

12. Myers SS, Zanobetti A, Kloog I, et al. Increasing CO2 threatens human nutrition. Nature 2014; 510: 139–142.

13. Zhu C, Kobayashi K, Loladze I, et al. Carbon dioxide (CO2) levels this century will alter the protein, micronutrients, and vitamin content of rice grains with potential health consequences for the poorest rice-dependent countries. Science Advances 2018; 4: eaaq1012. doi: 10.1126/sciadv.aaq1012.

14. Smith MR, Myers SS. Impact of anthropogenic CO 2 emissions on global human nutrition. Nature Climate Change 2018; 8: 834–839.

15. Kattwinkel M, Kühne J-V, Foit K, Liess M. Climate change, agricultural insecticide exposure, and risk for freshwater communities. Ecological Applications 2011; 21: 2068–2081.

16. Food and Agriculture Organization of the United Nations. Pesticides Use, Pesticides Trade, and Pesticides Indicators: Global, Regional, and Country Trends, 1990–2020. Rome: FAO, 2022. Available at: https://doi.org/10.4060/cc0918en. Accessed February 23, 2023.

17. Centers for Disease Control and Prevention. Ammonia. Available at: https://www. cdc.gov/niosh/topics/ammonia/default.html. Accessed February 23, 2023.

18. Pozzer A, Tsimpidi AP, Karydis VA, et al. Impact of agricultural emission reductions on fine-particulate matter and public health. Atmospheric Chemistry and Physics 2017; 17: 12813–12826.

19. Ramankutty N, Evan AT, Monfreda C, Foley JA. Farming the planet: 1. Geographic distribution of global agricultural lands in the year 2000. Global Biogeochemistry Cycles 2008; 22: GB1003. doi:10.1029/2007GB002952.

20. U.S. Environmental Protection Agency. Renewable Fuel Standard Program. Available at: https://www.epa.gov/renewable-fuel-standard-program. Accessed February 23, 2023.

21. U.S. Department of Agriculture, Economic Research Service. Feedgrains Sector at a Glance: Corn and Other Feed Grains. Available at: https://www.ers.usda.gov/topics/crops/corn-and-other-feedgrains/feedgrains-sector-at-a-glance/. Accessed February 23, 2023.

22. U.S.Department of Agriculture, Economic Research Service. U.S. Bioenergy Statistics. Available at: https://www.ers.usda.gov/data-products/u-s-bioenergy-statistics/. Accessed February 23, 2023.

23. U.S. Energy Information Administration. How Much Ethanol is in Gasoline, and How Does It Affect Fuel Economy? Available at: https://www.eia.gov/tools/faqs/faq.php?id=27&t=4. Accessed February 23, 2023.

24. U.S. Department of Energy, Alternative Fuels Data Center. Ethanol Fuel Basics. Available at: https://afdc.energy.gov/fuels/ethanol_fuel_basics.html. Accessed February 23, 2023.

25. Kim S, Dale BE. Global potential bioethanol production from wasted crops and crop residues. Biomass Bioenergy 2004; 26: 361–375.

26. Flugge M, Lewandrowski J, Rosenfeld J, et al. A Life-cycle Analysis of the Greenhouse Gas Emissions of Corn-based Ethanol. Report prepared by ICF under USDA Contract No. AG-3142-D-16-0243. January 12, 2017. Available at: https://digitalcommons.unl.edu/usdaarsfacpub/1617/. Accessed February 23, 2023.

27. Searchinger T, Heimlich R, Houghton RA, et al. Use of U.S. Croplands for Biofuels Increases Greenhouse Gases Through Emissions From Land-use Change. 2008. Available at: http://econpapers.repec.org/paper/isugenres/12881.htm. Accessed February 23, 2023.

28. Nickerson C, Borchers A. How is Land in the United States Used: A Focus on Agricultural Land. U.S. Department of Agriculture, Economic Research Service. March 1, 2012. Available at: https://www.ers.usda.gov/amber-waves/2012/march/data-feature-how-is-land-used/. Accessed February 23, 2023.

29. Bhutada G. From Feed to Fuel: This is How Corn is Used Around the World. World Economic Forum, Sustainable Development. June 2, 2021. Available at: https://www.weforum.org/agenda/2021/06/corn-industries-sustainability-food-prices/. Accessed February 23, 2023.

30. Hristov AN. Technical note: Contribution of ammonia emitted from livestock to atmospheric fine particulate matter (PM2.5) in the United States. Journal of Dairy Science 2011; 94: 3130–3136. doi: 10.3168/jds.2010-3681.

31. Hill J, Goodkind A, Tessum C, et al. Air-quality-related health damages of maize. Nature Sustainability 2019; 2: 397–403.

32. Ribaudo M. Reducing Agriculture's Nitrogen Footprint. U.S. Department of Agriculture, Economic Research Service. September 1, 2011. Available at: https://www.ers.usda.gov/amber-waves/2011/september/nitrogen-footprint/. Accessed February 23, 2023.

33. Fields S. Global nitrogen: Cycling out of control. Environmental Health Perspectives 2004; 112: A556–A563.

34. Mueller JT, Gasteyer S. The widespread and unjust drinking water and clean water crisis in the United States. Nature Communications 2021; 12: 3544. https://doi.org/10.1038/s41467-021-23898-z.

35. Townsend AR, Howarth RW, Bazzaz FA, et al. Human health effects of a changing global nitrogen cycle. Frontiers in Ecology and the Environment 2003; 1: 240–246.

36. Knobeloch L, Salna B, Hogan A, et al. Blue babies and nitrate-contaminated well water. Environmental Health Perspectives 2000; 108: 675–678.

37. Ward MH, Jones RR, Brender JD, et al. Drinking water nitrate and human health: An updated review. International Journal of Environmental Research and Public Health 2018; 15: 1557. https://doi.org/10.3390%2Fijerph15071557.

38. Food and Agriculture Organization of the United Nations. The State of Food and Agriculture 2008. Biofuels: Prospects, Risks, and Opportunities. Rome: FAO, 2008. Available at: http://www.fao.org/docrep/011/i0100e/i0100e00.htm. Accessed February 23, 2023.

39. Scully MJ, Norris GA, Falconi TMA, MacIntosh DL. Carbon intensity of corn ethanol in the United States: State of the science. Environmental Research Letters 2021; 16: 043001. doi: 10.1088/1748-9326/abde08.

40. Hill J, Nelson E, Tilman D, et al. Environmental, economic, and energetic costs and benefits of biodiesel and ethanol biofuels. Proceedings of the National Academy of Sciences USA 2006; 103: 11206–11210.

41. Lark TJ, Hendricks NP, Smith A, Gibbs HK. Environmental outcomes of the US Renewable Fuel Standard. Proceedings of the National Academy of Sciences USA 2022; 119: e2101084119. https://doi.org/10.1073/pnas.2101084119.

42. Lewandrowski J, Rosenfeld J, Pape D, et al. The greenhouse gas benefits of corn ethanol: Assessing recent evidence. Biofuels 2020; 11: 361–375.

43. Malins, C. Biofuel to the Fire: The Impact of Continued Expansion of Palm and Soy Oil Demand Through Biofuel Policy. Rainforest Foundation Norway, 2020. Available at: https://dv719tqmsuwvb.cloudfront.net/documents/RF_report_biofuel_0320_en g_SP_update.pdf. Accessed February 23, 2023.

44. Scovronick N, Wilkinson P. Health impacts of liquid biofuel production and use: A review. Global Environmental Change 2014; 24: 155–164.

45. Niven RK. Ethanol in gasoline: Environmental impacts and sustainability review article. Renewable and Sustainable Energy Review 2005; 9: 535–555.

46. Williams PRD, Cushing CA, Sheehan PJ. Data available for evaluating the risks and benefits of MTBE and ethanol as alternative fuel oxygenates. Risk Analysis 2003; 23: 1085–1115.

47. U.S. Energy Information Administration (EIA). Frequently Asked Questions (FAQs). Available at: https://www.eia.gov/tools/faqs/faq.php. Accessed February 23, 2023.

48. Mueller S, Unnasch S, Keesom B, et al. The Impact of Higher Ethanol Blend Levels on Vehicle Emissions in Five Global Cities. University of Illinois Chicago, November 2018. Available at: https://erc.uic.edu/wp-content/uploads/sites/633/2020/03/ UIC5cities_HEALTH_Nov12_Final.pdf. Accessed February 23, 2023.

49. Morefield PE, LeDuc SD, Clark CM, Iovanna R. Grasslands, wetlands, and agriculture: The fate of land expiring from the Conservation Reserve Program in the Midwestern United States. Environmental Research Letters 2016; 11: 094005.

50. Fargione J, Hill J, Tilman D, et al. Land clearing and the biofuel carbon debt. Science 2008; 319: 1235–1238.

51. Larson JA, English BC, Ugarte DG, et al. Economic and environmental impacts of the corn grain ethanol industry on the United States agricultural sector. Journal of Soil and Water Conservation 2010; 65: 267–279.

52. Mitchell DA. A Note on Rising Food Prices. July 2008. Available at: http://papers.ssrn. com/abstract=1233058. Accessed February 23, 2023.

53. National Research Council. Renewable Fuel Standard: Potential Economic and Environmental Effects of U.S. Biofuel Policy. The National Academies Press, 2011. Available at: https://nap.nationalacademies.org/resource/13105/Renewable-Fuel-Standard-Final.pdf. Accessed February 23, 2023.

54. Rosegrant M. Biofuels and Grain Prices: Impacts and Policy Responses. 2008. Available at: http://large.stanford.edu/courses/2011/ph240/chan1/docs/rosegrant2 0080507.pdf. Accessed February 23, 2023.

55. U.S. Bureau of Labor Statistics. Consumer Expenditures—2021. September 8, 2022. Available at: https://www.bls.gov/news.release/cesan.nr0.htm. Accessed February 23, 2023.

56. UN World Food Program. How High Food Prices Affect the World's Poor. September 4, 2012. Available at: https://reliefweb.int/report/world/how-high-food-prices-aff ect-world%E2%80%99s-poor. Accessed February 23, 2023.

57. Agriculture and Economic Development Analysis Division. The State of Food Insecurity in the World 2013: The Multiple Dimensions of Food Security. Rome: Food and Agriculture Organization of the United Nations, 2013. Available at: https://www. fao.org/3/i3434e/i3434e.pdf. Accessed February 23, 2023.

58. Devarenne S, DeSimone B. History of the United States Farm Bill. Available at: https://www.loc.gov/ghe/cascade/index.html?appid=1821e70c01de48ae899a7 ff708d6ad8b&bookmark=What%20is%20the%20Farm%20Bil. Accessed August 25, 2023.

59. U.S.Department of Agriculture, Economic Research Service. Agriculture Improvement Act of 2018: Highlights and Implications. Available at: https://www. ers.usda.gov/agriculture-improvement-act-of-2018-highlights-and-implications/. Accessed February 23, 2023.

60. Laborde D, Mamun A, Martin, et al. Agricultural subsidies and global greenhouse gas emissions. Nature Communications 2021; 12: 2601. https://doi.org/10.1038/s41 467-021-22703-1.

61. Springmann M, Freund F. Options for reforming agricultural subsidies from health, climate, and economic perspectives. Nature Communications 2022; 13: 82. https:// doi.org/10.1038/s41467-021-27645-2.

62. Walls HL, Johnston D, Tak M, et al. The impact of agricultural input subsidies on food and nutrition security: A systematic review. Food Security 2018; 10: 1425–1436.

63. Merem EC, Twusami Y, Wesley J, et al. Assessing the environmental impacts of farm subsidy in North Texas region. International Journal of Agriculture and Forestry 2022; 12: 9–28.

64. Alston JM, Sumner DA, Vosti SA. Are Agricultural Policies Making Us Fat? Likely Links Between Agricultural Policies and Human Nutrition and Obesity, and Their Policy Implications. International Association of Agricultural Economists, 2005. Contributed Paper prepared for the 26th Triennial Conference of the International Association of Agricultural Economists, Queensland, Australia. August 12–18, 2006. Available at: http://ideas.repec.org/p/ags/iaae06/25343.html. Accessed February 23, 2023.

65. Farnese PL. Remembering the farmer in the agriculture policy and obesity debate. Food and Drug Law Journal 2010; 65: 391–401, iii.

66. Mekonnen MM, Hoekstra AY. A global assessment of the water footprint of farm an-imal products. Ecosystems 2012; 15: 401–415.

67. Gerber PJ, Steinfeld H, Henderson B, et al. Tackling Climate Change Through Livestock: A Global Assessment of Emissions and Mitigation Opportunities. Rome: Food and Agriculture Organization of the United Nations, 2013. Available at: http://www.fao.org/docrep/018/i3437e/i3437e.pdf. Accessed February 23, 2023.

68. Horrigan L, Lawrence RS, Walker P. How sustainable agriculture can address the environmental and human health harms of industrial agriculture. Environmental Health Perspectives 2002; 110: 445–456.
69. Franck C, Grandi SM, Eisenberg MJ. Agricultural subsidies and the American obesity epidemic. American Journal of Preventive Medicine 2013; 45: 327–333.
70. Bazzano LA. The high cost of not consuming fruits and vegetables. Journal of the American Dietetic Association 2006; 106: 1364–1368.
71. Jackson RJ, Minjares R, Naumoff KS, et al. Agriculture policy is health policy. Journal of Hunger & Environmental Nutrition 2009; 4: 393–408.
72. Stewart H, Hyman J, Dong D, Carlson A. The more that households prioritise healthy eating, the better they can afford to consume a sufficient quantity and variety of fruits and vegetables. Public Health Nutrition 2021; 24: 1841–1850.
73. Mozaffarian D, Griffin T, Mande J. The 2018 Farm Bill: Implications and opportunities for public health. Journal of the American Medical Association 2019; 321: 835–836.
74. Gibbs J, Cappuccio FP. Plant-based dietary patterns for human and planetary health. Nutrients 2022; 14: 1614. doi: 10.3390/nu14081614.
75. Satija A, Bhupathiraju SN, Rimm EB, et al. Plant-based dietary patterns and incidence of type 2 diabetes in US men and women: Results from three prospective cohort studies. PLoS Medicine 2016; 13: e1002039. https://doi.org/10.1371/journal.pmed.1002039.
76. Willett W, Rockström J, Loken B, et al. Food in the Anthropocene: The EAT–Lancet Commission on healthy diets from sustainable food systems. Lancet 2019; 393: P447–P492.
77. Springmann M, Clark M, Mason-D'Croz D, et al. Options for keeping the food system within environmental limits. Nature 2018; 562: 519–525.
78. Wisconsin Department of Health Services. Environmental Hazards: Concentrated Animal Feeding Operations (CAFOs) and Public Health. Available at: https://www.dhs.wisconsin.gov/environmental/cafo.htm. Accessed February 23, 2023.
79. Hribar C. Understanding Concentrated Animal Feeding Operations and Their Impact on Communities. National Association of Local Boards of Health, 2010. Available at: https://www.cdc.gov/nceh/ehs/docs/understanding_cafos_nalboh.pdf. Accessed February 23, 2023.
80. Food and Agriculture Organization of the United Nations. Meat Market Review: Overview of Global Meat Market Developments in 2018. Rome: FAO, 2019. Available at: https://www.fao.org/publications/card/en/c/CA3880EN/. Accessed February 23, 2023.
81. Griscom BW, Adams J, Ellis PW, Fargione J. Natural climate solutions. Proceedings of the National Academy of Sciences USA 2017; 114: 11645–11650.
82. McMichael AJ, Powles JW, Butler CD, Uauy R. Food, livestock production, energy, climate change, and health. Lancet 2007; 370: 1253–1263.
83. Godfray HCJ, Beddington JR, Crute IR, et al. Food security: The challenge of feeding 9 billion people. Science 2010; 327: 812–818.
84. Steinfeld H, Gerber P, Wassenaar T, et al. Livestock's Long Shadow: Environmental Issues and Options. 2006. Available at: https://www.europarl.europa.eu/climatechange/doc/FAO%20report%20executive%20summary.pdf. Accessed February 23, 2023.
85. Food and Agriculture Organization of the United Nations. Climate-Smart Agriculture Sourcebook. FAO, 2013. Available at: http://www.fao.org/docrep/018/i3325e/i3325e00.htm. Accessed February 23, 2023.

86. LeZaks D, Ellerton M. The Regenerative Agriculture and Human Health Nexus: Insights from Field to Body. Croatan Institute, September 2021. Available at: https://croataninstitute.org/wp-content/uploads/2021/09/RegenAg_HumanHealth_2021.pdf. Accessed February 23, 2023.

87. Baslam M, Garmendia I, Goicoechea N. Enhanced accumulation of vitamins, nutraceuticals, and minerals in lettuces associated with arbuscular mycorrhizal fungi (AMF): A question of interest for both vegetables and humans. Agriculture 2013; 3: 188–209.

88. Mukherjee A, Omondi EC, Hepperly PR, et al. Impacts of organic and conventional management on the nutritional level of vegetables. Sustainability 2020; 12: 8965. https://doi.org/10.3390/su12218965.

89. Poore J, Nemecek T. Reducing food's environmental impacts through producers and consumers. Science 2018; 360: 987–992.

Further Reading

Mozaffarian D, Griffin T, Mande J. The 2018 Farm Bill: Implications and opportunities for public health. Journal of the American Medical Association 2019; 321: 835–836.

This paper outlines important health considerations linked to the largest agriculture and nutrition policy in the United States, the "Farm Bill."

Poore J, Nemecek T. Reducing food's environmental impacts through producers and consumers. Science 2018; 360: 987–992.

This study consolidates data that quantify the environmental impact of food produced on about 38,000 farms globally. The authors analyze the environmental cost of producing various foods, accounting for variability in production methods and suggesting that a small number of producers create the most environmental impact.

LeZaks D, Ellerton M. The Regenerative Agriculture and Human Health Nexus: Insights from Field to Body. Croatan Institute. September 2021. Available at: https://croataninstitute.org/wp-content/uploads/2021/09/RegenAg_HumanHealth_2021.pdf. Accessed February 23, 2023.

This report summarizes data linking agricultural practices and soil health to human health. The relationship between these factors is complex and much is unknown. The authors assert that there is growing evidence that food produced in healthy soil will be important for food-system transformation via regenerative agriculture—with broad climate, environment, and health impacts.

Roundtable on Environmental Health Sciences, Research, and Medicine; Board on Population Health and Public Health Practice; Institute of Medicine. The Nexus of Biofuels, Climate Change, and Human Health: Workshop Summary. Washington, DC: National Academies Press, 2014. Available at: http://www.ncbi.nlm.nih.gov/books/NBK196451/. Accessed February 15, 2023.

This report examines air, water, land use, food, and social impacts of biomass feedstock as an energy source and the science and health policy implications of using different types (and generations) of biofuels as an energy source.

Roberts S, Forster T, Laestadius L, et al. Growing Healthy Food and Farm Policy: The Impact of Farm Bill Policies on Public Health. Baltimore: Johns Hopkins Center for a Livable Future, 2012. Available at: https://clf.jhsph.edu/sites/default/files/2019-04/growing-healthy-food-farm-policy-report.pdf. Accessed February 15, 2023.

This report examines the health impact of the 2008 U.S. Farm Bill, although it does not include the Supplemental Nutrition Assistance Program, the Conservation Stewardship Program, or commodity subsidies. Although the Farm Bill has since been updated, this paper is a good example of how general U.S. agriculture and nutrition policies in the Farm Bill can shape public health.

PART IV

DEVELOPING AND IMPLEMENTING ACTIONS FOR ADAPTATION

15

Developing and Implementing
Interventions for Health Adaptation

Kristie L. Ebi and Peter Berry

Adaptation in human systems is "the process of adjustment to actual or expected climate and its effects, in order to moderate harm or exploit beneficial opportunities."[1] It consists of measures to prepare for and cope with climate change. Adaptation focuses on decreasing exposure to hazards, reducing vulnerabilities, and increasing capabilities to manage the consequences of climate change. The health consequences of climate change require new adaptation measures within and outside the health sector that are sustainable and resilient on a large scale.[2-5]

Health adaptation aims to reduce the consequences of climate change for population health and health systems. Health adaptation measures include, but are not limited, to the following:

- Research, such as investigating the impact of extreme heat on chronic noncommunicable diseases (such as chronic lung disease and diabetes), on occupational illnesses and injuries, and on mental health and well-being
- Improved surveillance for vectorborne and waterborne diseases
- Assessment of risk, such as by developing detailed maps of flood-prone areas and socioeconomic inequities and vulnerabilities
- Early warning and response systems for infectious diseases, heat waves, increased ozone levels, and extreme weather events
- Greening communities to reduce the urban heat island effect
- Proper siting and design of healthcare facilities to withstand extreme weather impacts
- Developing programs to support food security, livelihoods, and cultures in Indigenous Communities.

Health adaptation measures often need to be disease-specific. For example, adaptation measures to reduce childhood asthma include vulnerability assessments, improved ventilation and heating, enhanced community education, and use of models to forecast increased ambient levels of ozone.[6] Adaptation measures for

Kristie L. Ebi and Peter Berry, *Developing and Implementing Interventions for Health Adaptation* In: *Climate Change and Public Health*. Edited by: Barry S. Levy and Jonathan A. Patz, Oxford University Press. © Oxford University Press 2024.
DOI: 10.1093/oso/9780197683293.003.0015

changes in the geographic range and seasonality of vectorborne diseases can include improved surveillance and monitoring, integrated vector management programs, and early warning and response systems. In addition, adaptation measures should be appropriately designed and prioritized for specific settings. Prioritization requires an understanding of the magnitude and pattern of current and future health risks, the vulnerabilities of populations and health systems, and the effectiveness of policies and programs to address these risks and reduce inequities. (See Table 15-1.)

Health adaptation can be encouraged and facilitated in a variety of ways, such as (a) providing funding, data, and technical support for developing climate change and health vulnerability and adaptation assessments; (b) supporting communities of practice and peer-to-peer networks in sharing experiences about developing health adaptation plans and engaging communities; and (c) facilitating community engagement that co-creates adaptation measures, especially with those at risk of the health consequences of climate change. Effective health adaptation measures, in both public and private sectors, draw from expertise in public health; healthcare planning; environmental engineering; and decision-making in disaster management, energy, transportation, agriculture, water and sanitation, and other sectors.

Public health agencies have extensive experience in recognizing, preventing, and controlling many climate-related hazardous exposures, such as air pollution, and vectorborne and other diseases. But many facilities and systems are not prepared to address the growing challenges posed by climate change, such as the structural threats to healthcare facilities from storms and floods and the increased incidence of heat-related disorders, vectorborne diseases, waterborne diseases, and mental disorders associated with these impacts.[4,7,8] The health impacts of climate change are already contributing to surges in demand and resource utilization that can adversely affect patient care.[9] Because climate change is ongoing, adaptation measures will need to increase and evolve to address the health challenges and the related inequity and injustice consequences.[10]

NEED FOR HEALTH EQUITY APPROACHES

As climate change increases, associated health consequences are becoming more difficult to manage. For example, it is often necessary to simultaneously address heat-related disorders, vectorborne diseases, reduced access to freshwater, and decreased food security. Climate change is also exerting disproportionate impacts on higher-risk populations, such as women, ethnic and religious minorities, Indigenous Peoples, older people, youth, and low-income

Table 15-1 Examples demonstrating the range of available adaptation options to manage climate-sensitive health risks

Climate-sensitive health risk	Illustrative adaptation measures
Heat-related morbidity and mortality	Education to increase awareness of heat-related hazards
	Heat action plans, including early warning and response plans for heat waves
	Modifications of the built environment to address increased heat
Injuries, illnesses, and deaths from extreme weather and climate events	Explicit incorporation of health into risk management plans for disasters
	Stress testing of disaster response plans to ensure effectiveness
	Early warning and response systems
	Incorporation of mental health impacts into disaster risk management
Respiratory disorders associated with poor air quality	Early warning and response systems
	Education to increase awareness of the hazards of ozone, wildfire smoke, and aeroallergens
Vectorborne and waterborne diseases	Integrated vector management
	Integrated disease surveillance systems
	Early warning and response systems
	Improved water, sanitation, and hygiene systems
Undernutrition	Integrated surveillance and monitoring systems
	Early warning and response systems

Source: Smith KR, Woodward A, Campbell-Lendrum D, et al. Human health: Impacts, adaptation, and co-benefits. In CB Field, VR Barros, DJ Dokken DJ, et al. (eds). Climate change 2014: Impacts, adaptation, and vulnerability. Part A: Global and sectoral aspects. Contribution of Working Group II to the Fifth Assessment Report of the Intergovernmental Panel on Climate Change. Cambridge, United Kingdom: Cambridge University Press, 2014, pp. 709–754.

people, thereby widening health disparities (See Box 1-1 in Chapter 1).[1,11,12] The health of poor and marginalized people is often adversely impacted when extreme weather events damage healthcare facilities and/or injure healthcare workers. In low-income countries and communities, damage to schools from extreme weather events can further disadvantage students with inadequate education. For example, floods in Cambodia during 2013 and 2014 adversely affected more than 450,000 students—155 schools closed for up to 9 weeks and 40,000 textbooks were destroyed.[13] Therefore, equity needs to be considered in developing and implementing adaptation measures to ensure that historically marginalized people and those who are most vulnerable are protected.[14,15] (See Box 15-1 and Chapter 20.)

STATUS OF HEALTH ADAPTATION

Many individuals, communities, and health systems are not adequately prepared for the health impacts of climate change, creating an *adaptation gap*[4,11,16,17] For example, only 47 (52%) of 91 countries recently surveyed had a national health and climate change plan or strategy.[18] Major reasons for an adaptation gap are inadequate human and financial resources and inadequate research on health adaptation, including on the feasibility and effectiveness of measures.[17,18] There has been insufficient funding for health adaptation—less than 0.5% of global adaptation funding has gone to the health sector.[11,17] As of mid-2021, only 4% of subnational adaptation measures in the United States, 5% in Canada, and 7% in Mexico focused on public health and safety. Few studies have assessed the effectiveness of adaptation measures.[19,20]

Limits to Adaptation

The increasing and changing nature of climate-related hazards—together with institutional, technological, economic, and other factors—indicates that there are *adaptation limits* at which health systems can no longer avoid intolerable risks, even with the implementation of adaptation strategies.[21] *Soft adaptation limits* arise from the lack of implementation of interventions to ensure adequate functioning of health systems as the climate continues to change, such as access to safe water and improved sanitation. *Hard adaptation limits* occur when it is not possible to maintain functioning with further climate change, such as when healthcare facilities must relocate because they were built on permafrost or in vulnerable coastal floodplains. *Intolerable adaptation limits* are the points where health systems must either transform or accept escalating losses and damage.

Box 15-1 Guiding Principles for Vancouver Coastal Health and Fraser Health Climate Change and Health Adaptation Framework

Kristie L. Ebi and Peter Berry

The *Vancouver Coastal Health and Fraser Health Climate Change and Health Adaptation Framework* outlines the roles and responsibilities within the jurisdictions for climate change and health adaptation and presents recommendations to reduce key risks to the population identified through a robust climate change and health vulnerability and adaptation assessment process. The framework includes health authority priorities for action and supports adaptation planning at the program level within and between each of the four partner organizations. The guiding principles for the Framework include the following:

- Engagement: Engagement with communities and allied organizations ensures a stronger understanding of vulnerability, builds capacity and relationships, and facilitates implementation.
- Equity: A commitment to understanding the disproportionate and intersectional nature of climate impacts, ensuring that actions for adaptation options account for and reduce existing inequities. A further commitment to cultural safety when building new relationships and taking action.
- Collaboration: Collaborative action involving multiple sectors enables objectives and actions that reflect the unique needs of a community and leverage its skills and resources.
- Impact: A commitment to meaningful, deliberate, and demonstrable change and a focus on implementation and impact.
- Transparency: A commitment to sharing activities and reporting publicly on the impact of the work.
- Low carbon resilience: Ensuring that greenhouse gas reduction co-benefits of actions are considered and maximized and that climate change adaptation and mitigation are considered in concert when appropriate.[1]

Box Reference

1. Vancouver Coastal Health and Fraser Health. Climate Change and Health Adaptation Framework. April 2022. Available at: http://www.vch.ca/Documents/HealthADAPT-Climate-Change-and-Health-Adaptation-Framework.pdf. Accessed August 1, 2022.

These limits fundamentally threaten core social objectives associated with health, security, and sustainability.[21]

Adaptation limits will depend on the rate and magnitude of increases in mean temperature, temperature variability, and changes in precipitation patterns. If global temperatures reach 2.0°C (3.6°F) above preindustrial levels by 2100, then health systems and other critical infrastructures will have time to prepare for many consequences; but with a global temperature increase of more than 2.0°C, the challenges to adaptation will be more significant and will require much greater investment.

This assessment was global. The extent to which impacts occur in a specific location will depend on whether national and local health authorities have the capacity to effectively prepare health systems for a much warmer future.

One of us (K.E.) reviewed scientific studies projecting how risks of climate-sensitive health outcomes could change with further increases in global mean surface temperature under three adaptation scenarios. The assessment considered six health risks: heat-related morbidity and mortality, ozone-related mortality, malaria incidence, and incidence of dengue and other diseases spread by *Aedes* species mosquitoes, Lyme disease, and West Nile virus disease.[22] The three adaptation scenarios were based on the shared socioeconomic pathways (SSPs). The first scenario, with low challenges to adaptation (SSP1), would be characterized by adaptation that is more proactive and effective than today's in preventing adverse climate-sensitive health outcomes.[22] This scenario would emphasize international cooperation toward achieving sustainable development and thus has the greatest potential to avoid significant increases in risks under all but the highest greenhouse gas (GHG) emission pathways that could cause warming of up to 4.0°C (7.2°F) in 2100. But even under this optimistic scenario, there would be some increase in the burden of climate-sensitive health outcomes, leading to more preventable injuries, illnesses, and deaths that health systems will need to manage. In the second scenario, with moderate challenges to adaptation (SSP2), there would be slow, but not always timely, improvements in health systems, resulting in some increases in the magnitude and burden of climate-sensitive health outcomes. In the scenario with high challenges to adaptation (SSP3), efforts to address health risks would prevent few avoidable climate-sensitive injuries, illnesses, and deaths.[22] This assessment was global. The extent to which impacts occur in a specific location will depend on whether national and local health authorities have the capacity to effectively prepare health systems for a much warmer future.

NATIONAL ADAPTATION PLANS

As part of the Cancun Adaptation Framework (2010) under the United Nations Framework Convention on Climate Change (UNFCCC), countries are required to develop national adaptation plans (NAPs) to identify people, infrastructure, and industries that are most vulnerable to the effects of climate change and develop ways for international donors to assist them. Health national adaptation plans (HNAPs) are developed by each ministry of health and are integrated into the NAP. A survey of 95 countries in 2021 revealed that 23 (59%) of 39 low- and middle-income countries (LMICs) had developed a HNAP.[18] High-income countries, which are primarily responsible for climate change, agreed under the UNFCCC to financially support adaptation in LMICs, which are suffering most from the consequences of climate change. The United Nations Environment Program estimates that between US\$160 and US\$340 billion per year across all sectors is required for adaptation in developing countries between now and 2030. In contrast, only about US\$28.6 billion was invested in adaptation in developing countries in 2020.[23] Investment in adaptation makes good sense from an economic perspective: the World Bank estimates that every US\$1 spent on adaptation yields, on average, US\$4 in benefits.

SCALING UP HEALTH ADAPTATION

Health officials in many countries, from the national to the local level, are preparing for future impacts of climate change.[4,11,18] Given the current adaptation gap and projected increase in health risks, large-scale measures need to be implemented to increase the resilience of health systems and empower individuals and communities to protect health. Adaptation strategies, policies, and programs aim to build climate-resilient and environmentally sustainable health systems.[24] These health systems provide sustainable healthcare for populations while anticipating, responding to, coping with, recovering from, and adapting to climate-related shocks and stresses in a manner that minimizes negative impacts on the environment and leverages opportunities to restore and improve it.[24,25] To build resilience, adaptation measures strengthen key components of health systems. Health systems include the people, infrastructure, institutions, and resources to both (a) protect and promote health and well-being (public health) and (b) protect access to and delivery of healthcare services. Critical components of health systems include the delivery of health and social services, health workers, technology and infrastructure (such as health information systems), leadership and governance, and financing and investment.[24] Strong actions from health sector managers can make available the needed resources and partnerships for research, assessments to understand risks and vulnerabilities, development of health adaptation plans, and provision of training and capacity building for staff members. (See Box 15-2.)

Box 15-2 Systems Approach to Climate Services for Health in Ethiopia

Kristie L. Ebi and Peter Berry

The Ethiopian Ministry of Health and the National Meteorological Services of Ethiopia collaborated to develop climate services for health. Because the WHO health systems framework does not include environmental metrics, especially those related to climate change information, the Ministry of Health and the Meteorological Services created the Climate-Health Working Group to provide leadership and coordinate the efforts of synergizing climate information and health information for policy and program development to protect health. The Working Group facilitated the exchange, analysis, and interpretation of data and information and developed training and capacity building programs for health workers and decision-makers. Meteorological Services provides seasonal forecasts for areas favorable for malaria-carrying mosquitoes, and the Ministry of Health provides routine malaria surveillance data from these areas. The Working Group facilitates integration and interpretation of these data to ensure evidence-based decisions to plan timely malaria prevention activities, including awareness campaigns. It also facilitates implementation of population health and treatment options in areas affected or forecast to be affected by malaria, including distributing mosquito nets, spraying some mosquito breeding sites, and providing medications.[1]

Box Reference

1. Manyuchi AE, Vogel C, Wright CY, Erasmus B. Systems approach to climate services for health. Climate Services 2021; 24: 100271. https://doi.org/10.1016/j.cliser.2021.100271.

For most public health agencies, the first step toward adapting to climate change is conducting a climate change and health vulnerability and adaptation (V&A) assessment of the current and projected health risks related to climate change.[1,26] V&A assessments establish a knowledge base of current and projected health risks for:

- Develop, implement, monitor, and evaluate the effectiveness of adaptation options
- Identify especially vulnerable populations, regions, and health system operations, services, and infrastructure
- Detail the capacity of systems, organizations, and communities to prepare for and manage changes in the magnitude and pattern of climate-related risks.

The process of conducting V&A assessments includes building partnerships within the health sector (including public health and healthcare) to ensure integration of climate change into relevant policies and programs, such as vector control or maternal and child health programs. The process also builds partnerships across other sectors, ranging from energy to environmental protection, to develop early warning and response systems and other measures and to ensure that health considerations are included in all policies. V&A assessments help to prioritize adaptation strategies, policies, and programs and plans for their implementation.[1]

Many countries use World Health Organization (WHO) guidelines and steps for conducting such assessments. Specific guidance is also available that focuses on assessing the climate resilience and environmental sustainability of health facilities and monitoring and evaluating efforts toward increasing the preparedness of health systems.[1,25] Tools can be used to tailor assessments for national and subnational health authorities, such as the BRACE program in the United States[27,28] and comparable programs in Canada.[29-31]

A V&A assessment is integral to developing the health component of a national adaptation plan (HNAP).[32] An HNAP builds on existing national efforts toward health adaptation to climate change, including assessments, policies, and programs, to ensure that health adaptation is integrated into national health planning strategies, processes, and monitoring systems. HNAPs maximize synergies across sectors, such as food, water, energy, and housing, by building health considerations into their adaptation planning and addressing the upstream drivers of health.

ECONOMIC COSTS

To inform adaptation plans and measures, health sector decision-makers, such as local public health officials, healthcare system planners, and health facility administrators, require information on the economic costs of climate change impacts on health, the costs of adaptation measures for protecting health taken in various sectors, and the efficiency of various adaptation measures.[33] WHO has developed guidance for performing studies on the economic costs of health impacts and the benefits of adaptation and GHG mitigation actions.[33,34]

Ineffective or inadequate adaptation measures lead to increased economic costs associated with impacts that in turn can erode the capacity of individuals, communities, and health systems to cope with climate-related shocks and stresses and prepare for future climate change. Increased morbidity and mortality due to a warming climate result in a loss of welfare to society and forgone income and/or wealth creation. Climate hazards also can affect labor productivity and key determinants of health, such as educational attainment, thereby impacting economic output.[35] And as recent extreme heat events, wildfires, severe windstorms, floods, and other events have demonstrated, illnesses and injuries from climate disasters can increase stresses and costs to health systems and affect economic productivity through increased sick leave and absenteeism.[36,37]

Estimates of current and projected economic costs of climate change on health vary greatly due to the paucity of data and differences in the methods used in studies. One study by the WHO estimated that the economic costs for a narrow range of direct climate change impacts on health would be $2 to $4 billion by 2030.[38] In Canada, between 2013 and 2018, fine particulate matter associated with wildfires resulted annually in an average of 54 to 240 premature deaths from short-term exposure and 570 to 2,500 premature deaths from long-term exposure. The acute health impacts cost an estimated $410 million to $1.8 billion per year, and the chronic health impacts cost an estimated $4.3 to $19 billion per year.[39] Between 2005 and 2015, wildfires in the western United States resulted in an annual average of $165 million in health-related costs.[40]

PUBLIC HEALTH AGENCIES

State and territorial public health agencies have vital roles to play in implementing climate change adaptation measures. These roles include the following:

- Assessing the impacts of climate change on health and well-being
- Conducting or coordinating surveillance of climate-sensitive health outcomes
- Implementing and evaluating adaptation measures
- Providing education on climate change and its health consequences
- Providing technical assistance to local or regional health agencies
- Coordinate with meteorological services and other agencies that can contribute to health adaptation.[41]

Healthcare (planning and delivery) decision-makers also have important roles to play in implementing climate change adaptation measures.

CIVIL SOCIETY ORGANIZATIONS

Civil society organizations, including health professional associations, play important roles in mobilizing and empowering the public and health sector workers to prepare for the health impacts of climate change and slow the rate of warming (Box 15-3). At the international, national, and local levels, health professionals are undertaking climate change and health research, building networks among health organizations, and training health sector workers.

Box 15-3 Combining Mitigation and Adaptation:
Climate-Proofing of Healthcare Facilities

Kristie L. Ebi and Peter Berry

Healthcare facilities and other infrastructure are sources of GHG emissions and are affected by extreme weather and climate events that can increase demand for healthcare while altering access through damage to power generation and transportation infrastructure.[1] Climate-proofing healthcare facilities can reduce vulnerability to potential power and water shortages; it includes increasing energy and water self-sufficiency, reducing operational costs, and increasing preparedness of healthcare for service delivery interruptions—all while reducing GHG emissions. These measures can be implemented by retrofitting nonstructural items, implementing passive and active design measures for building resilience, placing transmission lines underground, and using on-site energy generation with solar panels and batteries.[2]

An illustrative economic assessment was conducted for the 545-bed Federal Government Polyclinic Hospital of Islamabad. Retrofitting, designing, undergrounding, and installing photovoltaic solar panels were considered and implemented independently or in combinations. The lifetime of the investments was assumed to be 20 years, and a 5% discount rate was used. Retrofitting nonstructural items and design measures were found to be the cheapest interventions and the most profitable, generating high benefit-cost ratios and short payback periods.[3]

Box References

1. Health Care Without Harm. Safe haven in the storm: Protecting lives and margins with climate-smart health care. 2018. Available at: https://noh arm-uscanada.org/sites/default/files/documents-files/5146/Safe%20Haven. pdf. Accessed August 1, 2022.

2. World Health Organization. WHO Guidance for Developing Climate Resilient and Environmentally Sustainable Health Facilities. Geneva: World Health Organization, 2020. Available at: https://apps.who.int/iris/handle/ 10665/335909. Accessed August 1, 2022.

3. Japan International Cooperation Agency (JICA). Data collection survey on health facilities and equipment in the Islamic Republic of Pakistan: Final report. 2018. Available at: https://openjicareport.jica.go.jp/pdf/12322293. pdf. Accessed August 1, 2022.

Because health professionals are among the most trusted voices, they are critical for communicating the public health risks and the need for action and for advocating for measures to increase health system resiliency and green healthcare operations, including the reduction of GHG emissions.[42]

Concerns about health can be important drivers of climate action.[43,44] In 2021, the COP26 Health Programme at the UNFCCC Conference of the Parties in Glasgow, Scotland, included national commitments for building climate-resilient and low-carbon sustainable health systems and enabling health professionals to advocate for stronger measures to address climate change.[45] Seventy-three countries have committed to this program (called the Alliance for Transformative Action on Climate and Health), which is administered by the World Health Organization. In 2022, the United Nations General Assembly recognized, as a universal human right, access to a clean, healthy, and sustainable environment.[46] This recognition was incorporated into the agreements reached at COP27 in Sharm el-Sheikh, Egypt, later that year.

INDICATORS FOR ASSESSING HEALTH ADAPTATION PROGRESS

What constitutes successful adaptation will vary as climate-related exposures change with further warming and shifting degrees of inequities and vulnerability. A successful V&A assessment identifies indicators of the baseline rate of climate-sensitive health outcomes, the effectiveness of current climate-relevant policies and programs, and the process of adaptation, considering that adaptation strategies implemented today may not show effectiveness for several years. Monitoring, evaluation, and learning from evaluations can create a virtuous cycle of improvements to adaptation policies and programs.

The WHO developed a framework for measuring the climate resilience of health systems (Figure 15-1).[25] The framework can be used by health authorities to identify indicators that track the status and trends in the following:

- Upstream determinants of exposures to climate-related hazards and vulnerabilities
- Robustness of health system functions
- Overall capacity of a health system to manage climate-related shocks and stresses
- Extent to which resilience outcomes change following adaptation interventions.

Short- and long-term climate-related risks need to be monitored to ensure that interventions build flexibility and capacity to manage additional climate change–related risks over coming decades.

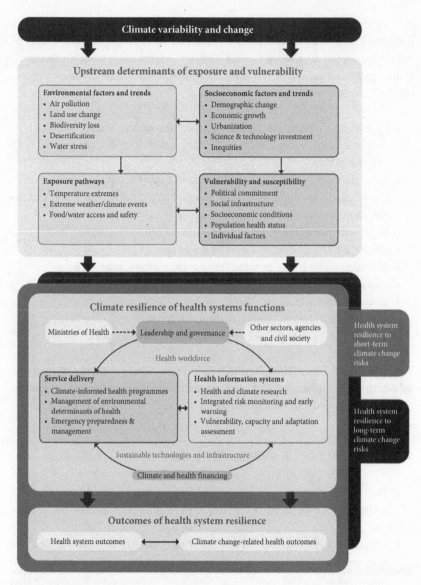

Figure 15-1 Framework for measuring the climate resilience of health systems. (*Source*: World Health Organization, 2022.)

Many health systems are tracking the incidence of climate-sensitive injuries, illnesses, and deaths. Other entities are tracking select indicators of exposure and vulnerability that health authorities should use in addition to identifying new data sources that might be needed. There is no universal set of indicators for adaptation and resilience building to be used by all health authorities. Health officials need to work with climate change programs in other sectors (such as environment, transportation, natural resources, and energy) to identify a set of indicators needed to achieve the national and regional resilience and sustainability goals. Indicators for a specific jurisdiction should ideally meet the following criteria:

- Specific, based on an association between climate and health
- Actionable, related to climate and health conditions that are amenable to adaptive actions
- Measurable, based on timely and unbiased data of acceptable quality
- Understandable, applicable, and acceptable to stakeholders and potential users
- Representative of the issues and areas of concern
- Consistent and comparable over time and space
- Robust and unaffected by minor changes in method, scale, or data
- Capable of being used at different scales
- Cost-effective
- Implemented at an acceptable cost–benefit ratio
- Sustainable—able to provide data for the next 20 to 30 years.[25]

The availability, accessibility, quality, and resource requirements of data for monitoring indicators vary. Different approaches to using indicators may be required based on available human and financial resources at local to national scales.

SUMMARY POINTS

- The increasing scale and pace of climate change impacts require rapid scaling up of policies and programs to protect health and close the adaptation gap.
- Effective actions to protect health include assessing risks and vulnerabilities to current and future climate change, identifying populations that suffer disproportionately, and developing needed interventions while empowering civil society organizations and the public.
- The increasing complexity and severity of health risks requires that officials in the health sector work closely with those in other sectors to develop climate-resilient, low-carbon health systems.
- Success in protecting health from climate change requires health authorities to regularly monitor progress in adaptation and mitigation.
- Adequate investment in adaptation is critical to ensuring that individuals, communities, and health systems can effectively respond to the challenges of climate change.

References

1. Matthews JBR. Annex I: Glossary. In: V Masson-Delmotte, P Zhai, H-O Pörtner, et al., editors. Global Warming of 1.5°C: An IPCC Special Report on the Impacts of Global Warming of 1.5°C Above Preindustrial Levels and Related Global Greenhouse Gas Emission Pathways, in the Context of Strengthening the Global Response to the Threat of Climate Change, Sustainable Development, and Efforts to Eradicate Poverty. Geneva: Intergovernmental Panel on Climate Change, 2018.

2. Ebi KL, Hess JJ. Health risks due to climate change: Inequity in causes and consequences. Health Affairs 2020; 39: 2056–2062.

3. Vicedo-Cabrera AM, Scovronick N, Sera F, et al. The burden of heat-related mortality attributable to recent human-induced climate change. Nature Climate Change 2021; 11: 492–500.

4. Cissé G, McLeman R, Adams H, et al. Health, wellbeing, and the changing structure of communities. In H-O Pörtner, DC Roberts, M Tignor M, et al. (eds.). Climate Change 2022: Impacts, Adaptation, and Vulnerability. Contribution of Working Group II to the Sixth Assessment Report of the Intergovernmental Panel on Climate Change. Cambridge: Cambridge University Press, 2022, pp. 1041–1170. doi: 10.1017/9781009325844.009.

5. New M, Reckien D, Viner D, et al. Decision making options for managing risk. In H-O Pörtner, DC Roberts, M Tignor M, et al. (eds.). Climate Change 2022: Impacts, Adaptation, and Vulnerability. Contribution of Working Group II to the Sixth Assessment Report of the Intergovernmental Panel on Climate Change. Cambridge: Cambridge University Press, 2022, pp. 2539–2654. doi: 10.1017/9781009325844.026.

6. Hu Y, Cheng J, Liu S, et al. Evaluation of climate change adaptation measures for childhood asthma: A systematic review of epidemiological evidence. Science of the Total Environment 2022; 839: 156291. doi: 10.1016/j.scitotenv.2022.156291.

7. Frumkin H, Hess J, Luber G, et al. Climate change: The public health response. American Journal of Public Health 2008; 98: 435–445.

8. Ebi KL, Otmani del Barrio M. Lessons learned on health adaptation to climate variability and change: Experiences across low- and middle-income countries. Environmental Health Perspectives 2017; 125: 065001. https://doi.org/10.1289/EHP405.

9. Theron E, Bills CB, Calyello Hynes EJ, et al. Climate change and emergency care in Africa: A scoping review. African Journal of Emergency Medicine 2022; 12: 121–128.

10. Eriksen SH, Aldunce P, Bahinipati CS, et al. When not every response to climate change is a good one: Identifying principles for sustainable adaptation. Climate and Development 2011; 3: 7–20. doi: 10.3763/cdev.2010.0060.

11. Romanello M, McGushin A, Napoli CD, et al. The 2021 report of the Lancet Countdown on health and climate change: Code red for a healthy future. Lancet 2021; 398: 1619–1662.

12. Gasparri G, Tcholakov Y, Gepp S, et al. Integrating youth perspectives: Adopting a human rights and public health approach to climate action. International Journal of Environmental Research and Public Health 2022; 19: 4840. doi: 10.3390/ijerph19084840.

13. United Nations Children's Fund (UNICEF). It Is Getting Hot: Call for Education Systems to Respond to the Climate Crisis. Bangkok: UNICEF East Asia and Pacific Regional Office, 2019. Available at: https://www.unicef.org/eap/reports/it-getting-hot. Accessed August 1, 2022.

14. Juhola S, Glaas E, Linner BO, Neset TS. Redefining maladaptation. Environmental Science & Policy 2016; 55: 135–140.

15. Schnitter R, Moores E, Berry P, et al. Climate change and health equity. In P Berry, R Schnitter (eds.). Health of Canadians in a Changing Climate: Advancing Our Knowledge for Action. Ottawa: Government of Canada, 2022, pp. 614–667. Available at: https://changingclimate.ca/site/assets/uploads/sites/5/2022/02/CCHA-REPORT-EN.pdf. Accessed August 30, 2023.

16. Ebi KL, Berry P, Bowen KJ, et al. Health system adaptation to climate variability and change. Rotterdam: Global Center on Adaptation, 2019. Available at: https://gca.org/reports/health-system-adaptation-to-climate-variability-and-change/.   Accessed August 1, 2022.

17. United Nations Environment Programme (UNEP). The Adaptation Gap Report 2018: Health. Nairobi: UNEP, 2018. Available at: https://www.unep.org/resources/adaptation-gap-report-2018. Accessed August 1, 2022.

18. World Health Organization. 2021 WHO health and climate change global survey report. Geneva: World Health Organization, 2021. Available at: https://www.who.int/publications/i/item/9789240038509. Accessed August 1, 2022.

19. Carbon Disclosure Project. Available at: https://data.cdp.net/d/feaz-9v5k/visualizat ion. Accessed August 1, 2022.

20. Scheelbeek PFD, Dangour AD, Jarmul S, et al. The effects on public health of climate change adaptation responses: A systematic review of evidence from low- and middle-income countries. Environmental Research Letters 2021; 16: 073001. doi: 10.1088/1748-9326/ac092c.

21. Klein RJT, Midgley GF, Preston BL, et al., Adaptation Opportunities, Constraints, and Limits. In CB Field, VR Barros, DJ Dokken, et al. (eds.). Climate Change 2014: Impacts, Adaptation, and Vulnerability. Part A: Global and Sectoral Aspects. Contribution of Working Group II to the Fifth Assessment Report of the Intergovernmental Panel on Climate Change. Cambridge: Cambridge University Press, 2014, pp. 899–943.

22. Ebi KL, Boyer C, Ogden N, et al. Burning embers: Synthesis of the health risks of climate change. Environmental Research Letters 2021; 16. doi: 10.1088/1748-9326/abeadd.

23. United Nations Environment Programme (2022). Adaptation Gap Report 2022: Too Little, Too Slow – Climate adaptation failure puts world at risk. Nairobi. https://www.unep.org/adaptation-gap-report-2022

24. World Health Organization. WHO Guidance for Climate-Resilient and Environmentally Sustainable Health Care Facilities. Geneva: World Health Organization, 2020. Available at: https://www.who.int/publications/i/item/978924 0012226. Accessed August 1, 2022.

25. World Health Organization. Measuring the Climate Resilience of Health Systems. Geneva: World Health Organization, 2022. Available at: https://apps.who.int/iris/han dle/10665/354542. Accessed August 1, 2022.

26. Berry P, Enright PM, Shumake-Guillemot J, et al. Assessing health vulnerability and adaptation to climate change: A review of international progress. International Journal of Environmental Research and Public Health 2018 15: 1–25. Available at: https://www.mdpi.com/1660-4601/15/12/2626. Accessed August 1, 2022.

27. Centers for Disease Control and Prevention. CDC's Building Resilience Against Climate Change (BRACE) framework. Last reviewed: October 31, 2022. Available at: https://www.cdc.gov/climateandhealth/BRACE.htm. Accessed January 6, 2023.

28. Guenther R, Balbus J. Primary Protection: Enhancing Health Care Resilience for a Changing Climate. Washington DC: United States Department of Health and Human Services, 2014. Available at: https://toolkit.climate.gov/sites/default/files/SCRH CFI%20Best%20Practices%20Report%20final2%202014%20Web.pdf.   Accessed August 1, 2022.

29. Ebi KL, Paterson J, Yusa A, et al. Ontario Climate Change and Health Vulnerability and Adaptation Assessment Guidelines. Toronto: Ministry of Health and Long-Term Care, 2016. Available at: https://www.health.gov.on.ca/en/common/ministry/publi cations/reports/climate_change_toolkit/climate_change_health_va_guidelines.pdf. Accessed August 1, 2022.

30. Lower Mainland Facilities Management. Moving towards climate resilient health facilities for Vancouver Coastal Health. October 2018. Available at: https://etccdi. pacificclimate.org/sites/default/files/publications/VCH_ClimateReport_Final.pdf. Accessed August 1, 2022.

31. Health Canada. Climate Change and Health Vulnerability and Adaptation Assessments: Workbook for the Canadian Health Sector. Ottawa: Government of Canada, 2022. Available at: https://www.canada.ca/en/health-canada/programs/hea lth-adapt.html#a4. Accessed August 1, 2022.

32. World Health Organization. WHO Guidance to Protect Health from Climate Change Through Adaptation Planning. Geneva: World Health Organization, 2014. Available at: https://www.who.int/publications/i/item/9789241508001. Accessed August 1, 2022.

33. World Health Organization Regional Office for Europe. Climate Change and Health: A Tool to Estimate Health and Adaptation Costs. Copenhagen: World Health Organization Regional Office for Europe, 2013. Available at: https://apps.who.int/ iris/handle/10665/329517. Accessed August 2, 2022.

34. World Health Organization. Economic Report: A Framework for the Quantification and Economic Evaluation of Health Outcomes Originating from Health and Non-health Climate Change Mitigation and Adaptation Action. Geneva: World Health Organization, 2023. Available at: https://www.who.int/publications/i/item/978924 0057906. Accessed September 12, 2023.

35. Ebi KL, Hess JJ, Watkiss P. Health risks and costs of climate variability and change. In CN Mock, R Nugent, O Kobusingye, KR Smith (eds.). Disease Control Priorities (3rd edition), Volume 7: Injury Prevention and Environmental Health. Washington, DC: World Bank, 2017, pp. 153–169.

36. Campbell-Lendrum D, Villalobas PE, Kendrovski V. Estimating the cost of health adaptation [Webinar Presentation]. World Health Organization. Available at: https:// unfccc.int/sites/default/files/resource/CGE_wenbinar%236_presentation.pdf. Accessed August 1, 2022.

37. Sanchez Martinez G, Berry P. The adaptation health gap: A global overview. In H Neufeldt, GS Martinez, A Olhoff, et al. (eds.). The Adaptation Gap Health Report 2018. Nairobi: UNEP, 2018. Available at: https://backend.orbit.dtu.dk/ws/files/ 163240215/AGR_Final_version.pdf. Accessed August 1, 2022.

38. World Health Organization. Factsheet: Climate change and health. October 30, 2021. Available at: https://www.who.int/news-room/fact-sheets/detail/climate-change-and-health. Accessed August 1, 2022.

39. Egyed M, Blagden P, Plummer D, et al. Air quality. In P Berry and R Schnitter (eds.). Health of Canadians in a Changing Climate: Advancing Our Knowledge for Action. Ottawa: Government of Canada, 2022. Available at: https://changingclimate.ca/hea lth-in-a-changing-climate/chapter/5-0/. Accessed August 1, 2022.

40. Jones B, Berrens RP. Application of an original wildfire smoke health cost benefits transfer protocol to the Western US, 2005–2015. Environmental Management 2017; 60: 809–822.

41. Errett NA, Dolan K, Hartwell C, et al. Adapting by their bootstraps: State and territorial public health agencies struggle to meet the mounting challenge of climate change. American Journal of Public Health 2022; 112: 1379–1381.

42. Naidu-Ghelani R. Who do you have faith in? The world's most trusted professions. Ipsos; October 2, 2019. Available at: https://www.ipsos.com/en/who-do-you-have-faith-worlds-most-trusted-professions. Accessed August 1, 2022.
43. Climate for health. Moving forward toolkit. https://climateforhealth.org/moving-forward-toolkit/. Accessed August 1, 2022.
44. Rudolph L, Harrison C. A Physician's Guide to Climate Change, Health, and Equity. Oakland, CA: Public Health Institute, 2016. Available at: https://climatehealthconn ect.org/resources/physicians-guide-climate-change-health-equity/. Accessed August 1, 2022.
45. World Health Organization. COP26 health programme. 2022. Available at: https://www.who.int/initiatives/alliance-for-transformative-action-on-climate-and-health/cop26-health-programme. Accessed August 1, 2022.
46. United Nations (UN). UN General Assembly declares access to clean and healthy environment a universal human right. Available at: https://news.un.org/en/story/2022/07/1123482. Accessed January 2, 2023.

Further Reading

Huang C, Vaneckova P, Wang X, et al. Constraints and barriers to public health adaptation to climate change: A review of the literature. American Journal of Preventive Medicine 2011; 40: 183–190.

This article provides an overview of the constraints and barriers to public health adaptation.

Mallen E, Joseph HA, McLaughlin M, et al. Overcoming barriers to successful climate and health adaptation practice: Notes from the field. International Journal of Environmental Research and Public Health 2022; 19: 7169. doi: 10.3390/ijerph19127169.

This paper explores the barriers to and enablers of several successful adaptation projects.

Ebi KL, Hess JJ, Isaksen TB. Using uncertain climate and development information in health adaptation planning. Current Environmental Health Reports 2016; 3: 99–105.

This paper asserts that to balance the potential bias of overestimating the extent to which climate change could alter the magnitude and pattern of health outcomes, decision-makers need projections of other drivers of health outcomes that are recognized determinants of some disease burdens.

Austin SE, Ford JD, Berrang-Ford L, et al. Enabling local public health adaptation to climate change. Social Science and Medicine 2019; 220: 236–244.

This paper examines how federal and regional governments can contribute to enabling and supporting public health adaptation to climate change at the local level in federal systems. The authors propose 10 specific measures that upper-level governments can take to build local public health authorities' capacity for adaptation.

Araos M, Austin SE, Berrang-Ford L, Ford JD. Public health adaptation to climate change in large cities: A global baseline. International Journal of Health Services 2016; 46: 53–78.

In this article, the authors develop and apply systematic methods to assess the state of public health adaptation in 401 urban areas globally with more than 1 million people, providing the first global baseline for urban public health adaptation.

16

Planning Healthy and Sustainable Built Environments

Jason Vargo

The *built environment* consists of human-made surroundings, ranging from buildings and parks to neighborhoods and entire cities, that are the setting for daily activities. When you woke up this morning, many decisions about your built environment were already made for you—a long time ago. These decisions influence, if not determine, your options for transportation, energy use, the type of home you live in, and even how long you are likely to live. A legacy of policies and the status quo create preferential access—or discriminatory exclusion—to live and work in certain built environments. But decisions about your built environment and policies limiting or granting access to them are not permanent. Especially at the local and regional scales, working to address and live with climate change involves smart decisions to improve our built environments, construct new ones, and expand access to healthy and resilient built environments for all.

Consider how you traveled to school, work, or elsewhere today. Probably part of your travel was by walking, even just to and from a car. Driving a car is, by far, the most carbon-intensive part—if not also the most dangerous—of our routine trips; driving less reduces greenhouse gas (GHG) emissions. Active transportation—getting around under our own power, such as by walking or bicycling—decreases GHG emissions and promotes exercise. Increasing physical activity helps to prevent many diseases and improves health.[1,2] (See Box 13-1 in Chapter 13.)

If you were motivated to routinely travel only by walking or bicycling, could you? Many people live in places where driving is the only option to get around because the built environment makes healthy behaviors like walking and bicycling impractical, impossible, or unsafe. In addition, designs and policies that promote highway use and convenient free parking also promote energy-intensive transportation, also known as *auto-dependence*.

It may seem counterintuitive to describe driving as "energy-intensive" because it can feel so effortless. But consider that making a 5-mile trip in a car that gets 30 miles per gallon (mpg) uses about 5,250 calories of fuel (or approximately

Jason Vargo, *Planning Healthy and Sustainable Built Environments* In: *Climate Change and Public Health*.
Edited by: Barry S. Levy and Jonathan A. Patz, Oxford University Press. © Oxford University Press 2024.
DOI: 10.1093/oso/9780197683293.003.0016

1,500 for an electric vehicle). This same trip with a bicycle, ridden at a leisurely pace, would require only about 220 calories. Although the trip by car may save 15 minutes, it is more than 20 times as energy-intensive.

The design of a built environment can lock us into using "a gallon of gas to get a gallon of milk,"[3] or it can offer choices that are better for one's health and the environment. Design of the built environment is one of the most effective, efficient, lasting, and equitable approaches for promoting healthy behaviors. In this age of apps and other technologies, it can be difficult to remember that simple designs can solve many problems better than any app. Design does not need to be turned on or charged up. There is no plug-in to download. No server farms support it. And there is often no learning curve or barrier to its use. Since designs are always "on" and people are always interacting with them, they must be carefully developed, appropriate, and versatile.

The design of our homes, neighborhoods, and communities—especially in urban and suburban areas—has tremendous implications for energy use, land use, environmental degradation, and our health. The amount of construction in the next 40 years will equal the amount from ancient times to the present.[4] Will built environments continue to be designed and constructed in ways that exacerbate climate change and encourage unhealthy behaviors, or will they be designed and constructed in ways that mitigate climate change and promote health?

CURRENT STATUS OF POLICIES AND ACTIONS

Transport in the Built Environment

Strategies that promote short trips by walking, cycling, and walking to transit can help to mitigate climate change, especially if these trips replace car trips and decrease fuel consumption. Transportation accounts for 23% of total GHG emissions, of which 70% come from road vehicles.[5] These percentages are increasing, especially in middle-income countries, where increasing urban sprawl (expansion of lower-density development away from central urban areas into previously remote and rural areas) requires more people to use personal vehicles for transportation.[6]

Three features of the built environment are key to reducing car travel and increasing active transportation: population density, connectivity, and human-scale design of facilities for active transportation.[7]

Population density relates to the number of people living in a specific area. *Land-use intensity* describes the number and variety of uses (such as residential or commercial) in a specific area; it is related to the number and types of activities present in an area. Proximity of various types of land use helps to create environments that promote walking, bicycling, and access to transit.

Connectivity, the second factor that facilitates active transportation, describes the number of route options available for making a single trip. Connectivity can be improved by increasing the density of road intersections, sometimes by creating special routes and paths for nonmotorized travel. One way to think about connectivity is to examine the area one could cover by traveling a specific distance in any direction (Figure 16-1). Increased connectivity of a neighborhood dramatically increases the number of locations one can visit and the number of trips one can complete during a 5-minute walk. Cul-de-sacs, which

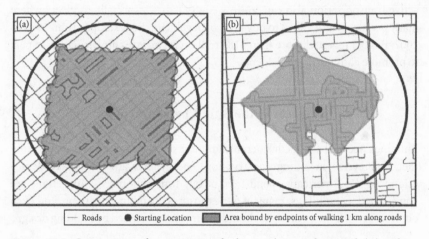

| — Roads | ● Starting Location | ▩ Area bound by endpoints of walking 1 km along roads |

Figure 16-1 Comparison of connectivity of a dense urban road network (A) and a lower-density suburban road network (B). Both A and B are schematic drawings of neighborhoods surrounded by a circle with a 1-kilometer radius. However, that is where the similarity between the neighborhoods ends. Neighborhood A has a dense urban road network, whereas Neighborhood B has a lower-density road network. Therefore, a person living at the center of the circle in Neighborhood A could accomplish more daily tasks by walking 1 kilometer (and have more options of paths to take in accomplishing these tasks) than a person living at the center of the circle in Neighborhood B. The two panels illustrate the concept of connectivity, which is the ratio of (a) the shaded area in each of the panels to (b) the area within the circle in each of the panels. This ratio is much higher in Panel A than in Panel B, illustrating that Neighborhood A has much higher connectivity than Neighborhood B. Assessing a location's connectivity using only the circle may misrepresent the number of destinations that one can easily walk to. (*Source*: Oliver LN, Schuurman N, Hall AW. Comparing circular and network buffers to examine the influence of land use on walking for leisure and errands. International Journal of Health Geographics 2007; 6: 41.)

are typically present in U.S. suburbs, severely limit connectivity and require residents to travel further to reach roads that connect to their daily destinations.

Human-scale design of facilities influences how people feel about modes of active transportation. Sidewalks and bicycle paths improve safety but also encourage people to choose active transportation. Street design can improve pedestrians' and bicyclists' perceptions of safety by reducing speeds on roads, narrowing streets at crossings, and creating barriers between automobiles and lanes for nonmotorized travel. Some cities, such as Copenhagen, illustrate best practices for standardizing active transportation infrastructure and safety.[8]

Encouraging active transportation, which uses less fuel and emits less carbon dioxide, helps to slow climate change. At the same time, active transportation can improve physical and mental health. Local actions to encourage and increase active transportation also can address equity issues, especially in places where many people cannot afford to buy their own vehicles. For example, from 1999 to 2002, Bogotá invested more than $180 million on bicycle infrastructure, in part to promote equity among residents.[9]

Spaces in the Built Environment

Buildings account for a considerable amount of GHG emissions, considering their full cycle—including construction, energy use, and ultimately demolition.[10,11] We Americans spend 90% of our time indoors.[12] Our actions are constrained by buildings, many of our exposures are defined by them, and even our interactions with other people and nature are influenced by building design.

Homes

The Inflation Reduction Act, which was passed in 2022, has provided homeowners with unprecedented incentives to add solar power and improve the energy efficiency of home heating and cooling systems. Nevertheless, decisions about where to place homes and about density of neighborhoods have consequences for climate change and public health. In addition, the ability to be very selective in where to reside is a privilege of those with relatively substantial resources. Separated and segregated suburban patterns of settlement in which Americans live affect the use of natural landscapes, which are important carbon sinks. The type of residential development in a region profoundly influences ecological preservation.[13]

The suburban style of development in the United States is characterized by larger separated homes on large individual lots, often built on greenfields (undeveloped land used for agriculture or landscape design, or left to evolve naturally).[14] Larger homes that do not share walls with other homes require more energy to heat and cool.[15] Other characteristics of homes, such as appropriate orientation and arrangement of buildings could decrease energy use by one-third throughout the United States.[16]

Homes built on large lots are often set further back from roads, therefore requiring longer driveways and facilitating larger lawns. Yards maintained with gas mowers and carbon-intensive fertilizers require more use of fossil fuels that contribute to climate change and pollute air and water—creating substantial environmental impacts. For example, residential fertilizers are generally applied at up to triple the concentrations used in large-scale farming.[17] Their use is built into the style of development.

Work

In workplaces, energy is typically used to power lights, computers, and other machines. However, workplaces offer some of the largest opportunities for conservation and capitalizing on economies of scale. For example, the tremendous size of many workplaces and their parent companies enables them, on a large scale, to recycle and purchase materials from nearby suppliers. Since many workplaces, such as offices and stores, turn over and renovate frequently, they provide many opportunities to improve internal components and systems to reduce energy use and improve worker productivity—such as with passive heating, cooling, and lighting strategies. Natural light, fresh air, and views of natural settings can also improve work and school settings and thereby help workers and students focus[18,19] and reduce stress.[20]

Businesses contribute to ancillary energy use and GHG emissions as employees drive to work and customers travel to stores to make purchases. The location of businesses affects fuel usage. *Transit-oriented development* (TOD)—the construction of mixed land uses centered around transit stops—is an example of how new construction of homes and workplaces co-located with transit access can help reduce car travel (Chapter 13).

"Third Places" and Public Spaces

Many of us seek *third places* in which to spend time, engage with others, and participate in our communities. These places traditionally include places of worship, community centers, hair salons, and coffee shops and bars. The most important third places are owned by families and operated by people who know many people in their neighborhoods.[21] Social interactions facilitated by third places foster community[22,23] and encourage creativity.[24] The value of these connections and interactions is termed *social capital*. Places with more social capital have better education, health, equality, and safety. The structure of communities profoundly influences their amount of social capital.[25] Strong social connections, trust among community members, and trust in public institutions are important in developing resilience to climate change, which often requires collective action.[26]

Open green spaces, such as public parks, offer excellent opportunities to create versatile third places while addressing climate change and improving health. Parks often offer natural landscapes and public access, improving environmental quality and equitably offering space for physical activity and social interaction.[27-29] Community centers, sometimes in or adjacent to these spaces, often provide health services and may serve as places of shelter in emergencies, especially for marginalized populations.[30] Because they set aside land from potential construction, parks can also guide surrounding development and influence population density. As landscapes, parks help manage stormwater and act as carbon sinks—and they can be managed with energy-efficient and nutrient-efficient practices. Parks offer activities for people of all ages and cultures.[31,32] They enable children to run and play on playgrounds, teenagers and adults to participate in sports, and older people to take walks or meet friends. Exposure to nature is another benefit of parks, which offer the "win-win-win" of environmental, personal, and social benefits. Small *pocket parks* can offer access for nearby residents. Larger regional parks can offer more opportunities for biodiversity and contact with nature. Pocket parks can be more easily incorporated with development plans, while regional parks often need to be protected from development.

COVID-19

The COVID-19 pandemic profoundly impacted our relationships with all of these spaces. It increased the desirability of larger homes in suburban settings. Families with the ability to work from home sought more space for offices and yards for recreation. Many office spaces are now less densely utilized and some are no longer necessary. Since the pandemic, transit ridership has remained low, more commuters continue to work from home and use private vehicles, and serious injuries and fatalities from crashes have increased. At the same time, some cities began to close streets to automobiles so that residents could spend more time outside, socialize safely, and exercise more. Parking spaces were converted for outdoor dining and greenspace. And people used public spaces, such as parks, beaches, and trails, more frequently.

A VISION FOR THE FUTURE

Cities and Climate Change

Cities are focal points for addressing global climate change and related health impacts. As the foci of economic activity and dense populations, cities produce most GHG emissions while also being the locations where many of the health impacts of climate change are experienced. Cities, which are the sites of the most severe health impacts of heat waves, can be leaders in reducing GHG emissions.[33-35]

Cities benefit from actions to mitigate climate change. The largest opportunities are in fast-growing cities, where low-carbon adaptive measures can be incorporated into construction. Ultimately, co-benefits to health, energy savings, and security will often exceed the up-front costs of investments in mitigating climate change.

Human-Scale Design

In many places, the design and construction of built environments cater to automobiles rather than people. Consider how the size of storefront signs corresponds to the width of adjacent roads and the speed of the traffic. Signs in older, traditional business districts, where there is regular pedestrian traffic, are much smaller than those set back from the road and read by drivers traveling at 55 miles per hour. Human-centered design accommodates life for people—not cars. The resultant built environment is usually of higher quality and more enjoyable, and it yields lower GHG emissions.

New York City has produced comprehensive design guidelines for increasing physical activity that include recommendations on constructing buildings, integrating physical activity into urban public spaces, improving streets, and promoting sustainable development.[36] Guidelines and standards for building design aim to reduce energy use and create co-benefits related to physical activity, stormwater management, energy efficiency, and urban heat. Recommendations for indoor spaces include designing attractive and accessible stairways to promote physical activity. Such guidelines and standards focus on design at a human scale.

The details of urban design are crucial to ensuring that changes in the urban environment promote health co-benefits. Design of physical environments must consider density, proximity, and connectivity and make places safe and attractive to encourage social and commercial activities. A high-quality physical environment that provides human-centered spaces for people, away from noise, dirt, and vehicular traffic, can dramatically increase optional outdoor activities and, in turn, invite more social interaction.[37] Details are important. For example, sidewalks should be designed to facilitate active transportation and pedestrian safety.[38]

There are many opportunities to integrate urban infrastructure with human-scale design. New York City's guidelines promote high-performance (low-waste, energy-efficient) infrastructure that also enhance the experience of pedestrians while managing stormwater and reducing the urban heat island effect. Figure 16-2 provides a cross-section of a street with several best management practices. Nearly every design strategy potentially has multiple environmental, health, and economic benefits.[38]

Efficiency of Urban Structure

Optimal organization of people into cities can reduce per-capita energy consumption. Infrastructure investments, such as those for sewers and roads, are considerable expenses for municipalities. Density around these investments makes them operationally efficient and can increase tax revenues to make them affordable. Compact urban centers can help decrease per-capita GHG emissions. Figure 16-3 shows the average energy use among three types of homes of increasing density—single-family detached, single-family attached, and multifamily. Suburban areas typically have larger homes and more cars. They require people to drive longer distances than residents of compact developments and urban centers. Moving a household from a less-dense suburban area to a more-dense urban area reduces energy use—more than technological improvements can achieve. For the average household in the United States, moving from a traditional suburb to a green urban location could reduce energy use by 75%.[39] In the past several years, some major U.S. cities and California have passed laws allowing single-family zoning to be removed from local codes.[40,41]

While the effect of urban structure on energy varies from place to place, city residents generally use less energy and create lower GHG emissions than suburban residents.[42] Personal and household behaviors and actions have only a modest effect on these emissions. Within large metropolitan areas in the United States, household emissions are more closely associated with population density than they are with household size or income.[43]

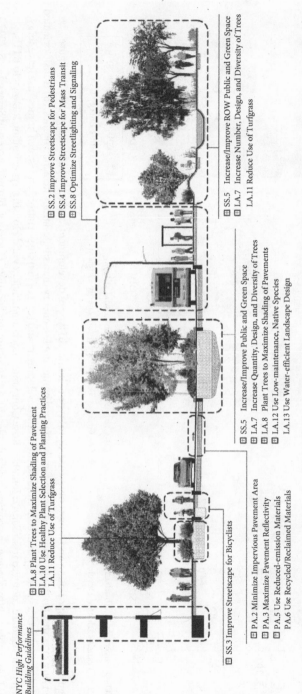

Figure 16-2 Selected New York City high-performance public infrastructure guidelines that have health co-benefits. These guidelines facilitate implementation of sustainable practices that can conserve energy, improve air and water quality, and make cities more beautiful and more livable. Many practices for streets and public right-of-way accomplish several of these goals while reducing costs. This figure is a cross-section demonstrating structures that promote both a sustainable urban environment and public health. For example, integrating vegetation into street design can reduce sewer overflows and decrease the urban heat island effect. (*Source:* New York City Department of Design and Construction. High Performance Infrastructure Guidelines, October 2005. Available at: http://www.nyc.gov/html/ddc/downloads/pdf/hpig.pdf. Accessed September 8, 2022.)

NYC High Performance Building Guidelines

LA.8 Plant Trees to Maximize Shading of Pavement
LA.10 Use Healthy Plant Selection and Planting Practices
LA.11 Reduce Use of Turfgrass

SS.2 Improve Streetscape for Pedestrians
SS.4 Improve Streetscape for Mass Transit
SS.8 Optimize Streetlighting and Signaling

SS.5 Increase/Improve ROW Public and Green Space
LA.7 Increase Number, Design, and Diversity of Trees
LA.11 Reduce Use of Turfgrass

SS.5 Increase/Improve Public and Green Space
LA.7 Increase Quantity, Design, and Diversity of Trees
LA.8 Plant Trees to Maximize Shading of Pavements
LA.12 Use Low-maintenance, Native Species
LA.13 Use Water-efficient Landscape Design

PA.2 Minimize Impervious Pavement Area
PA.3 Maximize Pavement Reflectivity
PA.5 Use Reduced-emission Materials
PA.6 Use Recycled/Reclaimed Materials

SS.3 Improve Streetscape for Bicyclists

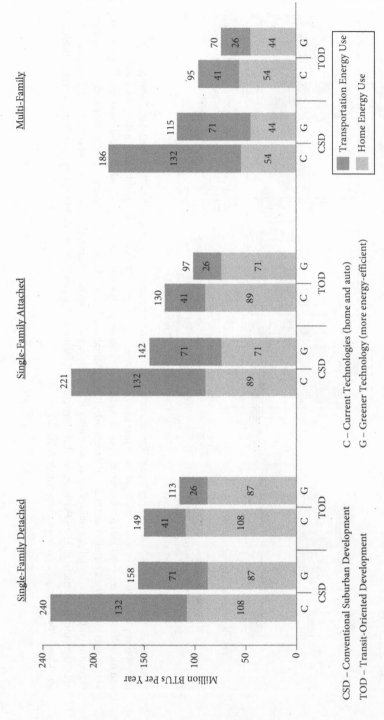

Figure 16-3 Transportation and home energy use, by (a) type of home, (b) type of development, and (c) type of energy use. The figure demonstrates that (a) multifamily and single-family attached homes use less energy than single-family detached homes; (b) transit-oriented development uses less energy than conventional suburban development; and (c) "green transportation" and "green homes" use less energy than their standard counterparts. (*Source*: Jonathan Rose Companies. Location Efficiency and Housing Type: Boiling It Down to BTUs, 2011. Available at: https://www.epa.gov/sites/default/files/2014-03/documents/location_efficiency_btu.pdf. Accessed September 8, 2022.)

Using denser, more diverse, and compact development as a means of mitigating climate change also offers greater health co-benefits than many technology-based solutions, including reducing fossil fuel use (and related GHG emissions) in transportation. National policies affecting fuel blends and efficiency of vehicles can reduce GHG emissions and improve air quality, but if people continue to drive long distances—even if they do so in more efficient vehicles—they will still be less physically active than if they lived in cities and walked to work. Encouraging dense, vibrant, transit-friendly community development can reduce vehicle emissions while providing additional health co-benefits. The magnitude of these benefits, especially those related to physical activity, are often much greater than the health benefits of reducing pollution alone.[44] In modeled analyses, active-transportation scenarios result in greater health co-benefits than other options, including extensive vehicle electrification.[45] A study found that, among various options considered, bicycling generated the greatest health benefit per ton of carbon reduction.[46] (See Chapter 13.)

POLICIES AND ACTIONS FOR THE FUTURE

Guiding the construction of the built environment involves many stakeholders and levels of administration and influence. Creating and changing the built environment is influenced by various policies and approaches operating at different scales.

Laws, processes, and policies affect the construction of buildings—probably the most detailed level at which the built environment can be influenced. Building codes and architectural standards improve the safety and health of people inside cities and surrounding communities. Building codes that emphasize reduced energy use and encourage physical activity and nutrition act to mitigate climate change while providing health co-benefits, such as by encouraging stair use and providing space for employees to shower after an active commute.[47]

Voluntary green building standards, most notably Leadership in Energy & Environmental Design (LEED, a green building certification program that recognizes best-in-class building strategies and practices), have raised much awareness about energy use in construction and demolition of buildings and in their everyday operation.[48]

Updating existing buildings to the newest codes is accomplished through building permit processes. For example, as homes are updated, they are often required to update electrical systems. These updates may also include making improvements in energy efficiency. Berkeley, California, was the first city to pass legislation to prohibit natural-gas connections to newly constructed houses to ensure that, as housing availability increases, residents of new houses are not locked into a lifetime of fossil fuel use. In addition, funding from the State of California has helped to retrofit existing low-income houses with energy-efficient upgrades and solar energy systems.[49-51]

The traditional suburb divides a large piece of land into smaller parcels, with one use and low density. Changing zoning can improve the land use in suburbs, but site planning is necessary to correct the connectivity of streets and provision of greenspace. In 1938, the American Housing Authority (AHA) revised its subdivision guidelines, decreasing connectivity and reinforcing a hierarchical street network.[52] This change required more driving by residents and funneled traffic onto larger, faster-moving roadways. Many believe that the AHA adversely impacted local character.[53] This example highlights the importance of understanding the multiple levels of influence on construction patterns and urban development.

Zoning is a policy mechanism by which uses are assigned to areas. Zoning codes are local but apply to larger areas than single pieces of land. Zoning determines the densities permitted on pieces of land in a given area. Zoning codes that permit multiple uses are key to facilitating transit and active transportation. Zoning may also dictate the amount of land that may be built up and the amount to remain natural or vegetated. This can be important for managing stormwater as well as for reducing the urban heat island effect. Zoning that sets the minimum lot size high often permits increased lot width and larger setbacks of homes from roads, making homes further apart, reducing density, and creating longer trips. In 2019, Minneapolis became the first major city in the United States to abolish zoning that permits only single-family houses.[40]

At the municipal level, a *climate action plan* (CAP) describes the amounts of GHG emissions for the city, by sector and end use. For example, in the commercial sector, there may be GHG emissions separately included for construction of buildings, maintenance of buildings, and building operations. The CAP may set reduction goals for the city and suggest strategies, such as mandates, financial incentives, and other policies, for achieving these goals. In many cases, CAPs focus on reducing energy and related emissions; CAPs have given less attention to the impact of urban growth on local climate or potential health co-benefits. Ideally, CAPs are integrated with other planning and activities, such as for transportation, housing, land use, and construction.

Growth management policies are often used at a regional level to promote or restrict development. Rapid development in a region, especially at a low density, requires major investment in new infrastructure and fosters its inefficient use. Some regions use *capital improvement plans* (CIPs) to ensure that development does not outpace the ability of local governments to provide services. CIPs stress efficient use of public funds for schools, parks, sewers, and other uses.

Another growth management tool is the *urban growth boundary* (UGB). In Portland, Oregon, for several years, a UGB implemented a legislated border around the metropolitan area, which limited the supply of developable land to preserve surrounding forest and farmland.[54] While the UGB in Portland encouraged density, it had the unintended consequence of inflating land prices, therefore making purchase of land unaffordable for many more people.

Transferrable development rights represent conservation tools for directing people and development to urban/town centers while preserving surrounding lands. A market for transferable development rights can be established in a region, allowing property owners in bucolic fringe areas to sell their rights to property owners in the dense urban/town centers. This decreases development far from these centers and increases the density at which downtown developers can build, thereby helping to preserve greenspace outside of cities and creating more vibrant urban areas where transit and other infrastructure can be developed and used efficiently.

CASE STUDIES

The Climate Action Plan in Portland, Oregon

Portland has been praised for its progressive planning, promotion of bicycling, and protection of environmental quality. In 1993, it drafted and adopted the first city-level CAP in the United States. Since then, it has been the only city to reduce per-capita and overall GHG emissions. Portland's CAP has emphasized that actions taken to mitigate climate change are aligned with many other municipal interests and that the many benefits of climate action "promote economic and environmental goals and enhance livability."[55]

The built environment of Portland contributed to the early success of its CAP. Its UGB served as a growth management tool, promoting concentration of people in the urban core, which helped support investments in increased transit use. By 1997, the use of transit by people in the region had increased by 30%, and their commutes by car to the central business district had decreased by 15%.[56] By 2000, Portland's per-capita GHG emissions were 7% below 1990 levels.[57]

In 2001, Portland, the county seat of Multnomah County, extended its CAP from the city limits to the entire county, acknowledging the importance of acting at multiple scales to address climate change. In 2013, the county's carbon emissions were 14% below 1990 levels, representing a 35% reduction per person.[58] The CAP also engaged community partners, focusing on new buildings being constructed throughout the city—rather than focusing only on municipal buildings. It helped to create the Green Investment Fund, which distributed more than $800,000 to new housing and commercial projects.[59]

Between 2009 and 2010, Portland's per-capita energy use decreased by 3%. From 2001 to 2010, bicycling in Portland grew five-fold.[60] Portland has one of the highest rates (6.5%) of bicycle commuting among large U.S. cities—about six times the national average.[60] And Multnomah County residents report obesity and physical inactivity rates 2–7% below national rates.[61] (The COVID-19 pandemic substantially reduced all types of commuting, including by bicycle.[62])

Portland's CAP has been both visionary and ambitious. It initially established a goal of reducing carbon emissions 20% below 1990 levels by 2010, and it recently revised this goal—to 40% below 1990 levels by 2030 and 80% by 2050.[58] In 2010, Portland reported that it had reduced emissions 6% below the 1990 level; during the same period, emissions for the United States *increased* about 13%.[63] In 2018, Portland's total emissions were almost 20% below 1990 levels, even while its population and its number of jobs had grown more than 20% since 1990.[64]

Despite rapid population growth during recent decades, Portland has reduced GHG emissions by dramatically reducing per-capita emissions. Its plan explicitly establishes objectives concerning urban design and transportation, aiming to "create vibrant neighborhoods where 80% of residents can easily walk or bicycle to meet all basic daily non-work needs and have safe pedestrian or bicycle access to transit." The plan also aims to decrease per-capita vehicle miles traveled by 30%, compared to 2008 levels.[58] Without focusing on the built environment by investing in transit and active transportation and by promoting green buildings, Portland cannot achieve these objectives.

California's SB 1000: Embedding Environmental Justice in Planning

In 2016, California passed a law entitled "Environmental Justice in Local Land Use Planning" (SB 1000).[65] A central goal of the law is to create healthier communities by protecting them from polluting land uses and prioritizing the needs of disadvantaged communities. The law is designed to ensure that the official long-term plans for low-income communities and communities of color (also known as Comprehensive Plans or General Plans) actively seek to correct historic inequities. Because these communities often experience the health impacts of climate change earlier and more extremely than other communities, this is an important strategy for sustainability. The law requires cities and counties to integrate environmental justice throughout their General Plans, helping to facilitate transparency and public engagement in local government. Plans must reduce unique and compounded health risks in disadvantaged communities, such as by improving air quality. The state has created a tool to facilitate planning, one that identifies disadvantaged communities and documents baseline conditions.

It is difficult or impossible for the environmental justice elements of plans to succeed if they do not prioritize improvements and programs that address the needs of disadvantaged communities. Ensuring that the goals identified in the environmental justice elements of plans align with the needs and priorities of disadvantaged communities requires close partnership between government and these communities. Therefore, SB 1000 also places high value on promoting community engagement in the public decision-making process. In the best examples of such processes, community advisory boards will be formed or a community representative will serve as an environmental justice consultant on the general plan advisory committee. Through early and ongoing proactive engagement, communities will have input into the goals and plans that affect the design and inclusivity of built environments. Community engagement also builds trust. Further information on SB 1000 and its implementation is available.[66] (See also Chapter 20.)

The Mariposa Healthy Living Initiative in Denver, Colorado

In the early 2000s, the development of a new TOD infrastructure was planned for the La Alma/Lincoln Park neighborhood near downtown Denver, one of the city's oldest neighborhoods. As part the TOD, the Denver Housing Authority (DHA) redeveloped South Lincoln Homes, a public housing project adjacent to the proposed transit station. The historical and cultural importance of the neighborhood as well as substantial health and wealth inequalities in the area led the DHA to consider a more comprehensive approach to improve the overall quality of life for residents there. In 2009, the DHA began the Mariposa Healthy Living Initiative, in which it used physical, mental, and community health to measure success.[67]

As a first step, a Health Impact Assessment (HIA) was done to determine the baseline status of the community. The HIA identified several needs and opportunities to improve health. The community had high levels of obesity (55%) and inactivity (72%). At the same time, most residents (65%) did not own a car and most (54%) relied on public transportation. Safety risks, limited access to healthcare, and inadequate opportunities for youth were also important concerns. Community engagement was important for confirming the findings of the HIA, identifying new issues, and developing solutions. Activities such as stakeholder interviews, a pedestrian audit, youth visioning sessions, and outreach to targeted populations helped to identify other local concerns and generated community-driven design elements, including a central plaza, parks and connective greenspace, and community gardens.[67]

The HIA and community input helped shape the overall direction and specific elements of a redevelopment plan. Priority goals include increasing physical activity, improving pedestrian and bicycle opportunities, increasing mobility and traffic safety, improving access to healthy foods, increasing safety and security, and improving access to healthcare. There have since been specific projects and programs to help accomplish these goals, including[67]

- Increasing mobility, with neighborhood traffic calming, a new bicycle lane, access to a bicycle share system, and a free shuttle bus to Denver Health Medical Center
- Creating a safe, attractive public realm, with new street tree plantings, building murals and other public art that celebrate the neighborhood's cultural diversity, public plazas, and a community garden
- Programs to support health, including walking groups, health classes, and job training opportunities.

To help evaluate progress, the Mariposa Healthy Living Tool was created.[67] It assesses health impacts of the built and social environments in six dimensions: healthy housing, sustainable and safe transportation, environmental stewardship, social cohesion, public infrastructure, and a healthy economy. As examples, social cohesion indicators include the proportion of the population living within a half-mile of gathering spaces, and environmental stewardship indicators include per-capita miles driven daily.

Results show positive trends for many indicators. Average transit commute time has decreased for neighborhood residents—lower than the city average. Residents have increased access to community space and increased access to foods from community gardens and a community facility. The percentage of residents with access to nearby open space has increased from 26% to 32%. And the crime rate has decreased more than 30%.[67]

In contrast to the previous two case studies, the Mariposa Healthy Living Initiative did not begin with an interest in climate change. This case study demonstrates that a project with the goal of improving communities and reducing inequalities can incorporate many strategies to address climate change and promote health. As the initiative reduces poverty, it also increases its capacity to adapt to climate change and improve community resilience.

SUMMARY POINTS

- Neighborhoods, communities, towns, and cities have important roles to play in addressing climate change.

- Strategies to reduce GHG emissions, conserve natural resources, and build climate resilience equitably offer significant health co-benefits.
- Addressing the design of the built environment at the scale at which people regularly experience it can protect the environment and promote healthy lifestyles.
- Built environments that reduce GHG emissions can help make green and healthy choices routine.
- The three case studies in this chapter demonstrate how communities, focusing on the built environment, can address climate change and improve health for all.

References

1. U.S. Department of Health and Human Services. Physical Activity and Health: A Report of the Surgeon General. Atlanta: U.S. Department of Health and Human Services, 1996. Available at: www.cdc.gov/nccdphp/sgr/pdf/sgrfull.pdf. Accessed September 8, 2022.
2. Powell KE, Blair SN. The public health burdens of sedentary living habits: Theoretical but realistic estimates. Medicine & Science in Sports & Exercise 1994; 26: 851–856.
3. Cabanatuan M. Transportation boosts costs of living in suburbs. SFGate, February 29, 2012. Available at: http://www.sfgate.com/bayarea/article/Transportation-boosts-cost-of-living-in-suburbs-3368916.php#src=fb. Accessed September 8, 2022.
4. United Nations Department of Economic and Social Affairs Population Division. Population distribution, urbanization, internal migration and development: An international perspective. January 1, 2011. Available at: https://www.un.org/development/desa/pd/content/population-distribution-urbanization-internal-migration-and-development-international. Accessed September 8, 2022.
5. Jaramillo P, Riberio SK, Newman P, et al. Transport. In PR Shukla PR, J Skea, R Slade, et al. (eds.). Climate Change 2022: Mitigation of Climate Change. Contribution of Working Group III to the Sixth Assessment Report of the Intergovernmental Panel on Climate Change. Cambridge: Cambridge University Press, 2022, p. 98. Available at: https://www.ipcc.ch/report/ar6/wg3/. Accessed September 8, 2022.
6. Short J, Van Dender K, Crist P. Transport policy and climate change. In: S Daniel, JS Cannon (eds.). Reducing Climate Impacts in the Transportation Sector. Dordrecht: Springer, 2009, pp. 35–48.
7. Frank LD. Health and Community Design: The Impact of the Built Environment on Physical Activity. Washington, DC: Island Press, 2003.
8. National Association of City Transportation Officials. Full speed ahead: NACTO releases second edition of Urban Bikeway Design Guide. September 6, 2012. Available at: http://nacto.org/wp-content/uploads/2012/09/NACTOUrbanBikewayDesignGuide_PressRelease_SecondEditon_090612.pdf. Accessed September 8, 2022.
9. Cervero R. Progressive transport and the poor: Bogota's bold steps forward. ACCESS Magazine, 2005. Available at: http://escholarship.org/uc/item/3mj7r62w. Accessed September 8, 2022.

10. Bulkeley H, Betsill MM. Cities and Climate Change: Urban Sustainability and the Global Environment Governance. London: Routledge, 2004.

11. Lucon O, Ürge-Vorsatz D. Buildings. In O Edenhofer, R Pichs-Madruga, Y Sokona, et al. (eds.). Contribution of Working Group III to the Fifth Assessment Report of the Intergovernmental Panel on Climate Change. Cambridge: Cambridge University Press, 2014, p. 22.

12. United States Environmental Protection Agency. Report to Congress on Indoor Air Quality: Volume 2. Washington, DC: U.S. EPA, 1989.

13. Vargo J, Habeeb D, Stone B. The importance of land cover change across urban–rural typologies for climate modeling. Journal of Environmental Management 2013; 114: 243–252.

14. Randolph J. Environmental Land Use Planning and Management. Washington, DC: Island Press, 2004.

15. Jonathan Rose Companies. Location efficiency and housing type: Boiling it down to BTUs. 2011. Available at: https://www.epa.gov/sites/default/files/2014-03/docume nts/location_efficiency_btu.pdf. Accessed September 8, 2022.

16. Thomas R. Sustainable Urban Design: An Environmental Approach. London: Spon Press, 2003.

17. Robbins P. Lawn People: How Grasses, Weeds, and Chemicals Make Us Who We Are. Philadelphia: Temple University Press, 2007.

18. Taylor AF, Kuo FE, Sullivan WC. Coping with ADD: The surprising connection to green play settings. Environment and Behavior 2001; 33: 54–77.

19. Wells NM. At home with nature: Effects of "greenness" on children's cognitive functioning. Environment and Behavior 2000; 32: 775–795.

20. Ulrich RS, Simons RF, Losito BD, et al. Stress recovery during exposure to natural and urban environments. Journal of Environmental Psychology 1991; 11: 201–230.

21. Oldenburg R. The Great Good Place: Cafés, Coffee Shops, Bookstores, Bars, Hair Salons, and Other Hangouts at the Heart of a Community. New York: Marlowe, 1999.

22. Burls A. People and green spaces: Promoting public health and mental well-being through ecotherapy. Journal of Public Mental Health 2007; 6: 24–39.

23. Health Council of the Netherlands and Dutch Advisory Council for Research on Spatial Planning. Nature and Health: The Influence of Nature on Social, Psychological, and Physical Well-Being. The Hague: Health Council of the Netherlands and RMNO, 2004.

24. Kaplan R. The Experience of Nature: A Psychological Perspective. New York: Cambridge University Press, 1989.

25. Putnam RD. Bowling Alone: The Collapse and Revival of American Community. New York: Simon & Schuster, 2001.

26. Carmen E, Fazey I, Ross H, et al. Building community resilience in a context of climate change: The role of social capital. Ambio 2022: 51; 1371–1387.

27. Ravenscroft N, Markwell S. Ethnicity and the integration and exclusion of young people through urban park and recreation provision. Managing Leisure 2000; 5: 135–150.

28. Kweon B-S, Sullivan WC, Wiley AR. Green common spaces and the social integration of inner-city older adults. Environment and Behavior 1998; 30: 832–858.

29. Ferris J, Norman C, Sempik J. People, land, and sustainability: Community gardens and the social dimension of sustainable development. Social Policy and Administration 2001; 35: 559–568.

30. National LGBT Health Education Center. Emergency preparedness and lesbian, gay, bisexual & transgender (LGBT) people: What health centers need to know. Available at: https://www.lgbtqiahealtheducation.org/wp-content/uploads/Emergency-Prepa redness-for-LGBT-People-Final.pdf. Accessed September 8, 2022.
31. Gobster PH. Urban parks as green walls or green magnets? Interracial relations in neighborhood boundary parks. Landscape and Urban Planning 1998; 41: 43–55.
32. Lubben JE. Assessing social networks among elderly populations. Family & Community Health 1988; 11: 42–52.
33. Bloomberg MR, Aggarwala RT. Think locally, act globally. American Journal of Preventive Medicine 2008; 35: 414–423.
34. Younger M, Morrow-Almeida HR, Vindigni SM, Dannenberg AL. The built environment, climate change, and health. American Journal of Preventive Medicine 2008; 35: 517–526.
35. Harlan SL, Ruddell DM. Climate change and health in cities: Impacts of heat and air pollution and potential co-benefits from mitigation and adaptation. Current Opinion in Environmental Sustainability 2011; 3: 126–134.
36. Center for Active Design. Active design: Shaping the sidewalk experience. 2013. Available at: https://nacto.org/docs/usdg/active_design_shaping_the_sidewalk_ex perience_nycdot.pdf. Accessed September 8, 2022.
37. Gehl J. Cities for People. Washington, DC: Island Press, 2010.
38. New York City Department of Design and Construction. High performance infrastructure guidelines. October 2005. Available at: www.nyc.gov/html/ddc/downloads/ pdf/hpig.pdf. Accessed September 8, 2022.
39. Calthorpe P. Urbanism in the Age of Climate Change. Washington, DC: Island Press, 2010.
40. Trickey E. How Minneapolis freed itself from the stranglehold of single-family homes. POLITICO Magazine, July 11, 2019. Available at: https://www.politico.com/ magazine/story/2019/07/11/housing-crisis-single-family-homes-policy-227265/. Accessed September 8, 2022.
41. SB 9: The California HOME Act. Available at: https://focus.senate.ca.gov/sb9. Accessed September 8, 2022.
42. Jones C, Kammen DM. Spatial distribution of U.S. household carbon footprints reveals suburbanization undermines greenhouse gas benefits of urban population density. Environmental Science & Technology 2014; 48: 895–902.
43. Jones CM, Kammen DM. Quantifying carbon footprint reduction opportunities for U.S. households and communities. Environmental Science & Technology 2011; 45: 4088–4095.
44. Woodcock J, Givoni M, Morgan AS. Health impact modelling of active travel visions for England and Wales using an Integrated Transport and Health Impact Modelling Tool (ITHIM). PLoS ONE 2013; 8: e51462.
45. Maizlish N, Rudolph L, Jiang C. Health benefits of strategies for carbon mitigation in US transportation, 2017–2050. American Journal of Public Health 2022; 112: 426–433.
46. Maizlish N, Linesch NJ, Woodcock J. Health and greenhouse gas mitigation benefits of ambitious expansion of cycling, walking, and transit in California. Journal of Transport & Health 2017; 6: 490–500.

47. Kahn EB, Ramsey LT, Heath GW, Howze EH. Increasing physical activity: A report on recommendations of the Task Force on Community Preventive Services. Morbidity and Mortality Weekly Report 2001; 50: 1–16.
48. Owen D. Green Metropolis: Why Living Smaller, Living Closer, and Driving Less Are the Keys to Sustainability. New York: Riverhead Books, 2009.
49. City of Berkeley, California. Berkeley Municipal Code, Ch. 12.80 – Prohibition of natural gas infrastructure in new buildings. Available at: https://berkeley.municipal.codes/BMC/12.80. Accessed September 8, 2022.
50. The Guardian. Berkeley became first US city to ban natural gas. Here's what that may mean for the future. Politico, July 23, 2019. Available at: https://www.politico.com/magazine/story/2019/07/11/housing-crisis-single-family-homes-policy-227265/. Accessed September 8, 2022.
51. California Department of Community Services & Development. Low-income weatherization program fact sheet. February 2022. Available at: https://www.csd.ca.gov/Shared Documents/LIWP-Fact-Sheet.pdf. Accessed September 8, 2022.
52. Scott M. American City Planning Since 1890: A History Commemorating the Fiftieth Anniversary of the American Institute of Planners. Berkeley, CA: University of California Press, 1971.
53. Gallagher L. The End of the Suburbs: Where the American Dream is Moving. New York: Portfolio Hardcover, 2013.
54. Urban growth boundary. February 2014. Available at: www.oregonmetro.gov/urban-growth-boundary. Accessed September 8, 2022.
55. City of Portland. Global warming reduction strategy. 1993. Available at: https://www.portland.gov/sites/default/files/2019-08/global-warming-reduction-strategy-nov-1993.pdf. Accessed September 8, 2022.
56. Healthy Klamath. Portland's local action plan to reduce carbon dioxide emissions. 2001. Available at: www.healthyklamath.org/index.php?controller=index&module=PromisePractice&action=view&pid=155. Accessed September 8, 2022.
57. Slavin MI. Sustainability in America's Cities: Creating the Green Metropolis. Washington: Island Press, 2011.
58. City of Portland and Multnomah County. Local strategies to address climate change: Climate action plan summary. June 2015. Available at: https://www.portland.gov/sites/default/files/2019-07/cap-summary-june30-2015_web.pdf. Accessed December 29, 2022.
59. Boswell MR, Greve AI, Seale TL. Local Climate Action Planning. Washington, DC: Island Press, 2012.
60. League of American Bicyclists. Bicycling and walking in the United States: 2018 benchmarking report. Available at: https://bikeleague.org/sites/default/files/Benchmarking_Report-Sept_03_2019_Web.pdf. Accessed December 27, 2022.
61. University of Wisconsin Population Health Institute. County health rankings & roadmaps: 2013 Rankings: Wisconsin. 2013. Available at: https://uwphi.pophealth.wisc.edu/wp-content/uploads/sites/316/2017/11/WCHR_2013_Rankings.pdf. Accessed September 8, 2022.
62. League of American Bicyclists. First federal data on pandemic-era bike commuting. Available at: https://data.bikeleague.org/first-federal-data-on-pandemic-era-bike-commuting/. Accessed December 28, 2022.
63. City of Portland Bureau of Planning and Sustainability. City of Portland and Multnomah County climate action plan 2009: Year two progress report, April 2012.

2012. Available at: https://www.portland.gov/sites/default/files/2019-08/cap_progr ess-rept2012_web.pdf. Accessed September 8, 2022.

64. Multnomah County Office of Sustainability & City of Portland's Bureau of Planning and Sustainability. Climate action plan final progress report 2020. 2020. Available at: https://www.multco.us/sustainability/news/climate-action-plan-final-progress-report-2020. Accessed December 29, 2022.

65. State of California Department of Justice. SB 1000 – Environmental justice in local land use planning. Available at: https://oag.ca.gov/environment/sb1000. Accessed September 8, 2022.

66. California Environmental Justice Alliance. SB 1000 Toolkit: Planning for healthy communities. Available at: https://caleja.org/sb-1000-toolkit/. Accessed September 8, 2022.

67. Mithūn EnviroHealth Consulting and Denver Housing Authority. The Mariposa Healthy Living Initiative, 2012. 2012. Available at: https://www.gih.org/wp-content/uploads/2019/04/Mariposa-Healthy-Living-Initiative-2012.pdf. Accessed September 8, 2022.

Further Reading

Calthorpe P. Urbanism in the Age of Climate Change. Washington, DC: Island Press, 2010.

This work describes what urbanism is—"qualities, not quantities; diversity, not size; intensity, not density; connectivity, not just location"—and how it connects to global climate change.

New York City Active Design Group. Active design guidelines. Available at: http://www.nyc.gov/html/ddc/html/design/active_design.shtml.

These guidelines provide excellent visual and evidence-based guidance on the most effective and appropriate design guidelines for indoor and outdoor urban spaces.

Dannenburg A, Frumkin H, Jackson R (eds.). Making Healthy Places: Designing and Building for Health, Well-Being, and Sustainability (2nd ed.). Washington, DC: Island Press, 2022.

This introductory text covers the many ways that the built environment affects health and strategies for improving the places where we live.

United Nations Human Settlements Programme (UN-Habitat). World Cities Report 2022: Envisaging the Future of Cities. Nairobi: UN-Habitat, 2022. Available at: https://unhabitat.org/sites/default/files/2022/06/wcr_2022.pdf.

This document describes the direction and vision for cities globally and how the interlinked mega-trends of urbanization and climate change present opportunities for synergistic actions. Chapters 5, 6, and 7 of the report discuss urban planning, climate change, and health equity.

17

Promoting Health Through Nature-Based Climate Solutions

Howard Frumkin, Brendan Shane, and Taj Schottland

Nature-based climate solutions are strategies that are inspired by, supported by, or copied from nature. They tend to be less expensive and more environmentally friendly than engineered solutions. For example, shade trees are a nature-based approach to cooling a building, while air conditioning is an engineered approach. A nature-based solution for flood protection might rely on intact river deltas, while an engineered approach might rely on seawalls.

Nature-based climate solutions provide co-benefits, including health co-benefits. For example, planting street trees to cool a neighborhood on hot days—an adaptation to climate change—can also provide shade, beautify the neighborhood, reduce residents' stress, promote physical activity, and help manage stormwater.

Green infrastructure is a closely related concept. Defined narrowly, it refers to particular stormwater management techniques or to parks and greenspace; defined broadly, it refers to a set of concepts and strategies in urban ecology. Green infrastructure and nature-based climate solutions overlap; bioswales to capture stormwater and tree canopy to cool a neighborhood are examples of both.

The following sections address nature-based solutions in cities and in rural areas, and the challenges of implementing these solutions.

NATURE-BASED CLIMATE SOLUTIONS IN CITIES

Worldwide, urban centers are growing rapidly, with profound implications for climate and health. In 2023, approximately 56% of the global population resided in urban areas, but by 2050, this percentage is expected to reach almost 70%.[1] In contrast, the global rural population is projected to peak before 2030 and then decline. Many urban areas are located in at-risk settings, such as on coastlines and along rivers. Cities are characterized by impervious surfaces such as roadways and parking lots, a paucity of vegetation, pollution sources, and other attributes that exacerbate the adverse health effects of climate change from

Howard Frumkin, Brendan Shane, and Taj Schottland, *Promoting Nature-Based Climate Solutions* In: *Climate Change and Public Health*. Edited by: Barry S. Levy and Jonathan A. Patz, Oxford University Press. © Oxford University Press 2024.
DOI: 10.1093/oso/9780197683293.003.0017

extreme heat, flooding, contaminated water, polluted air, and other hazards. Adequately functioning natural systems in urban areas can reduce these hazards, but often these systems are compromised.

There is a growing global movement to "re-green" cities by using nature-based solutions. For example, Toronto is implementing a $1.25 billion flood mitigation project to protect the port district by "rewilding" a river and creating new parks and wildlife habitat that can naturally absorb flood waters before they threaten critical infrastructure.[2] While such large-scale projects attract much attention, nature-based solutions at the neighborhood scale can be just as important. They take many forms, such as rain gardens (depressed areas that collect rain water from a roof, driveway, or street and allow it to soak into the ground) and bioswales (channels designed to concentrate and convey stormwater runoff while removing debris and pollution), urban forests, and river daylighting (uncovering previously covered or buried waterways to restore natural drainage and other ecological functions). (See Figure 17-1.)

In low- and middle-income countries (LMICs), nature-based climate solutions can be an essential strategy for equitably reducing risk and improving health. In LMICs, where cities are growing rapidly and the cost of engineered solutions can be prohibitive, nature-based climate solutions offer substantial economic and practical advantages. People living in informal settlements and other underresourced parts of LMIC cities are especially vulnerable to heat,[3] floods, [4] water contamination, [5] and other climate-related threats to health and

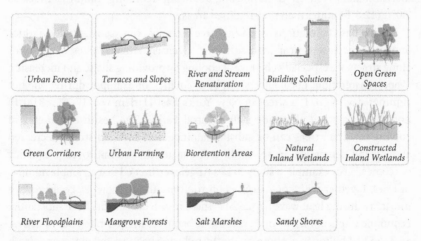

Figure 17-1 A wide variety of nature-based solutions provide not only resilience to climate-related challenges, such as extreme heat and flooding, but also health and social co-benefits. (*Source*: World Bank. A Catalogue of Nature-based Solutions for Urban Resilience. Washington, DC: World Bank, 2021.)

well-being.[6] For example, a study in two areas of Nepal—a hillside section of the capital city, Kathmandu, and a rural part of Gandaki province—found that people in informal settlements were vulnerable due to poor baseline health, poor-quality housing, poor water supplies, exposure to flooding, and insect infestations.[7]

The following section describes four common urban nature-based climate solutions and their health benefits:

- Keeping cities cool
- Managing stormwater to improve water quality and reduce flooding
- Building social cohesion, a foundation of community resilience
- Improving air quality and sequestering greenhouse gas (GHG) emissions.

Keeping Cities Cool

Cities are often hotter than surrounding areas, a phenomenon known as the *urban heat island effect* (Chapter 4). Contributing factors include (a) dark surfaces that absorb heat and re-radiate it at night, when cities ordinarily cool down; (b) local generation of heat by vehicles, machinery, and other sources; and (c) absence of vegetation that would provide cooling through evapotranspiration.[8]

Heat levels can vary greatly. among neighborhoods within a city because of variation in features such as paved surfaces and tree canopy, with resulting health disparities. In the United States, neighborhoods historically subjected to discriminatory restrictions in housing financing (redlining) can be as much as 7.0°C (12.6°F) hotter than non-redlined areas.[9]

Nature-based solutions such as tree canopy can substantially reduce temperatures at and near land surfaces.[10] Densely situated trees in parks or along streets are especially effective at cooling by providing shade and increasing evapotranspiration. Local effects can be dramatic; for example, during one hot summer day in 2021, a street in New York's East Harlem was 17.2°C (31.0°F) hotter than a street alongside leafy Central Park.[11] The health co-benefits of cooling by urban trees can be substantial, including reduced stress, decreased anxiety and depression, better cardiovascular health, and improved pregnancy outcomes.[12] For example, a study in Los Angeles estimated that an ample number of trees, together with cool surfaces on rooftops and pavements, could reduce mortality during heat waves by 25%.[13] These health co-benefits have substantial economic implications; by one estimate, urban trees in the United States deliver $5.3 to $12.1 billion in economic value by reducing heat-related illness and death as well as electricity consumption.[14]

But urban tree cover has been declining globally. An estimated 40,000 hectares (about 99,000 acres) of urban tree cover were lost to impervious surface annually between 2012 and 2017.[15] However, many cities are embarking on ambitious tree-planting projects through campaigns such as the Trees in Cities Challenge, a United Nations-sponsored global program in which mayors commit to making their cities greener, more sustainable, and more resilient.

Vegetation on building roofs and walls (sometimes called *green roofs* and *green facades*) can also cool buildings and nearby neighborhoods (Figure 17-2).[16] A study in Germany found that building exteriors covered with climbing plants were up to 15.5°C (27.9°F) cooler than bare walls, making interior walls almost 2.0°C (3.6°F) cooler.[17] Nearby bodies of water can also cool neighborhoods— another nature-based solution for extreme heat. But as water absorbs heat during the day, it can emit heat at night; therefore, the overall cooling effect may be minimal.

Managing Stormwater in Cities

When rainfall exceeds the capacity of stormwater systems, flooding can threaten human life and health and damage critical infrastructure, including

Figure 17-2 Vegetated surfaces on building walls and roofs, as shown here, keep the interiors of buildings and their neighborhoods cooler during hot weather. Buildings with vegetated surfaces can also be sites for urban agriculture, providing food for their residents and adding aesthetic appeal and charm to neighborhoods. (Copyright iStock/k_samurkas.)

hospitals and emergency services. Even modest rainfall events can lead to significant stormwater runoff that carries pollutants into rivers. Sewer systems that carry stormwater and sewage in the same pipes can be overwhelmed by heavy rainfall, sending a mixture of raw sewage and polluted stormwater runoff into streets and waterways—and causing waterborne disease (Chapter 7).

Globally, many cities are utilizing *green stormwater infrastructure* (GSI) to reduce flooding and improve water quality in a cost-effective manner.[18] Different types of GSI are suited to different landscape settings and different primary goals, but each is designed to capture or slow stormwater runoff, thereby preventing the rainwater from overwhelming sewer systems and carrying pollutants into waterways. Examples of GSI include bioswales, renaturalized waterways to channel stormwater, bioretention facilities (either natural or constructed) to hold and treat stormwater, green roofs, wetlands (either constructed or natural), and outfall retrofits (enabling reduced flow, stormwater retention, sedimentation, and filtration at points where stormwater pipes empty into waterways by restoring streambeds, enhancing wetlands, constructing obstacles to flow, and other nature-based interventions).

GSI delivers co-benefits, including reduction of extreme heat, improved mental and physical health, and enhanced habitat for urban wildlife. For example, Rodney Cook Sr. Park in Atlanta's Vine City neighborhood is a recreational and social asset which also serves as a major stormwater retention facility during heavy rainfall events, thereby preventing the flooding that had frequently occurred prior to the redesign of the park (Figure 17-3).[19] Similarly, Hunter Point South Park in Queens, New York, on the banks of the East River, is a community recreational and social asset that provides a buffer against storm surges that could threaten nearby neighborhoods. This park's nature-based features include a constructed tidal marshland, a green roof on the café that catches rain for reuse, and a soccer field designed to capture and drain water.[20]

GSI can provide additional co-benefits. For example, in Philadelphia, in and around areas where new GSI was installed, there was a reduction in crime, including narcotics possession, narcotics manufacturing, and burglaries.[21] But GSI can have unintended adverse consequences. In some circumstances, wetlands can not only fail to treat stormwater, but can also be a source of pollution with nutrients (nitrogen and phosphorus). Careful design and monitoring of GSI as well as *adaptive management*—a structured, intentional approach to managing complex systems that calls for decisions and adjustments based on emerging information—can help prevent such adverse consequences.

Urban trees can also assist with stormwater management by intercepting falling rain, stabilizing soil, absorbing water from soil, and returning water to the air through evapotranspiration. Trees also improve urban water quality by removing nutrient pollutants and some metals from runoff.

Figure 17-3 Rodney Cook Sr. Park in Atlanta, a 16-acre park in an African American neighborhood, which has had recurrent flooding after heavy rainfalls, was designed to capture and store up to 10 million gallons of stormwater. It is also a vibrant community asset. (Courtesy of Atlanta Area Parks/Photograph by Lang Binkins.)

Building Social Cohesion

Green infrastructure can promote *social capital*, the bonds of trust and reciprocity that link people with each other and with their communities. Parks and greenspace can help bring people together, strengthening bonds within defined social groups and between groups that differ by race, ethnicity, national origin, and other factors.[22,23] An urban community with a park or greenspace is likely to be a stronger, more connected community than one without these natural amenities. Not only is social capital a strong predictor of health, but when a disaster such as a wildfire, flood, or heat wave strikes, social capital plays a defining role in the immediate response and in post-disaster recovery. In a disaster, neighbors are often the "first responders" for each other, providing rescue, first aid, food, shelter, and emotional support. During the recovery phase, social capital is a strong predictor of restored community function. Accordingly, an urban park may not only serve as a stormwater retention area or a flood barrier, but it can also build climate resilience and promote health by building social capital.

Improving Air Quality and Sequestering Greenhouse Gas Emissions

Emerging evidence suggests that urban forests can both sequester carbon and, by providing shade and decreasing the need for cooling, reduce GHG emissions.

Annually, urban forests in the United States sequester an estimated 36.7 million tons of carbon, and their cooling benefits reduce energy usage by $5.4 billion.[24] Urban forests, which can be as dense as forests in rural areas, can reduce airborne particulate matter and other pollutants. Urban trees in the United States remove an estimated 822,000 tons of air pollutants per year, preventing 670 deaths and 575,000 cases of associated respiratory illness.[24] But this contribution of trees to urban air quality is generally only incremental and depends on tree species, topography, and street design.

Urban Inequities

The health co-benefits of nature-based solutions in cities are not equitably distributed. In the United States, parks that serve mainly nonwhite populations are, on average, half the size of parks in majority white neighborhoods, yet they serve almost five times as many people. Similar disparities exist by income. Parks that serve low-income households are, on average, one-fourth the size of parks that serve high-income households.[25]

A similar pattern manifests globally. Cities in the Global North offer their residents about three times more greenspace than do cities in the Global South. Within Global South cities, greenspace inequity—the disparity in greenspace availability between wealthy and poor neighborhoods—is about double that in the Global North.[26] For example, a study in Cali, Colombia, found that low-income neighborhoods have less tree canopy cover and fewer parks than higher-income neighborhoods.[27] Addressing these inequities within and between cities is essential as part of a larger moral mandate to address climate inequities.

NATURE-BASED CLIMATE SOLUTIONS IN RURAL AND WILDLAND AREAS

Rural and wildland lands produce food, provide *ecosystem services* (Box 17-1), support biodiversity, and improve health and well-being. They also have the potential to capture and store massive amounts of atmospheric carbon. But in recent years there has been unprecedented conversion and degradation of rural lands, resulting in significant increases in global GHG emissions and forfeiting important human benefits. The Intergovernmental Panel on Climate Change (IPCC) has stated that "reducing and reversing land degradation, at scales from individual farms to entire watersheds, can provide cost effective, immediate, and long-term benefits to communities and support several Sustainable Development Goals (SDGs) with co-benefits for adaptation and mitigation."[28]

Box 17-1 Ecosystem Services and Nature's Contributions to People

---

Howard Frumkin, Brendan Shane, and Taj Schottland

Nature provides an abundance of goods and services, known as *ecosystem services*, that fulfill human needs—a concept that is fundamental to nature-based climate solutions. The Millennium Ecosystem Assessment, a United Nations initiative to assess human impacts on natural systems published in 2005, grouped ecosystem services into the following four categories[1]:

- Provisioning services fulfill such needs as food and potable water. For example, intact marine ecosystems provide seafood, and intact forests provide wood for use in building homes.
- Regulating services maintain the conditions that make life possible. For example, intact forests help conserve biodiversity, filter water, stabilize soil, and improve air quality.
- Supporting services underlie and facilitate natural processes. Examples include the cycles of water, nitrogen, and phosphorus; the creation of soil; and photosynthesis.
- Cultural services consist of nonmaterial benefits that the natural world provides, such as spiritual and recreational benefits.

Critics have pointed out that this ecosystem services framework relies heavily on Western rationalist ways of thinking and, in response, have urged inclusion of non-Western approaches to valuing the gifts of nature. Critics also have urged considering not only one-directional transactions, but also benefits co-produced by nature and people. In addition, critics have urged greater emphasis on equity and justice. These critiques have led to a broader reframing of ecosystem services as *nature's contributions to people*.[2]

Health has not typically been considered an ecosystem service, although all four of the above categories of ecosystem services provide health benefits.[3,4] As this chapter describes, nature-based climate solutions protect people from disasters, infectious diseases, extreme heat, and many other health threats and also help reduce GHG levels in the atmosphere—all of which are ecosystem services.

Box References

1. Millennium Ecosystem Assessment Program. Ecosystems and Human Well-Being: Synthesis. Washington, DC: Island Press; 2005.

2. Díaz S, Pascual U, Stenseke M, et al. Assessing nature's contributions to people. Science 2018; 359: 270–272. doi 10.1126/science.aap8826.

3. Lindgren E, Elmqvist T. Ecosystem Services and Human Health. Oxford Research Encyclopedia, Environmental Science. Oxford: Oxford University Press; 2017.

4. Hernández-Blanco M, Costanza R, Chen H, et al. Ecosystem health, ecosystem services, and the well-being of humans and the rest of nature. Global Change Biology 2022; 28: 5027–5040. https://doi.org/10.1111/gcb.16281.

Nature-based climate solutions in rural areas, including regenerative agriculture, sustainable ecosystem management, and conservation and restoration of diverse ecosystems, yield numerous health co-benefits. One is more reliable food and water supplies. Another is disaster resilience. Still another is the wide-ranging health and recreational benefits of contact with natural settings.

The following section describes several pathways to health benefits from rural and wildland nature-based climates solutions: sequestering carbon in forests, coastal (blue carbon) ecosystems, and croplands; building disaster resilience; and controlling infectious diseases.

Sequestering Carbon

Carbon dioxide removal from the atmosphere is essential for achieving the goal of the 2015 Paris Agreement to limit global warming to well under 2.0°C (3.6°F) above preindustrial levels. Forests, grasslands, wetlands, and agricultural land can sequester and store vast amounts of carbon. *Sequestration* (also called *carbon dioxide removal* or *negative emissions*) is the process of carbon uptake in biomass or other pools, expressed as the quantity of carbon taken up per unit time. *Storage* is the amount of carbon residing in a specific pool, such as a forest or a wetland, usually expressed as the quantity of carbon per unit area. Technological solutions for carbon sequestration are under development, but nature-based solutions are already in place and offer some relative advantages: wide availability, potential for large-scale carbon sequestration, and many environmental, economic, and health co-benefits.

The magnitude of potential carbon sequestration by natural systems is large, although it varies by region and land type. While several land types—grasslands, wetlands, forests, and croplands—all have considerable potential to store carbon, forests account for the largest global pool because of the amount of land they cover. A 2017 analysis considered a range of strategies, and confirmed that reforestation, conservation of existing forests, and natural forest management

are among the highest-yield mitigation strategies.[29] A 2022 analysis estimated
that biomass and soils could capture and store 287 petagrams of carbon (PgC)
above and beyond what lands are currently storing, even after accounting for the
need to safeguard lands currently utilized for food production and human hab-
itation.[30] To put this in perspective, this quantity exceeds the estimated 250 PgC
ceiling of total emissions after 2021 needed to keep the global temperature rise to
below 2.0°C (3.6°F) by the year 2100.

Nature-based climate solutions protect and restore natural forests, grasslands,
and wetland ecosystems and improve management of agricultural lands. Land management decisions are typically complex, with many competing social, economic,
and environmental considerations. Balancing these competing demands is critical, given the high potential of land-based carbon sequestration and the urgency
of implementing solutions. One proposed decision framework uses a hierarchy
of nature-based climate solutions based on magnitude, immediacy, and cost-
effectiveness of carbon sequestration and co-benefits for biodiversity and human
health and well-being.[31] Under this approach, protecting existing natural carbon
sinks emerges as a top priority, followed by management of landscapes to enhance
carbon storage, and by reforestation and other longer-term restoration efforts.

Nature-based climate solutions for carbon sequestration can be selected and
implemented in ways that optimize other benefits using frameworks such as the
Sustainable Development Goals[32] and the concept of Nature's Contributions to
People (Box 17-1). While all nature-based approaches to carbon capture deliver
some benefits to people and sustainable societies, approaches such as wetland
restoration and soil carbon sequestration tend to yield the most positive impacts. [33] Other approaches, such as afforestation (planting on land not previously
forested) and producing some forms of biofuels, can have adverse effects because
they compete for land use with food production and other beneficial land uses.

Forests
Of the various land-based strategies for climate change mitigation, forestry
receives the most attention—and with good reason. Forests, when preserved,
managed, and restored, are the largest nature-based contributor to carbon
sequestration, with many co-benefits.[29] One benefit is prevention of infectious diseases, as discussed below. Another is cooling. For example, a study
in Indonesia found that forest loss increased local temperatures, causing
unsafe working conditions, heat-related morbidity, and increased mortality.[34] Forests provide recreational opportunities. And time spent in forests
contributes to stress reduction, improved cognition, and better mental
health.

Forests can also promote local prosperity through tourist economies. For example, a study of 34 low-income countries in Africa, Asia, and Latin America

found that households near protected areas (usually forests) with tourism activity had less poverty, more robust child growth, and less childhood stunting compared to distant households.[35]

Forests provide additional benefits. They are sources of medications, such as xylitol (from birch trees), which is used in the prevention of dental caries, and paclitaxel (from Pacific yew trees), which is used in cancer chemotherapy. Forests protect soil, provide storm protection, and supply clean drinking water—all benefits that protect human health.

Ensuring the multiple benefits of forests requires thoughtful management and planning. Poorly planned or executed forest projects can be counterproductive—as evidenced by low tree survival, reduced biodiversity, diversion of important land resources, economic failure, and other poor outcomes. Recommended practices include preserving existing forests first, managing to maximize biodiversity, engaging local communities, incorporating indigenous knowledge, and planning for economic sustainability.[36] (Afforestation and reforestation have limits; Box 17-2 describes the controversy over Earth's capacity for trees.)

Coastal (Blue Carbon) Ecosystems

Blue carbon refers to carbon captured by ocean and coastal ecosystems. Like other natural landscapes, coastal ecosystems including mangroves, seagrass, and tidal marshes, store significant amounts of carbon (Figure 17-4). It is estimated that more than 30 petagrams of carbon dioxide equivalents are stored in up to 185 million hectares (457 million acres) of coastal areas globally.[37] However, over the past 50 to 100 years, an estimated 25%, and perhaps as much as 50%, of blue carbon ecosystems have been lost; this loss continues at up to 3% a year in some coastal ecosystems.[38] Restoring degraded blue carbon ecosystems could sequester 0.8 petagrams of carbon dioxide equivalents per year—equivalent to 3% of global emissions.[37] Blue carbon ecosystems also provide other economic, social, and health benefits by supporting critical habitat and fisheries and protecting coastal communities from erosion, storm surges, and pollution.[39]

Croplands

Of all ice-free land on Earth, 49% is used for cropland and pasture.[28] Great expanses of cultivated soils and grasslands store vast amounts of carbon and could sequester an estimated 63 petagrams of carbon dioxide equivalents (more than 25% of the carbon storage potential of the Earth's biomass).[40]

Regenerative agriculture (also known as *conservation agriculture*) refers to a range of farming practices including reduced tillage, use of cover crops, crop rotation, integration of livestock grazing with crops, and reduced use of inputs such as fertilizers and pesticides. A principal benefit of regenerative agriculture is carbon sequestration. Improving soil health—the ability of soil to carry out its

Box 17-2 How Many Trees Can the World Support?

Howard Frumkin, Brendan Shane, and Taj Schottland

Trees represent one of the most powerful nature-based climate solutions. They not only sequester carbon, but they also provide shade and reduce local temperatures on hot days, help manage stormwater and stabilize soil, support biodiversity, yield numerous useful products, and promote human health. Wouldn't we want to plant as many trees as possible everywhere?

Many people think so. Globally, the number of tree-planting organizations nearly tripled between 1990 and 2020. A study found that these organizations had planted almost 1.4 billion trees in tropical nations from the 1960s to the 2010s.[1] High-profile regional initiatives have included:

- The Green Belt Movement in East Africa, for which Kenya's Wangari Maathai received the 2004 Nobel Peace Prize
- China's Great Green Wall, which aims to halt the expansion of the Gobi Desert
- Conservation International, which is reforesting 30,000 hectares (74,132 acres) in the Brazilian Amazon.

Ambitious campaigns have also focused on tree planting on a global scale. In 2006, building on Africa's Green Belt Movement, the United Nations Environment Program launched its Billion Tree Campaign. Within just a few years, more than 12 billion trees had been planted by the 193 participating nations, led by China with 2.8 billion and India with 2.1 billion, followed by Ethiopia, Mexico, and Turkey. This inspired Felix Finkbeiner, a 9-year-old student in Germany, to propose planting a *trillion* trees globally; this goal is being pursued by the youth-run Plant-for-the-Planet Foundation in its Trillion Tree Campaign. In 2011, Germany and the International Union for Conservation of Nature launched the Bonn Challenge, with the aim of restoring 350 million hectares (865 million acres) of forest globally by 2030.

Indeed, there is an enormous global need and capacity for more trees. In 2015, it was estimated that there were approximately 3 trillion trees globally, which represented a 50% reduction since the start of human agriculture.[2] A follow-up 2019 report estimated that an additional 900 million hectares (about 2.2 billion acres)—about the land area of Canada—could accommodate trees.[3]

But soon after these proposals were made, many objections arose. Some scientists pointed out that ecosystems targeted for afforestation, such as grasslands and savannas, are already carbon-rich ecosystems and are often ecologically unsuitable for planting trees. Ecologists pointed out that large-scale tree planting could reduce biodiversity and create unsustainable ecosystems.

Still others pointed out that, in some places, newly planted trees were promptly harvested for wood products, thus undermining the purposes for which they were planted. In addition, in other places, seedling mortality was high due to water scarcity, severe weather, lack of resources for maintenance, and other factors. An additional concern is that climate change means that trees planted in an appropriate setting now might be unlikely to survive when climate changes in the future.[4-7]

Trees are indeed an effective nature-based climate solution, one that can deliver wide-ranging co-benefits. But the practical challenges of large-scale tree-planting exemplify the complexity of these solutions, the need for tradeoffs, and the importance of careful systems-based science in guiding policy.[8]

Box References

1. Martin MP, Woodbury DJ, Doroski DA, et al. People plant trees for utility more often than for biodiversity or carbon. Biological Conservation 2021; 261: 109224. doi: 10.1016/j.biocon.2021.109224.

2. Crowther TW, Glick HB, Covey KR, et al. Mapping tree density at a global scale. Nature 2015; 525: 201–205. doi 10.1038/nature14967.

3. Bastin J-F, Finegold Y, Garcia C, et al. The global tree restoration potential. Science 2019; 365: 76–79.

4. Veldman JW, Overbeck GE, Negreiros D, et al. Where tree planting and forest expansion are bad for biodiversity and ecosystem services. BioScience 2015; 65: 1011–1018. doi: 10.1093/biosci/biv118.

5. Veldman JW, Silveira FAO, Fleischman FD, et al. Grassy biomes: An inconvenient reality for large-scale forest restoration? A comment on the essay by Chazdon and Laestadius. American Journal of Botany 2017; 104: 649–651. doi: https://doi.org/10.3732/ajb.1600427.

6. Lewis SL, Wheeler CE, Mitchard ETA, Koch A. Restoring natural forests is the best way to remove atmospheric carbon. Nature 2019; 568: 25–28. doi: 10.1038/d41586-019-01026-8.

7. St. George Z. Can planting a trillion new trees save the world? New York Times 2022 13 July.

8. Holl KD, Brancalion PHS. Tree planting is not a simple solution. Science 2020; 368: 580–581. doi: 10.1126/science.aba8232.

ecological functions, marked by biological, chemical, and water balance—also enhances the capacity of soil to store carbon.[41] (Both organic carbon, in such forms as lignin, cellulose, humus, proteins, and carbohydrates, and inorganic carbon, mostly in the form of carbonates, are stored in the soil.) As the concentration of carbon in soil increases, soil can hold more water, thereby reducing the

Figure 17-4 Mangrove swamps, like this one, store significant amounts of carbon and help protect against floods. (Photograph by Matt Tilghman/Shutterstock.com.)

risk of flooding and better supporting crops.[42] In addition, as the concentration of carbon in soil increases, soil fertility is improved.[42] There is some evidence that food grown through regenerative farming techniques is more nutritious—with improved micronutrient and phytochemical levels in food crops and improved lipid profiles in livestock—compared to conventionally grown food.[43] (See Chapter 14.)

Building Disaster Resilience

Weather-related disasters, supercharged by a warming climate, threaten human health and economic well-being. Over the past 50 years, weather-related disasters have caused $4.3 trillion in losses and led to over 2 million deaths (90% in LMICs).[44] In rural areas, nature-based climate solutions offer protection by stabilizing steep slopes in mountainous areas and by reducing the impacts of droughts, wildfires, and flooding. Examples of drought resilience include using abandoned rice fields in Japan to recharge groundwater supplies,[45] protecting and restoring riparian forests and streambanks in Greece to reduce water runoff,[45] and clearing invasive trees in South Africa to enhance stream flow during dry periods.[46] Examples of flood resilience include dune restoration to reduce coastal erosion and protect built infrastructure from storm surges, restoration of floodplains to store and convey water, and protection of upland forests to help slow and retain runoff. [47] Mangrove forests and coral reefs are also effective in flood protection.

Controlling Infectious Diseases

In addition to storing carbon, forests help to control infectious diseases. There is some evidence that exposure to forests improves immune function, as measured by the number and activity of natural killer (NK) lymphocytes and by levels of cytotoxic effector molecules.[48] There is also some evidence that phytoncides— volatile organic chemicals produced by trees and other vegetation—may have direct antimicrobial activity.[49] In addition, ecological mechanisms, such as an intact, biodiverse forest ecosystem, may limit the profusion of pathogenic microorganisms.[50]

Land use change—especially the conversion of forests to other uses, such as agriculture and grazing—has long been associated with the risk of infectious diseases.[51] This relationship is complex, involving the extent and pattern of de- forestation; human migration, travel, settlement, and activity patterns; post- deforestation land use and management; and other factors.[52] Deforestation may perturb stable relationships in complex multihost disease systems, changing the balance of pathogen species and/or transmission rates.[53] Increased human con- tact with wildlife species in the context of deforestation may lead to increased spillover of zoonoses to humans.

Intact forest ecosystems generally seem to reduce the risk of infectious diseases, as illustrated by the following:

- Malaria: Deforestation, especially the loss of primary forests in tropical regions, has been linked with malaria outbreaks, although not all studies support this conclusion. Studies in Asia and Latin America suggest that de- forestation shifts the balance of mosquito species, favoring those species that serve as malaria vectors.[54] On the other hand, deforestation is some- times associated with an improved local standard of living, which can re- duce malaria risk.

- Ebola: Studies of Ebola outbreaks in West and Central Africa have found that most index cases—resulting from spillover from animals to humans— and subsequent outbreaks occurred where there had been recent deforesta- tion and forest fragmentation.[55]

- Hendra virus: Deforestation across Southeast Asia and Oceania has disrupted natural habitats and winter food resources for black flying foxes (fruit bats), forcing them to seek out food in human-dominated areas of Australia, which, in turn, has led to outbreaks of bat-borne Hendra virus disease in horses and people. In addition to being driven to roost closer to humans, these displaced bats have been shown to carry higher levels of Hendra virus due to stress on their immune systems following the destruc- tion of their natural habitats.[56]

- Diarrheal disease: Deforestation increases the incidence of childhood diarrhea,[57] possibly reflecting the fact that forests improve the quantity and quality of water in rivers and streams.

IMPLEMENTING NATURE-BASED CLIMATE SOLUTIONS

Nature-based climate solutions can be challenging to implement and, when implemented poorly, may not yield the anticipated benefits and co-benefits, including for health—or may actually do harm.

Confronting Tradeoffs

Implementing nature-based climate solutions can sometimes require difficult tradeoffs between costs and benefits. For example, extensive afforestation in settings such as grasslands can disrupt stable ecosystems, reduce biodiversity, and reduce resilience to climate change. Similarly, afforestation on agricultural lands can compete with food production, as can the reallocation of agricultural lands from food crop to biofuel production.[58] When governments offer financial incentives to scale up nature-based climate solutions, local land rights can be compromised, leading to land grabs by governments and private investors. These examples remind us that nature-based climate solutions need to be evaluated holistically, with consideration of the full range of costs and benefits, to support wise policy decisions.

Using Policy Levers to Advance Nature-Based Climate Solutions

Nature-based climate solutions are best implemented as part of a comprehensive approach to climate change mitigation and adaptation. In urban areas, this can mean weaving them into regional planning efforts conducted by metropolitan planning organizations, creating dedicated structures (such as chief resilience officers), and implementing specific policies (such as municipal tree ordinances). The American Planning Association has published *Planning for Urban Heat Resilience*, a report that recommends strategies, including peri-urban land conservation, park and open space protection, tree planting, green roofs and walls, and green stormwater infrastructure. It also recommends that planning be supported by enforceable regulations to guide city development, zoning, and streetscape design—regulations that reduce development that exacerbates the urban heat island effect.[59]

Because no city has a "Department of Green Infrastructure" to coordinate nature-based climate solutions, addressing a challenge such as extreme heat typically falls both within and outside the jurisdiction of multiple city agencies, leading to uncoordinated action—or none at all. To address these challenges, some cities and counties have created a chief heat officer position to oversee and coordinate preparedness and responses to extreme heat.

A growing number of policy incentives and requirements support the adoption of nature-based climate solutions to manage stormwater. For example, U.S. cities that have violated provisions of the Clean Water Act have been required to reduce pollution through consent decrees that prescribe discharge reductions. In some cities, including Chicago, Seattle, and Washington, DC, these consent decrees mandate the use of green infrastructure to reduce stormwater runoff and combined sewer overflows.

In rural areas and natural landscapes, other types of policies can facilitate the use of nature-based climate solutions for mitigation and adaptation and increase co-benefits. The IPCC has identified a wide range of policies that can support nature-based climate solutions, including the following:

- Financial support for conservation and restoration
- Designations of protected and management areas
- Resilience preparedness for storms, droughts, or fires
- Land use planning to protect natural land uses.

At the same time, the IPCC has also identified barriers to adoption of these policies, in particular the adverse impacts of social inequality and corruption in systems of governance and finance.[28]

Using Economic Levers

There are many ways to scale up investment in nature-based climate solutions. One of the more widespread innovations is *payment for ecosystem services*.[60] The concept is relatively simple: a willing payor incentivizes landowners to manage their land in a way that boosts certain ecosystem services, such as enhanced carbon storage and sequestration, reduced deforestation, or improved water quality. While the concept is simple, the implementing mechanisms to facilitate these transactions are often highly complex and varied. Generally, most mechanisms function at local or regional scales, but there are a few international examples, including Reducing Emissions from Deforestation and Forest Degradation in Developing Countries (REDD+), a framework established by the United Nations Framework Convention on Climate Change Conference of the Parties primarily to reduce emissions from deforestation and degradation.[60]

In the United States, payments for forest-based ecosystem services in 2019 were estimated to be $3.6 billion, including $176 million for carbon and $889 million for water.[61] While these payments provide much-needed capital to help scale nature-based climate solutions, the impact of financial incentives on the behavior of forest landowners is generally limited, and many of the financial incentives have been targeted to people who might have taken action without the incentives.[61]

Communicating Nature-Based Climate Solutions

The enormous scale and decentralized nature of nature-based climate solutions means that effective implementation rests on the actions of people globally, from elected leaders and government decision-makers to individual landowners. Effective communication is essential to build a broad understanding of practices and their benefits and limitations.

The label "natural" is often positively received by the public and policymakers, but different audiences can interpret very differently what solutions are "natural" and how they should be implemented.[62] The powerful narrative around expanding forestry for carbon capture, for example, may overlook or even undermine the importance of protecting other natural ecosystems and the interests of local communities.[62] In addition, "natural" framing may obscure risks and tradeoffs that come with nature-based climate solutions, which must be made explicit in public discourse to enable transparent and informed decision-making.[63] Another challenge arises when nature-based climate solutions are erroneously viewed as alternatives to reducing emissions. In fact, addressing the climate crisis requires *both* rapid phaseout of fossil fuels and nature-based measures.

Assuring Equity in Nature-Based Climate Action

Implementation of nature-based climate solutions is not inherently just or fair.[64] The impacts of structural racism, underinvestment, and marginalization of racial and ethnic minorities extend to greenspace and natural areas. Disinvested neighborhoods have more limited access to the natural settings that support human and environmental health.[65] New investment in nature-based climate solutions can worsen inequities by increasing segregation and displacement—a consequence sometimes termed *green gentrification*.[66] Especially in urban areas that are home to historically marginalized groups, implementing nature-based climate solutions should be undertaken as part of a comprehensive, deliberate approach to ensure climate action does not worsen inequities.[64] (See Chapter 20.)

Fortunately, an increasing number of examples of thoughtful, integrated planning for nature-based climate solutions are grounded in a commitment to equity. Early on, Burlington, Vermont, began its plan for nature-based climate solutions with strategies for equity and inclusion, which were designed to benefit those who were most vulnerable to the impacts of climate change and those who had historically not been engaged in local planning. In its plan, Burlington set universal goals for nature-based solutions, but designed implementation strategies based on how different groups were situated structurally, culturally, and geographically—a process known as *targeted universalism*.[67]

By responding to existing conditions, expanding nature-based solutions can remedy lack of access to nature and reduce inequities. In Washington, DC, the 11th Street Bridge Park project illustrates how early planning to engage members of historically underinvested communities succeeded in addressing the challenges of expanding greenspace in a rapidly gentrifying city. This project's equitable development plan outlined a comprehensive strategy for park development as part of a multisector approach that included affordable housing, workforce and small business development, and recognition of the arts and culture of neighboring communities.[68]

SUMMARY POINTS

Nature-based climate solutions:

- Represent a set of strategies and tools that can help reduce carbon emissions, remove carbon dioxide from the atmosphere, promote adaptation to climate change, and build resilience.
- Need to be implemented as part of an integrated set of climate actions.
- Aim to accomplish specific mitigation and adaptation goals effectively, economically, and safely.
- Deliver a wide range of co-benefits, including co-benefits for health and well-being.
- Can cause unintended and even harmful consequences, and therefore need to be carefully planned, implemented, and managed.

References

1. World Bank. Urban development. Available at: https://www.worldbank.org/en/topic/urbandevelopment/overview. Accessed August 2, 2023.
2. Bochove D. Is this the future of urban resilience? Bloomberg.com, July 27, 2022. Available at: https://www.bloomberg.com/news/features/2022-07-27/is-toronto-s-port-lands-flood-protection-project-the-future-of-urban-resilience. Accessed October 28, 2023.

3. Ramsay EE, Fleming GM, Faber PA, et al. Chronic heat stress in tropical urban informal settlements. iScience 2021; 24: 103248. https://doi.org/10.1016/j.isci.2021.103248.

4. Escobar Carías MS, Johnston DW, Knott R, Sweeney R. Flood disasters and health among the urban poor. Health Economics 2022; 31: 2072–2089. https://doi.org/10.1002/hec.4566.

5. Corburn J. Water and sanitation for all: Citizen science, health equity, and urban climate justice. Environment and Planning B: Urban Analytics and City Science 2022; 49: 2044–2053. https://doi.org//10.1177/23998083221094836.

6. Borg FH, Greibe Andersen J, Karekezi C, et al. Climate change and health in urban informal settlements in low- and middle-income countries: A scoping review of health impacts and adaptation strategies. Global Health Action 2021; 14: 1908064. doi: 10.1080/16549716.2021.1908064.

7. Giri M, Bista G, Singh PK, Pandey R. Climate change vulnerability assessment of urban informal settlers in Nepal, a least developed country. Journal of Cleaner Production 2021; 307: 127213. https://doi.org/10.1016/j.jclepro.2021.127213.

8. Masson V, Lemonsu A, Hidalgo J, Voogt J. Urban climates and climate change. Annual Review of Environment and Resources 2020; 45: 411–444. https://doi.org/10.1146/annurev-environ-012320-083623.

9. Hoffman JS, Shandas V, Pendleton N. The effects of historical housing policies on resident exposure to intra-urban heat: A study of 108 US urban areas. Climate 2020; 8: 12. doi: 10.3390/cli8010012.

10. Shao H, Kim G. A comprehensive review of different types of green infrastructure to mitigate urban heat islands: Progress, functions, and benefits. Land 2022; 11: 1792. https://doi.org/10.3390/land11101792.

11. Leland J. Why an East Harlem street is 31 degrees hotter than Central Park West. New York Times, August 20, 2021.

12. Sadeghi M, Chaston T, Hanigan I, et al. The health benefits of greening strategies to cool urban environments: A heat health impact method. Building and Environment 2021; 207: 108546. doi: 10.1016/j.buildenv.2021.108546.

13. Kalkstein LS, Eisenman DP, de Guzman EB, Sailor DJ. Increasing trees and high-albedo surfaces decreases heat impacts and mortality in Los Angeles, CA. International Journal of Biometeorology 2022; 66: 911–925. doi: 10.1007/s00484-022-02248-8.

14. McDonald RI, Kroeger T, Zhang P, Hamel P. The value of US urban tree cover for reducing heat-related health impacts and electricity consumption. Ecosystems 2020; 23: 137–150.

15. Nowak DJ, Greenfield EJ. The increase of impervious cover and decrease of tree cover within urban areas globally (2012–2017). Urban Forestry & Urban Greening 2020; 49: 126638. https://doi.org/10.1016/j.ufug.2020.126638.

16. Besir AB, Cuce E. Green roofs and facades: A comprehensive review. Renewable and Sustainable Energy Reviews 2018; 82: 915–939.

17. Hoelscher M-T, Nehls T, Jänicke B, Wessolek G. Quantifying cooling effects of facade greening: Shading, transpiration and insulation. Energy and Buildings 2016; 114: 283–290.

18. Jessup K, Parker SS, Randall JM, et al. Planting stormwater solutions: A methodology for siting nature-based solutions for pollution capture, habitat enhancement, and multiple health benefits. Urban Forestry & Urban Greening 2021; 64: 127300. https://doi.org/10.1016/j.ufug.2021.127300.

19. Bryant R. How a stormwater park is revitalizing a historic Atlanta neighborhood. Parks and Recreation Magazine, March 17, 2022. Available at: https://www.nrpa.org/

parks-recreation-magazine/2022/april/how-a-stormwater-park-is-revitalizing-a-historic-atlanta-neighborhood/. Accessed October 28, 2023.

20. Brown EN. USA: This public park is a model for urban design in the age of climate crisis. Fast Company, September 23, 2019. Available at: https://www.preventionweb.net/news/usa-public-park-model-urban-design-age-climate-crisis. Accessed August 2, 2023.

21. Kondo MC, Low SC, Henning J, Branas CC. The impact of green stormwater infrastructure installation on surrounding health and safety. American Journal of Public Health 2015; 105: e114–e21.

22. Powers SL, Webster N, Agans JP, et al. The power of parks: How interracial contact in urban parks can support prejudice reduction, interracial trust, and civic engagement for social justice. Cities 2022; 131: 104032.

23. Peters K, Elands B, Buijs A. Social interactions in urban parks: Stimulating social cohesion? Urban Forestry & Urban Greening 2010; 9: 93–100.

24. Nowak DJ, Greenfield EJ. US urban forest statistics, values, and projections. Journal of Forestry 2018; 116: 164–177.

25. Trust for Public Land. The Heat Is On. San Francisco: Trust for Public Land; 2022. Available at: https://www.tpl.org/the-heat-is-on. Accessed October 28, 2023.

26. Chen B, Wu S, Song Y, et al. Contrasting inequality in human exposure to greenspace between cities of Global North and Global South. Nature Communications 2022; 13: 4636. https://doi.org/10.1038/s41467-022-32258-4.

27. Shiraishi K. The inequity of distribution of urban forest and ecosystem services in Cali, Colombia. Urban Forestry & Urban Greening 2022; 67: 127446. https://doi.org/10.1016/j.ufug.2021.127446.

28. Shukla PR, Skea J, Calvo Buendia E, et al. (eds.). Climate change and land: An IPCC special report on climate change, desertification, land degradation, sustainable land management, food security, and greenhouse gas fluxes in terrestrial ecosystems. 2019. Available at: at https://www.ipcc.ch/report/srccl/.Accessed August 2, 2023.

29. Griscom BW, Adams J, Ellis PW, et al. Natural climate solutions. Proceedings of the National Academy of Sciences 2017; 114: 11645–11650. doi: 10.1073/pnas.1710465114.

30. Walker WS, Gorelik SR, Cook-Patton SC, Baccini A, Farina MK, Solvik KK, et al. The global potential for increased storage of carbon on land. Proceedings of the National Academy of Sciences. 2022; 119(23): e2111312119. doi: 10.1073/pnas.2111312119.

31. Cook-Patton SC, Drever CR, Griscom BW, et al. Protect, manage, and then restore lands for climate mitigation. Nature Climate Change 2021; 11: 1027–1034. doi: 10.1038/s41558-021-01198-0.

32. United Nations Department of Economic and Social Affairs. The Sustainable Development Goals Report 2022. New York: United Nations; 2022. Available at: https://unstats.un.org/sdgs/report/2022/. Accessed October 28, 2023.

33. Smith P, Adams J, Beerling DJ, et al. Land-management options for greenhouse gas removal and their impacts on ecosystem services and the Sustainable Development Goals. Annual Review of Environment and Resources 2019; 44: 255–286. doi: 10.1146/annurev-environ-101718-033129.

34. Wolff NH, Zeppetello LRV, Parsons LA, et al. The effect of deforestation and climate change on all-cause mortality and unsafe work conditions due to heat exposure in Berau, Indonesia: A modelling study. The Lancet Planetary Health 2021; 5. https://doi.org/10.1016/S2542-5196(21)00279-5.

35. Naidoo R, Gerkey D, Hole D, et al. Evaluating the impacts of protected areas on human well-being across the developing world. Science Advances 2019; 5: eaav3006. doi: 10.1126/sciadv.aav3006.

36. Di Sacco A, Hardwick KA, Blakesley D, et al. Ten golden rules for reforestation to optimize carbon sequestration, biodiversity recovery, and livelihood benefits. Global Change Biology 2021; 27: 1328–1348. https://doi.org/10.1111/gcb.15498.

37. Macreadie PI, Costa MDP, Atwood TB, et al. Blue carbon as a natural climate solution. Nature Reviews Earth & Environment 2021; 2: 826–839.

38. Pendleton L, Donato DC, Murray BC, et al. Estimating global "blue carbon" emissions from conversion and degradation of vegetated coastal ecosystems. PloS One 2012; 7: e43542. https://doi.org/10.1371/journal.pone.0043542.

39. Barbier EB, Hacker SD, Kennedy C, et al. The value of estuarine and coastal ecosystem services. Ecological Monographs 2011; 81: 169–193. https://doi.org/10.1890/10-1510.1.

40. Walker WS, Gorelik SR, Cook-Patton SC, et al. The global potential for increased storage of carbon on land. Proceedings of the National Academy of Sciences 2022; 119: e2111312119. https://doi.org/10.1073/pnas.2111312119.

41. Mbow C, Rosenzweig C, Barioni LG, et al. Food security. In: PR Shukla, J Skea, E Calvo Buendia, et al. (eds.). Climate Change and Land: An IPCC Special Report on Climate Change, Desertification, Land Degradation, Sustainable Land Management, Food Security, and Greenhouse Gas Fluxes in Terrestrial Ecosystems. Geneva: IPCC, 2019, pp. 437–550.

42. Chabbi A, Rumpel C, Hagedorn F, et al. Carbon storage in agricultural and forest soils. Frontiers in Environmental Science 2022; 10. https://doi.org/10.3389/fenvs.2022.848572.

43. Montgomery DR, Biklé A, Archuleta R, et al. Soil health and nutrient density: Preliminary comparison of regenerative and conventional farming. PeerJ 2022; 10: e12848. https://doi.org/10.7717/peerj.12848.

44. World Meteorological Organization. WMO Atlas of Mortality and Economic Losses from Weather, Climate, and Water Extremes (1970–2019) (WMO-No 1267), Updated Through 2021. Geneva: World Meteorological Organization, 2023. Available at: https://public.wmo.int/en/resources/atlas-of-mortality. Accessed October 28, 2023.

45. Organisation for Economic Co-operation and Development. Nature-Based Solutions for Adapting to Water-Related Climate Risks. Paris: OECD, 2020. Available at: https://www.oecd.org/environment/nature-based-solutions-for-adapting-to-water-related-climate-risks-2257873d-en.htm. Accessed October 28, 2023.

46. Holden PB, Rebelo AJ, Wolski P, et al. Nature-based solutions in mountain catchments reduce impact of anthropogenic climate change on drought streamflow. Communications Earth & Environment 2022; 3: 51. https://doi.org/10.1038/s43247-022-00379-9.

47. Bridges TS, King JK, Simm JD, et al. International guidelines on natural and nature-based features for flood risk management. Washington DC: U.S. Army Corps of Engineers, Engineer Research and Development Center, 2021. Available at: https://erdc-library.erdc.dren.mil/jspui/handle/11681/41946. Accessed October 28, 2023.

48. Chae Y, Lee S, Jo Y, et al. The effects of forest therapy on immune function. International Journal of Environmental Research and Public Health 2021; 18: 8440. doi: 10.3390/ijerph18168440.

49. Antonelli M, Donelli D, Barbieri G, et al. Forest volatile organic compounds and their effects on human health: A state-of-the-art review. International Journal of Environmental Research and Public Health 2020; 17. doi: 10.3390/ijerph17186506.

50. Keesing F, Belden LK, Daszak P, et al. Impacts of biodiversity on the emergence and transmission of infectious diseases. Nature 2010; 468: 647–652. https://doi.org/10.1038/nature09575.

51. Gottdenker NL, Streicker DG, Faust CL, Carroll CR. Anthropogenic land use change and infectious diseases: A review of the evidence. EcoHealth 2014; 11: 619–632.

52. Faust CL, McCallum HI, Bloomfield LSP, et al. Pathogen spillover during land conversion. Ecology Letters 2018; 21: 471–483.

53. Murray KA, Daszak P. Human ecology in pathogenic landscapes: two hypotheses on how land use change drives viral emergence. Current Opinion in Virology 2013; 3: 79–83.

54. Burkett-Cadena ND, Vittor AY. Deforestation and vector-borne disease: Forest conversion favors important mosquito vectors of human pathogens. Basic and Applied Ecology 2018; 26: 101–110.

55. Rulli MC, Santini M, Hayman DTS, D'Odorico P. The nexus between forest fragmentation in Africa and Ebola virus disease outbreaks. Scientific Reports 2017; 7: 41613. doi 10.1038/srep41613.

56. Eby P, Peel AJ, Hoegh A, et al. Pathogen spillover driven by rapid changes in bat ecology. Nature 2022. https://doi.org/10.1038/s41586-022-05506-2.

57. Herrera D, Ellis A, Fisher B, et al. Upstream watershed condition predicts rural children's health across 35 developing countries. Nature Communications 2017; 8: 811. doi: 10.1038/s41467-017-00775-2.

58. Hasegawa T, Sands RD, Brunelle T, et al. Food security under high bioenergy demand toward long-term climate goals. Climatic Change 2020; 163: 1587–1601. doi: 10.1007/s10584-020-02838-8.

59. Keith L, Meerow S. Planning for Urban Heat Resilience. Chicago: American Planning Association, 2022. Available at: https://www.planning.org/publications/report/9245695/. Accessed October 28, 2023.

60. Kaiser J, Haase D, Krueger T. Payments for ecosystem services: A review of definitions, the role of spatial scales, and critique. Ecology and Society 2021; 26. https://doi.org/10.5751/ES-12307-260212.

61. Frey GE, Kallayanamitra C, Wilkens P, James NA. Payments for forest-based ecosystem services in the United States: Magnitudes and trends. Ecosystem Services 2021; 52:101377. https://doi.org/10.1016/j.ecoser.2021.101377.

62. Seddon N, Smith A, Smith P, et al. Getting the message right on nature-based solutions to climate change. Global Change Biology 2021; 27: 1518–1546. https://doi.org/10.1111/gcb.15513.

63. Osaka S, Bellamy R, Castree N. Framing "nature-based" solutions to climate change. WIREs Climate Change 2021; 12: e729. https://doi.org/10.1002/wcc.729.

64. Haase A. The contribution of nature-based solutions to socially inclusive urban development– Some reflections from a social-environmental perspective. In: N Kabisch, H Korn, J Stadler, A Bonn (eds.). Nature-Based Solutions to Climate Change Adaptation in Urban Areas: Linkages Between Science, Policy, and Practice. Cham: Springer International Publishing, 2017, pp. 221–236.

65. Bikomeye JC, Namin S, Anyanwu C, et al. Resilience and equity in a time of crises: Investing in public urban greenspace is now more essential than ever in the

US and beyond. International Journal of Environmental Research and Public Health 2021; 18: 8420. doi: 10.3390/ijerph18168420.

66. Anguelovski I, Connolly JJT, Cole H, et al. Green gentrification in European and North American cities. Nature Communications 2022; 13: 3816. doi: 10.1038/s41467-022-31572-1.

67. City of Burlington. Nature-based climate solutions: An addendum to the Burlington Open Space Protection Plan 2022. Available at: https://www.burlingtonvt.gov/DPI/CB/Open-Space-Addendum/What-is-the-Open-Space-Climate-Change-Addendum. Accessed October 28, 2023.

68. 11th Street Bridge Park. 11th Street Bridge Park's Equitable Development Plan, 2018. Available at: https://bbardc.org/wp-content/uploads/2018/10/Equitable-Development-Plan_09.04.18.pdf. Accessed October 28, 2023.

Further Reading

Chausson A, Turner B, Seddon D, et al. Mapping the effectiveness of nature-based solutions for climate change adaptation. Global Change Biology 2020; 26: 6134–6155. doi 10.1111/gcb.15310.

Girardin CAJ, Jenkins S, Seddon N, et al. Nature-based solutions can help cool the planet – if we act now. Nature 2021; 593: 191–194. doi: 10.1038/d41586-021-01241-2.

Kabisch N, Korn H, Stadler J, Bonn A (eds.). Nature-Based Solutions to Climate Change Adaptation in Urban Areas: Linkages Between Science, Policy, and Practice. Cham: Springer International Publishing, 2017.

Seddon N, Chausson A, Berry P, et al. Understanding the value and limits of nature-based solutions to climate change and other global challenges. Philosophical Transactions of the Royal Society B: Biological Sciences 2020; 375: 20190120. doi: 10.1098/rstb.2019.0120.

World Bank. A catalogue of nature-based solutions for urban resilience. Washington, DC: World Bank, 2021. Available at: https://openknowledge.worldbank.org/handle/10986/36507. Accessed January 23, 2023.

These publications provide helpful overviews of nature-based climate solutions.

Shukla PR, Skea J, Calvo Buendia E, et al. (eds.). Climate change and land: An IPCC special report on climate change, desertification, land degradation, sustainable land management, food security, and greenhouse gas fluxes in terrestrial ecosystems. Intergovernmental Panel on Climate Change, 2019. Available at: https://www.ipcc.ch/srccl/. Accessed January 23, 2023.

This authoritative report is an excellent source on the role of land in the context of climate change.

Soz SA, Kryspin-Watson J, Stanton-Geddes Z. The Role of Green Infrastructure Solutions in Urban Flood Risk Management. Washington, DC: World Bank, 2016. Available at: https://openknowledge.worldbank.org/handle/10986/25112. Accessed January 23, 2023.

A useful resource on nature-based climate solutions for urban flood management.

Smith AC, Tasnim T, Irfanullah HM, et al. Nature-based solutions in Bangladesh: Evidence of effectiveness for addressing climate change and other Sustainable Development Goals. Frontiers in Environmental Science 2021; 9. doi: 10.3389/fenvs.2021.737659.

This publication provides an overall conceptual and practical picture of nature-based climate solutions in a setting with limited resources.

PART V

STRENGTHENING PUBLIC AND POLITICAL SUPPORT

18

Communicating the Health Relevance
of Climate Change

Mona Sarfaty and Edward Maibach

Although the current impacts and future risks of climate change to public health are increasingly well documented, they are not necessarily well known and understood by the public or policymakers. This chapter reviews the public's understanding of the health implications of climate change and how health professionals and others can effectively communicate about the health impacts of climate change and the health co-benefits of mitigation and adaptation measures.

From a public health perspective, communication ensures that the public, policymakers, and other key stakeholders understand how climate change affects human health. From a clinical perspective, communication helps patients and their families understand how climate change is affecting the health of specific people. And from a societal perspective, communication enables health professionals to advocate for changes in society that will protect people from climate change and promote better health, and for changes in the healthcare delivery system that will reduce emission of greenhouse gases and increase resilience of healthcare facilities to the impacts of climate change.[1]

GLOBAL WARMING'S SIX AMERICAS

Public understanding of climate change has been studied extensively. For example, the George Mason University Center for Climate Change Communication and the Yale Project on Climate Change Communication have been conducting a nationally representative survey of American adults twice per year since 2008. An important finding from this project has been the identification of six distinct segments of the U.S. population defined by their perceptions of global warming, called Global Warming's Six Americas.[2] People in each of these segments have a distinct pattern of climate change beliefs, behaviors, policy preferences, and levels of engagement, as described below.[3]

Mona Sarfaty and Edward Maibach, *Communicating the Health Relevance of Climate Change* In: *Climate Change and Public Health*. Edited by: Barry S. Levy and Jonathan A. Patz, Oxford University Press. © Oxford University Press 2024. DOI: 10.1093/oso/9780197683293.003.0018

The Alarmed: Members of this segment are highly engaged in global warming. They know that it is occurring, is caused by human activity, and is a very real threat. They are changing their behaviors and strongly support a range of public policies to address global warming.

The Concerned: Members of this segment are similar to members of the Alarmed segment, except they are less certain of their conclusions, less personally engaged, less likely to be changing their behaviors, and slightly less supportive of policies to address climate change. They accept the harmful nature of global warming but tend to see the harm as beginning one to two decades in the future.

The Cautious: Members of this segment also believe that global warming is happening, but they are less certain that human activity is the cause and that the consequences will be serious, and they do not perceive global warming as a personal threat. They tend not to be considering personal behavior changes, but they do have moderate levels of support for policies to address climate change.

The Disengaged: Members of this segment have not thought much about global warming. Their defining characteristic is that they respond "Don't know" to most survey questions about global warming. They are inclined to believe that, if global warming is real, it is likely to be harmful. They also show moderate levels of support for climate change policies. And they acknowledge that they could easily change their minds about global warming.

The Doubtful: These people tend to be skeptical that human-caused global warming is happening or that climate change poses a serious threat in the foreseeable future, but they do not hold strong views. And almost all of them believe that the United States is already doing enough to address global warming.

The Dismissive: Like the Alarmed, members of this segment are relatively highly engaged in global warming as an issue. They consider themselves to be well informed about it and yet they have concluded that it is not happening and, therefore, it is not a threat. As a result, they feel strongly that policies to address global warming are misguided at best.

Between 2014 and 2020, there were large shifts in the percentages of these six segments in the U.S. population, with an especially notable increase in the number of the Alarmed (Table 18-1). Similarly, between 2014 and 2020, there were large increases in the proportion of people "very concerned" about climate change in Germany (19%), the United Kingdom (18%), Australia (16%), South Korea (13%), Spain (10%), and Canada (7%).[4]

Table 18-1 Percentage of Global Warming's Six Americas segments in 2014 and 2021

Segment	2014	2021
The Alarmed	13%	33%
The Concerned	31%	25%
The Cautious	23%	17%
The Disengaged	7%	5%
The Doubtful	13%	10%
The Dismissive	13%	9%

Source: Yale University and George Mason University.

While there are only modest differences between these six segments in demographic characteristics, such as age and gender, there are large differences among them in political ideology and party affiliation. The Alarmed and the Concerned segments are overwhelmingly Democrats; in contrast, the Doubtful and the Dismissive segments are overwhelmingly Republicans. Therefore, political perceptions present a major challenge to public engagement in climate change.[5] Despite these challenges, segmentation analysis enables communication specialists to develop, test, and evaluate different communication approaches for different groups.[6]

In 2014, few of the people in *any* of the six segments were more than vaguely aware of the health threats posed by climate change, and none thought that the human health consequences of climate change were a major concern. When asked "When you think of global warming, what is the first word or phrase that comes to your mind?" none of the survey respondents gave an answer that indicated they thought about human health.[7] When asked "Worldwide, how many people do you think currently become injured or ill or die each year as a result of global warming?" Approximately 40% answered "Don't know"; about 40% responded "None" or "Hundreds"; 16% responded "Thousands"; and the remaining 4% responded "Millions."[8] Almost all the people who correctly responded "Thousands" or "Millions" were members of the Alarmed and the Concerned segments.

Between 2014 and 2020, there was a 20 percentage point increase in the proportion of Americans who said that global warming was increasing the prevalence in their communities of 10 specific threats to health (heat stroke, asthma and/or other lung diseases, diseases carried by insects, foodborne or waterborne illness due to bacteria or viruses, pollen-related allergies, severe anxiety, depression, and bodily harm from severe storms or hurricanes, wildfires, and flooding).[9] However, these increases were not evenly distributed across the population: the Alarmed were 21 percentage points more likely to say climate change was causing a threat to health in their community, the Concerned were 15 points more likely, the Cautious 14 points more likely, and the Disengaged 25 points more likely; the perceptions of the Doubtful and Dismissive did not shift.[8]

In 2021, the Pew Research Center found similar results in a survey of adults in 17 countries. It found that an average of 72% were somewhat or very concerned that climate change will harm them personally at some point in their lives.[4]

Although government agencies and nongovernmental organizations have long attempted to educate the public about climate change, much of their work focused on scientific aspects; harm to nonhuman forms of life, such as polar bears; and impacts to the environment, such as to glaciers.[10] Until recently, there has been relatively little news coverage of the relevance of climate change to human health. A study in 10 countries in both the Global North (including the United States) and the Global South found that, between 2006 and 2018, fewer than 1% of news stories about climate focused on health.[11] In addition, studies in the United States, Canada, and New Zealand found that reporting on the health aspects of climate change was often inaccurate.[12–14] News stories on the impacts of climate change on human health were typically embedded in stories about a specific heat wave, storm, flood, or fire, rather than reported in a broader context that explained the long-term consequences of climate change.

In addition, health organizations are increasingly educating the public and key stakeholders about the health impacts of climate change and ways of reducing these impacts. For example, in 2017, the Medical Society Consortium on Climate & Health, which represents 52 organizations with a total membership of more than 700,000 physicians in the United States, started communicating four messages:

- Climate change is harming Americans today and these impacts will increase unless actions are taken.
- Climate change places everyone at risk, but is harming some—including children, pregnant women, older people, people of color, and immigrants—more than others.
- The way to slow or stop these impacts is to decrease the use of fossil fuels and increase energy efficiency and use of clean energy.
- These changes in energy choices will improve air and water quality and yield rapid health benefits.[15]

Since 2019, when more than 180 U.S. health organizations declared that climate change is a "public health emergency,"[16] the news media has had increased interest in this subject. After the 26th Conference of the Parties to the United Nations Framework Convention on Climate Change (COP26) in 2021, which included a health program, a health pavilion, and national commitments to low-carbon health systems, the *New York Times* noted that reporting on climate change as a health issue was becoming more frequent.[17] This increased interest has created new opportunities for health professionals to share this information with the public, which has a strong interest in news related to health.[18]

MENTAL MODELS, FRAMES, AND EFFECTIVE COMMUNICATION

People organize their experience and understanding of the world with *mental models*—interconnected sets of associated ideas, beliefs, and feelings that are likely encoded in neural networks in the brain and are largely based on the stimulation of repeated exposures and activations. Few Americans have mental models of climate change that include associations with health. Rather, the most common concepts in Americans' mental models about climate change tend to be centered around vague concepts, such as heat, melting ice, and "bad for the planet"; apocalyptic concepts, such as "the end of everything"; and naysayer concepts, such as conspiracy theories, doubts about the science, and beliefs that the science is hyped and that climate change is part of a natural cycle.[19]

In contrast to mental models, which are attributes of people, *frames* are attributes of communication. Any given issue has many different facets and can be viewed from many different perspectives—or through many different frames.[10] When choosing which facet of an issue to discuss with others, a communicator is choosing a frame, or perspective, for the audience to focus on the issue. For example, communicators who tell stories about the harm of climate change to wildlife, such as polar bears, have chosen an environmental frame.

The choice of frames has consequences. For example, when media coverage of an issue primarily uses one frame, that frame will tend to strongly influence public understanding of that issue, and people will not likely consider other equally valid frames of the issue. These *framing effects* can be subtle and gradually adopted or powerful and rapidly adopted, especially for issues for which the public's mental models are not yet well defined. Either way, they are likely to be pervasive.

Based on people's preexisting values and interests, any given frame is likely to engage specific types of people, have little impact on others, and may even antagonize some. For example, information about climate change that is framed as an environmental problem is likely to engage people who see themselves as environmentalists (about one-third of Americans), but is likely to be dismissed by people who believe that environmentalists are misguided (another one-third of Americans).

The U.S. experience with environmental tobacco smoke (ETS) illustrates the impact that frames can have on attempts to change public policy and behavior. Prior to the mid-1980s, cigarette smoking was largely defined as a personal choice, with adverse health effects limited to *individual* smokers. Evidence of the adverse health effects of ETS on other people emerged and created an opportunity for health professionals to change the frame to one in which tobacco smoke was a threat to *everyone* exposed, especially people in enclosed spaces, such as airplanes and buildings. Over time, the new frame contributed to a change in mental models about smoking, which eventually heightened public acceptance of the need for public policies that protect nonsmokers from ETS.

The frames that have dominated U.S. public discourse on climate change have included *an environmental frame, a political frame,* and *an economic frame*— all of which have been highly polarizing—as well as *a scientific frame,* which resonates with few people. In addition, other frames that have more recently been introduced in U.S. public discourse include:

- <u>A national security frame</u>, as illustrated by military strategists who inform the public that climate change can lead to global instability, which will threaten U.S. national security
- <u>An energy frame</u>, as illustrated by entrepreneurs who highlight the benefits of clean renewable forms of energy in contrast to the costs of relying on fossil fuels
- <u>A moral frame</u>, as illustrated by civic and faith community leaders who assert that some (mostly high-income) countries are harming people in other (mostly low-income) countries and harming future generations in all countries
- <u>A stewardship frame</u>, as illustrated by leaders of the faith community who assert that people have responsibility to protect God's creation
- <u>A human health frame</u>, as illustrated by health professionals who describe the increases in morbidity and mortality due to climate change and the health benefits associated with measures to address climate change. Until recently, there has been relatively little effort to frame climate change as a human health issue.

Frames tend to be most influential in shaping public understanding when there is congruence between the message (the frame) and the messenger. People are more likely to accept information if they perceive its source as trustworthy. And different people trust different sources of information.[20] The most effective messengers are trusted authorities on the frame being presented.[20] Physicians and other health professionals are the most trusted group of professionals in most countries.[21] Americans, especially those who are politically conservative, rate their primary care physicians as among their most highly trusted sources of information about global warming.[22]

THE HEALTH FRAME

Most Americans place great value on health. Health is an integral part of the founding documents of the United States. The Preamble to the U.S. Constitution states that a principal purpose of government is to "promote the general welfare." The Declaration of Independence states that "life, liberty, and the pursuit of happiness" are national goals. Health seems implicit.

Public health—what we, as a society, do collectively to assure the conditions in which people can be healthy[23]—also resonates with most people. Assuring the conditions for people to be healthy includes preventing health threats related to climate change and protecting people from these threats.

A health frame is an effective means of helping most Americans better understand, consider, and respond to climate change.[24] Information about climate change framed around health elicits a more productive set of responses than information framed around the environment or national security—especially for people who otherwise would be unlikely to consider information about climate change.[25] Presenting information about the ways in which climate change harms health—and whose health is most likely to be harmed—has been shown to increase people's concern about and engagement with climate change, including, and perhaps especially, people who are politically conservative.[26] In addition, providing information about the health benefits of measures to address climate change enhances people's intentions to advocate for these measures.[27] Messages about the health benefits of clean energy and improved community design are those that are most compelling.[27] However, since most people are not aware of the specific health impacts caused by climate change, there is a need to provide them with information on the direct association between climate change and increased risk of specific illnesses and death.

People in all six segments of the U.S. population that were described earlier respond positively to the benefits or the concept of *co-benefits* associated with taking action to limit global warming.[24] For example, most people endorse the statement: "Taking actions to limit global warming—by making our energy sources cleaner and our cars and appliances more efficient, by making our cities and towns friendly to trains, buses, and bikers and walkers, and by improving the quality and safety of our food—will improve the health of almost every American."

FIVE KEY FACTS, THREE SIMPLE MESSAGES

How can health professionals inform people about the health relevance of climate change, thereby enhancing public engagement in responding to it? There are five key facts about global warming, which can be summarized in 10 words (Figure 18-1). People who know and accept these five key facts about global warming are significantly more likely to support a societal response to climate change and to personally take actions that encourage a societal response.[20] In addition, people who feel they have directly experienced climate change are more likely to hold firm convictions that it is real.[28,29]

Health professionals can play an important role by communicating these key concepts through the following three simple, important messages:

1. There is a scientific consensus about human-caused climate change. Most Americans are either unaware or do not accept that climate scientists have reached a consensus about the reality of human-caused climate change. As a result of a decades-long disinformation campaign,[30] many people believe there is disagreement among experts about human-caused climate change. These people are less likely to be convinced that climate change is real, human-caused, and serious and that it can be halted or reversed.[30]

IT'S REAL	Global warming is happening.
IT'S US	Human activity is the main cause.
EXPERTS AGREE	More than 99% of the world's climate experts are convinced, based on evidence, that human activity is warming the planet.
IT'S BAD	The human health impacts are serious—especially for children, older adults, people with chronic illnesses, and members of low-income communities and communities of color.
THERE'S HOPE	There are many actions we can take that will address climate change and improve our health—in equitable ways.

Figure 18-1 Five key facts about global warming (in 10 words). (*Source*: Center for Climate Change Communication, George Mason University.)

When people *are* told that there is a consensus among scientists about human-caused climate change, their understanding changes.[31] For example, a presentation of the following statement increases from about 60% to about 80% the proportion of people who believe there is a consensus: "Based on the evidence, more than 97% of climate experts are convinced that human-caused climate change is happening."[32] By presenting information about this consensus, rather than explaining the facts of human-caused climate change, health professionals can avoid conversations about areas of climate science with which they themselves may not be familiar.

2. <u>Climate change is harming people's health everywhere</u>. Health professionals are in a unique position to educate the public about the health relevance of climate change. They are trusted members of every community. Health is their area of expertise. They can convey the ways in which climate change is already causing health effects and how these effects are likely to worsen unless actions are taken to address climate change. They can communicate about the co-benefits to health resulting from actions that address climate change.[33,34]

People process threat information more easily when it is explained in a way that reflects their own experience or that of others in their community. Since personal and community experience varies, specific content of messages needs to be tailored for specific communities. For example, in communities where air quality is poor, relevant stories might refer to the way that more-severe heat waves due to climate change are contributing to poor air quality and resulting in increased occurrence of serious respiratory disease.

3. <u>People and communities can take actions that will limit climate change, protect their health from the consequences of climate change, and make their communities healthier places to live</u>. Focusing on solutions can bring people together, even when the underlying ways of thinking may differ.[35] The belief that taking action will make a difference can bolster individual self-efficacy and collective efficacy and motivate people to act. Conversely, the absence of belief in the efficacy of acting is associated with a sense of helplessness, denial, and avoidance.[36]

Invoking the value of protecting people from harm can help to engage people in responding to climate change. Most people feel that protective behavior is worthwhile and sensible. When people learn about potential harm, they are more likely to take effective action to reduce the risk of that harm.

The actions people take to protect themselves from health risks can result in a healthier community. For example, encouraging people to walk or bike rather than drive improves their health and reduces the use of fossil fuels. Buying locally grown produce helps reduce both fat intake and long-distance food transportation using fossil fuels. A plant-forward diet can reduce reliance on raising meat, especially red meat, which has the largest carbon footprint and is leading to forest destruction worldwide as other countries develop a greater taste for the meat-based diets of the Americas. The percentage of individuals who reject meat entirely and choose vegetarian diets has also increased. (See Chapter 14.)

KEY ELEMENTS OF COMMUNICATION

In communicating about climate change, health professionals and others should follow basic principles of climate change communication (Box 18-1). Contacting, convening, collaborating, and putting a human face on climate change are four key elements of getting the message out.

Box 18-1 Principles of Climate Change Communication

Howard Frumkin and Edward Maibach

1. People's prior beliefs and cultural frames shape their knowledge, attitudes, and behaviors, so effective communication should ideally be tailored accordingly.
2. Engaging people by listening is more effective than talking at them.
3. Simply providing scientific information is unlikely to engage people.
4. People who doubt or deny the reality of climate change may support energy efficiency, conservation, and similar measures for other reasons, such as clean air, economic savings, or health.
5. People value immediate benefits more than long-term benefits.
6 Simple clear messages tend to be more effective than complex or abstract ones.
7. Trusted messengers, especially if they are known to people, are most effective.
8. Repetition is an important element in effective communication.
9. Climate change is scary. It is helpful to acknowledge the current impacts and future risks of climate change to human health, but too much fear can be counterproductive. Strive to increase audience members' sense of agency by focusing on what they can do to support solutions.
10. People are most likely to adopt behaviors that are easy, fun, and popular.
11. Communication is most effective when reinforced by policies, environmental changes, and products or services that make it easy to perform recommended actions. The default should be the healthy option.

Contacting

Health professionals can contact those who should be conveying information about the health consequences of climate change: government officials, leaders in public health and safety organizations, representatives of nongovernmental organizations, news reporters, members of newspaper editorial boards, radio and TV weather forecasters, and policymakers. They can strengthen the knowledge base in professional organizations and networks. They can engage colleagues through presentations, meetings, and discussion groups. And they can engage those in distant locations via websites, webinars, and teleconferences. Clinical encounters represent another set of opportunities for sharing information.

Convening

Since protection of health requires cross-sectoral approaches, health professionals can convene stakeholders from multiple sectors to plan adaptation and mitigation measures. Stakeholders include (a) traditional partners, such as government agencies, hospitals and clinics, healthcare providers, and nongovernmental organizations and (b) nontraditional partners, such as agencies and organizations involved in land use, environmental protection, education, transportation, economic development, and social justice and the faith community.

Collaborating

Health professionals can build partnerships and coalitions to facilitate communication about climate change, such as by increasing media coverage. They can work with others to train health professionals and journalists, organize hearings about public policy, and advocate for specific measures.

Health professionals and others can utilize many different venues for communicating and educating about climate change. These opportunities include one-on-one conversations, small group discussions, community presentations, and massive open online courses.

Putting a Human Face on Climate Change

People tend to understand life through the stories of individuals and families, rather than through statistics. Stories of people who have experienced health effects from climate change can powerfully influence people's beliefs and actions, such as stories about older people who have died during heat waves, children with asthma that has been exacerbated by air pollution, people with increased allergy symptoms because of longer pollen seasons, and children malnourished because of drought induced by climate change.

MISINFORMATION AND DISINFORMATION

Despite careful messaging, various factors can interfere with the public's understanding of climate change. Communication has been undermined by widespread misinformation (incorrect statements) and disinformation (purposely misleading statements). Disinformation has been developed and spread intentionally to undermine the truth, often with the intention of creating doubt about existing consensus. Misinformation and disinformation have adversely affected efforts to address climate change by spreading unsound critiques of sound scientific conclusions and by creating rationales for delaying the implementation of necessary measures.[30]

Health professionals and others have many opportunities to correct misinformation and disinformation during their interactions with the public. Publications have exposed the strategies implemented by some corporations and others to delay policies that could adversely affect their profits. For example, the book *Merchants of Doubt* detailed how the strategy of sowing doubt about scientific research has been used effectively by the tobacco industry and its trade group to raise questions about the health hazards of cigarette smoking and by the fossil fuel industry to raise doubts about climate change.[30] Similarly, *Big Oil vs the World*, a documentary by the BBC, reported how this same strategy has been utilized by oil companies, the American Petroleum Institute, and their allies to exaggerate uncertainties in climate science and to suggest that the conclusions of climate scientists about global warming are uncertain.[37]

Disinformation campaigns have often featured exaggerated or manipulated frames that play on public fears, especially about jobs and the economy. Many representatives of the fossil fuel industry and the politicians who support them have falsely claimed that decreasing the use of fossil fuels will have adverse effects on jobs and the economy. In fact, the number of jobs created in the renewable energy sector is growing faster than the number in the fossil fuel energy sector, and the cost of renewable energy is now less than the cost of energy from fossil fuels. Renewable energy and low-carbon transportation, such as electric vehicles, are growth industries bringing new opportunities and many jobs to the economy. For example, investing in solar-photovoltaic equipment manufacturing creates 1.5 times as many jobs, and investing in wind power 1.2 times as many jobs, as the same amount invested in fossil fuel production.[38]

A four-step "debunking" approach called the "truth sandwich"—as presented in the *Debunking Handbook*—can be effective.[39] It consists of:

1. <u>Fact: State the fact that has been misrepresented</u>. This statement should be simple, concrete, and plausible, without the use of jargon. For example: "Based on the evidence, more than 97% of the world's climate scientists have concluded that human-caused climate change is occurring."
2. <u>Myth: Warn that you are going to state the myth and then state it—once only</u>. Mention the myth preceded by a warning, such as: "It is a myth that there is disagreement among climate scientists about the reality of human-caused climate change."
3. <u>Fallacy: Explain how the myth is misleading</u>. This can be done in many ways, as long as it makes clear—even to people paying relatively little attention—why the myth is wrong. For example: "For many years, big oil companies claimed that the science of climate change was not yet settled because they knew that lie was reassuring to the public. But they knew it was a lie and they told it anyway."
4. <u>Fact: Restate the fact</u>. Closing with a restatement of the fact makes it more likely that people will accept and remember the fact, helping to supplant the myth.

Since misinformation is hard to dislodge, preventing it from taking root in the first place can be an effective form of "belief inoculation," forewarning, and preemptive refutation. Effective measures include:[40]

- Warning people that they might be misled can prevent them from later relying on misinformation.
- General warnings—such as "The media sometimes does not check facts before publishing information that turns out to be inaccurate."—can make people more receptive to later corrections.
- Specific warnings that content may be false can reduce the likelihood that people will share misinformation or disinformation online.

SUMMARY POINTS

- Communication can be an indispensable strategy for educating patients, the public, and policymakers about the human health impacts of climate change and engaging them in solutions-oriented actions.

- Recently, there has been a large increase in public concern about climate change and growing awareness of its relevance to human health. However, most people continue to see climate change as primarily an environmental, scientific, or political problem.
- Presenting health-framed information about climate change and measures to address climate change is effective for engaging people across the political spectrum on the issue and building public support for measures to address climate change.
- There is growing evidence on how to effectively present health-framed information about climate change.
- While misinformation and disinformation abound, there are proven strategies for "debunking" and "inoculating" people against erroneous and misleading information.

References

1. Maibach E, Frumkin H, Ahdoot S. Health professionals and the climate crisis: Trusted voices, essential roles. World Medical and Health Policy 2021; 13: 137–145. https://doi.org/10.1002whm3.421.
2. Maibach E, Leiserowitz A, Roser-Renouf C, Mertz CK. Identifying like-minded audiences for global warming public engagement campaigns: An audience segmentation analysis and tool development. PLoS ONE 2011; 6: e17571.
3. Leiserowitz A, Roser-Renouf C, Marlon J, Maibach E. Global Warming's Six Americas: A review and recommendations for climate change communication. Current Opinion in Behavioral Sciences 2021; 42: 97–103. doi: 10.1016/j.cobeha.2021.04.007
4. Bell J, Poushter J, Fagan M, Huang C. In response to climate change, citizens in advanced economies are willing to alter how they live and work. Pew Research Center, September 14, 2021. Available at: https://www.pewresearch.org/global/2021/09/14/in-response-to-climate-change-citizens-in-advanced-economies-are-willing-to-alter-how-they-live-and-work/. Accessed November 19, 2022.
5. McCright AM, Dunlap RE. Anti-reflexivity: The American conservative movement's success in undermining climate science and policy. Theory Culture & Society 2010; 27: 100–133.
6. Roser-Renouf C, Stenhouse N, Rolfe-Redding J, et al. Engaging diverse audiences with climate change: Message strategies for global warming's six Americas. In A Hansen, R Cox (eds.). Handbook of Environment and Communication. New York: Routledge, 2015, pp. 368–386.
7. Maibach E, Kreslake J, Roser-Renouf C, et al. Do Americans understand that global warming is harmful to human health? Evidence from a national survey. Annals of Global Health 2015; 81: 396–409.
8. Roser-Renouf C, Maibach E, Leiserowitz A, et al. Understanding the health harms of climate change: A Six Americas analysis. George Mason University Center for Climate Change Communication, 2020. Available at: https://www.climatechangeco

mmunication.org/all/understanding-the-health-harms-of-climate-change-a-six-americas-analysis/. Accessed December 29, 2022.

9. Kotcher J, Maibach E, Rosenthal S, et al. Americans increasingly understand that climate change harms human health. George Mason University Center for Climate Change Communication, 2020. Available at: https://www.climatechangecommun ication.org/all/americans-increasingly-understand-that-climate-change-harms-human-health/. Accessed December 29, 2022.

10. Nisbet MC. Communicating climate change: Why frames matter to public engagement. Environment 2009; 51: 12–23.

11. Hase V, Mahl D, Schafer MS, Keller TR. Climate change in news media across the globe: An automated analysis of issue attention and themes in climate change coverage in 10 countries (2006–2018). Global Environmental Change 2021; 70: 102353. https://doi.org/10.1016/j.gloenvcha.2021.102353.

12. Weathers M, Kendal B. Developments in the framing of climate change as a public health issue in US newspapers. Environmental Communication 2016; 10: 593–611.

13. King N, Bishop-Williams KE, Beauchamp S, et al. How do Canadian media report climate change impacts on health? A newspaper review. Climatic Change 2019; 152: 581–596.

14. Harrison S, Macmillan A, Rudd C. Framing climate change and health: New Zealand's online news media. Health Promotion International 2020; 35: 1320–1330.

15. Sarfaty M, Duritz N, Gould R, et al. Organizing to advance equitable climate and health solutions: The Medical Society Consortium on Climate and Health. Journal of Climate Change & Health 2022; 7: 100174. https://doi.org/10.1016/j.joclim.2022.100174.

16. U.S. Call to Action on Climate, Health, and Equity: A Policy Action Agenda. Available at: https://climatehealthaction.org/cta/climate-health-equity-policy/. Accessed August 24, 2023.

17. Choi-Schagrin W. Effort to reframe climate change as a health crisis gains steam. New York Times, November 4, 2021. Available at: https://www.nytimes.com/2021/11/04/climate/public-health-climate-change.html?login=smartlock&auth=login-smartlock. Accessed September 3, 2023.

18. Kennedy B, Funk C. Public interests in science, health, and other topics. Pew Research Center, December 11, 2015. Available at: https://www.pewresearch.org/science/2015/12/11/public-interest-in-science-health-and-other-topics/. Accessed December 29, 2022.

19. Smith N, Leiserowitz A. The rise of global warming skepticism: Exploring affective image associations in the United States over time. Risk Analysis 2012; 32: 1021–1032.

20. Maibach E, Uppalapati S, Orr M, Thaker J. Harnessing the power of communication and behavior science to enhance society's response to climate change. Annual Review of Earth and Planetary Science 2023; 51: 53–77.

21. Ipsos. Doctors and scientists are seen as the world's most trustworthy professions. 2022. Available at: https://www.ipsos.com/en/global-trustworthiness-index-2022. Accessed December 29, 2022.

22. Leiserowitz A, Maibach E, Rosenthal S, et al. Politics & Global Warming, April 2022. Yale Program on Climate Change Communication. New Haven, CT: Yale University and George Mason University, 2022. Available at: https://www.climatechangecommun ication.org/all/politics-global-warming-april-2022/. Accessed December 29, 2022.

23. National Research Council. The Future of the Public's Health in the 21st Century. Washington, DC: The National Academies Press, 2003.

24. Maibach EW, Nisbet M, Baldwin P, et al. Reframing climate change as a public health issue: An exploratory study of public reactions. BMC Public Health 2010; 10: 299. https://doi.org/10.1186/1471-2458-10-299.

25. Myers TA, Nisbet MC, Maibach EW, Leiserowitz AA. A public health frame arouses hopeful emotions about climate change. Climatic Change. 2012; 113: 1105–1112. doi: 10.1007/s10584-012-0513-6.

26. Kotcher J, Maibach E, Hassol S, Montoro M. How Americans respond to information about global warming's health impacts: Evidence from a national survey experiment. GeoHealth 2018; 2: 262–275. doi.org/10.1029/2018GH000154.

27. Kotcher J, Feldman L, Luong KT, et al. Advocacy messages about climate and health are more effective when they include information about risks, solutions and a normative appeal: Evidence from a conjoint experiment. Journal of Climate Change and Health 2021; 3: 100030. https://doi.org/10.1016/j.joclim.2021.100030.

28. Myers TA, Maibach EW, Roser-Renouf C, et al. The relationship between personal experience and belief in the reality of global warming. Nature Climate Change 2013; 3: 343–347.

29. Akerlof K, Maibach EW, Fitzgerald D, et al. Do people "personally experience" global warming, and if so how, and does it matter? Global Environmental Change 2013; 23: 81–91.

30. Oreskes N, Conway E. Merchants of Doubt. New York: Bloomsbury Press, 2010.

31. van der Linden SL, Leiserowitz AA, Feinberg GD, Maibach EW. The scientific consensus on climate change as a gateway belief: Experimental evidence. PLoS One 2015; 10: e0118489. doi: 10.1371/journal.pone.0118489.

32. Myers TA, Maibach EW, Peters E, Leiserowitz A. Simple messages help set the record straight about scientific agreement on human-caused climate change: The results of two experiments. PLoS One 2015; 10: e0120985. doi: 10.1371/journal.pone.0120985.

33. Sarfaty M. Climate change must be a central concern of all health professionals. Harvard Health Policy Review. June 11, 2022. Available at: http://www.huhpr.org/vol ume-21-issue-1-2/2022/6/10/climate-change-must-be-a-central-concern-of-all-hea lth-professionals-xeax3?rq=sarfaty. Accessed August 24, 2023.

34. Luong KT, Kotcher J, Miller J, et al. Prescription for healing the climate crisis: Insights on how to activate health professionals to advocate for climate and health solutions. Journal of Climate Change and Health 2021; 4. doi: 10.1016/j.joclim.2021.100082.

35. Johnson BB. Climate change communication: A provocative inquiry into motives, meanings, and means. Risk Analysis 2012; 32: 973–991.

36. Roser-Renouf C, Maibach EW, Leiserowitz A, Zhao X. The genesis of climate change activism: From key beliefs to political action. Climatic Change 2014; 125: 163–178. doi: 10.1007/s10584-014-1173-5.

37. BBC. Big Oil vs the World tells the 40 year story of how the oil industry delayed action on climate change. July 21, 2022. Available at: https://www.bbc.com/mediacen tre/2022/big-oil-vs-the-world. Accessed December 29, 2022.

38. Jaeger J, Walls G, Clarke E, et al. The green jobs advantage: How climate friendly investments are better job creators. World Resources Institute, October 2021. Available at: https://www.ituc-csi.org/IMG/pdf/the_green_jobs_advantage_-_wri_nce_and_ituc_working_paper.pdf. Accessed December 29, 2022.

39. Lewandowsky S, Cook J, Ecker U, et al. The Debunking Handbook. George Mason University, 2020. Available at: https://www.climatechangecommunication.org/wp-content/uploads/2020/10/DebunkingHandbook2020.pdf. Accessed December 29, 2022.
40. Cook J, Supran G, Lewandowsky S, et al. America Misled: How the Fossil Fuel Industry Deliberately Misled Americans about Climate Change. Fairfax, VA: George Mason University Center for Climate Change Communication, 2019. Available at: https://www.climatechangecommunication.org/america-misled/. Accessed December 29, 2022.

Further Reading

Centers for Disease Control and Prevention. Communicating the health effects of climate change: A toolkit for public health outreach. Available at: https://www.cdc.gov/climateandhealth/docs/ClimateandHealthPresentationGRANTEES-508.pdf. Accessed September 3, 2023.

This site provides a series of slides that can be used in presentations about the human health relevance of climate change and actions that can be taken to protect the public's health.

Gould R, Sarfaty M, Maibach E. The Health Promise of Climate Solutions. A Report of the Medical Society Consortium on Climate and Health. Fairfax, VA: George Mason University, 2022. Available at: https://test.ms2ch.org/wp-content/uploads/2022/10/The-Health-Promise-of-Climate-Solutions-5-22-1.pdf.

This work provides detailed, evidence-based guidance for making the case about the health benefits of measures to address climate change.

Maibach E, Uppalapati S, Orr M, Thaker J. Harnessing the power of communication and behavior science to enhance society's response to climate change. Annual Review of Earth and Planetary Science 2023; 51: 53–77.

This article provides detailed practical information about the principles of effective public communication about climate change and guidance for maximizing the effectiveness of recommendations concerning behavior.

Peters E, Salas R. Communicating statistics on the health effects of climate change. New England Journal of Medicine 2022; 387: 193–196.

This paper provides a practical summary of how to present statistical information about climate change as a health issue in a simple and compelling manner.

19

Building Movements to Address Climate Change

Teddie M. Potter, Julia Frost Nerbonne, and Vishnu Laalitha Surapaneni

Environmental health initiatives during the past several decades provide illustrative examples and helpful lessons for building and strengthening social movements. These movements include the Earth Day Movement, which played an important role in enacting federal environmental laws; the Environmental Justice Movement, which fostered an alliance between civil rights activists and environmentalists; and the Climate Justice Movement (Chapter 20). These movements have stimulated the genesis and growth of many grassroots climate organizations and initiatives. They illustrate the transformational potential of social movements and provide examples of best practices for building and strengthening them—such as by developing inspiring narratives, inclusive relationships, clear structures, and strategies for meaningful action.

Awareness of poor air quality and the health impacts of environmental degradation grew in the 1960s, which led Senator Gaylord Nelson to propose a "national day for the environment." He engaged college students and others in this vision, which led to the first Earth Day on April 22, 1970. By grassroots organizing, the Earth Day Movement mobilized more than 20 million people throughout the United States and influenced the enactment of several major federal environmental laws.[1]

The Environmental Justice Movement began in North Carolina in 1982, when a group of environmentalists, civil rights activists, and community leaders engaged in community protests against the proposed dumping of soil contaminated with polychlorinated biphenyls in a landfill in a low-income, African American community. When these key stakeholders organized and pooled their resources, they were able to create a movement that has engaged millions of people across the United States and beyond.[2]

The Planetary Health Movement evolved from analyzing and addressing the impacts of human disruptions to the Earth's natural systems that affect all forms

Teddie M. Potter, Julia Frost Nerbonne, and Vishnu Laalitha Surapaneni, *Building Movements to Address Climate Change*
In: *Climate Change and Public Health*. Edited by: Barry S. Levy and Jonathan A. Patz, Oxford University Press.
© Oxford University Press 2024. DOI: 10.1093/oso/9780197683293.003.0019

of life. It now includes an alliance of more than 340 universities, nongovern-mental organizations, research institutes, and government entities from more than 64 countries.[3]

This chapter focuses on how health professionals, climate scientists, and others can help build and strengthen social movements like those briefly described above.

A *social movement* is defined as "a loosely organized but sustained campaign in support of a social goal, typically either the implementation or the prevention of a change in society's structure or values."[4] Policies and actions to address climate change are more likely to succeed if they are associated with social movements that engage people, develop shared goals, pool their resources, and focus their energy.

Two complementary theories on movement building are the Resource Mobilization Theory (RMT) and the New Social Movement Theory (NSMT). In RMT, the success of a movement depends on the ability of its organizers to identify and mobilize available resources—such as knowledge, relationships, and financial support—to take advantage of a strategic opportunity. In NSMT, the success of a movement depends on developing transformational identities of movement participants and the stories or narratives that motivate them. RMT focuses on bringing about institutional, state-centered change; in contrast, NSMT focuses on change in cultural spheres and civil society.

THE MARSHALL GANZ APPROACH
TO MOVEMENT BUILDING

Many highly successful social movements have been based on the work of scholar and activist Marshall Ganz, Senior Lecturer in Leadership, Organizing, and Civil Society at Harvard's Kennedy School of Government. He has described five neces-sary elements for building social movements for transformative change.[5-8] These elements, which include aspects of both RMT and NSMT, are described below.

1. Developing a shared story: It is important develop a shared story that resonates with participants' values so that they are able to confront a shared challenge. In successful movements, stories inspire people to action and protect them from self-doubt and isolation. People join movements be-cause they have been motivated by personal experience. It is valuable to have participants share their own stories—each of them "a story of self"—about what motivated them to join others in the movement.

2. Building relationships: People move people. And people are moved by examples of people moving people. Successful social movements

intentionally create relationships. Among participants, "story of self" is integrated into a "story of us"—how people can become more powerful together. Individuals who try to take action by themselves often feel alienated, powerless, and ineffective.[9] In contrast, if people can find others who share their values, they can work together to build power. (Many instructional materials are available on how to build relationships, such as by conducting personal interactions, recruiting team members, and inviting others to join in a social movement.[10])

3. Shared structure: Successful movements provide explicit and transparent structures that facilitate collaboration, planning, strategizing, decision-making, and action. They also develop effective ways to identify, recruit, and train leaders. Clear norms, goals, and organizational structure are essential for a movement to be successful. ("Working in Groups: Start Here"[11] is a useful tool for setting up group structures.)

4. Shared strategy: Strategizing is determining how to aggregate participants' resources into power so that they can turn what they have into what they need to get what they want. Developing a shared strategy helps to focus on the "story of now," which leverages the power of a social movement to reach specific goals. Some questions that can be used to develop a shared strategy include: What is our goal? How can we identify the resources needed? How can we organize these resources effectively to achieve our goals?[7]

5. Shared action: Social movements are fundamentally action-oriented. To be effective, movements must turn stories, relationships, structure, and strategy into action. Actions can be measured or counted, such as counting the number of people coming to a meeting or a protest demonstration (Figure 19-1), signing on to letters or petitions, and voting. One success can lead to another, ultimately leading to transformational change.

One model of shared action is a three-legged stool, symbolizing (a) practical actions, such as installing solar panels to reduce greenhouse gas (GHG) emissions; (b) systemic actions on a larger scale, such as advocating for specific state or federal laws or regulations; and (c) building and maintaining relationships, which are essential for success.[12]

THE MANUEL PASTOR APPROACH
TO MOVEMENT BUILDING

A complementary approach to movement building, developed by Manuel Pastor and his colleagues at the Equity Research Institute at the University of Southern California,[13] focuses on the fundamental elements, implementation tools, and

Figure 19-1 Some of the more than 300,000 people who participated in the People's Climate March in New York in 2014. (Photograph by Jonathan A. Patz.)

scale of movement building, and stresses the role of strategic understanding of government and a need for economic system change. Like Marshall Ganz, they describe fundamental elements of a successful movement as being rooted in shared story, vision, and commitment built through transparent and strategic relationship building.

They are also committed to a theory of change that does not assume that status-quo economics will serve us in the future. Since social movements redistribute resources and power, they urge us to plan for prosperity after this redistribution. For example, social movements addressing climate change can demonstrate how the transition to renewable energy will strengthen the economy and provide opportunities for sustainable well-being.

Social movements are iterative and interactive. To be successful, they culti-vate and support identifying individuals and communities adversely affected by climate change and engage them in decision-making that affects them. Social movements need a solid base of data, research, and monitoring so that they can create and evaluate policies for mitigation, adaptation, and reparation.

Finally, social movements need to work with stakeholder organizations that have enough members and resources to challenge existing power. They can do this by broadening their range of issues and their geographic base and by converging with other social movements. Healthcare professionals can be key intermediaries, connecting those who address health issues to those who address worker rights, systemic racism, rights of Indigenous Peoples, and the climate crisis.

IMPORTANCE OF MAINTAINING RESILIENCE

Social movements and their members inevitably experience setbacks in their work, which can generate despair and disengagement. Movements need to build resilience to maintain long-term engagement of their members. This can be accomplished, in part, by focusing on both short-term objectives and long-term goals. Building and maintaining relationships is critically important for strengthening movements, fostering a sense of community, and reducing the likelihood of burnout.

CASE STUDIES

This section, which briefly describes two case studies of climate movements led and supported by health professionals, illustrates some elements of movement building strategy.

Stop Line 3 Movement: Enbridge Line 3 Tar Sands Oil Pipeline Resistance

The Enbridge Line 3 Replacement Project, also known as Line 3/93, daily carries 760,000 barrels of carbon-intensive tar sands oil from a mining site in Alberta through Minnesota to Wisconsin. Before construction, Line 3/93 was estimated to have a maximum annual climate impact equivalent to that of 50 coal-fired power plants or 38 million gasoline-powered cars.[14]

In Minnesota, members of Health Professionals for a Healthy Climate (HPHC) joined the Indigenous-led Stop Line 3 Movement. They co-authored a white paper on the health impacts of the construction and operation of Line 3/93,[15] submitted testimony during regulatory hearings, wrote op-eds in newspapers, and participated in social media campaigns and demonstrations against Line 3/93.

The actions of HPHC built on Marshall Ganz's teachings, including elements such as shared relationships and shared strategy, and helped to build the scope and scale of the movement described by Manuel Pastor. For example, HPHC contacted national and state health organizations involved in advocating for climate action. It hosted and participated in webinars on the health impacts of Line 3/93, participated in local and federal advocacy efforts, and organized the "National Day of Action to Stop Line 3" with Indigenous Communities and environmental justice organizations. (More information about its work and the Stop Line 3 Movement can be found at https://www. stopline3.org/#intro.)

Declaring Climate Change a Public Health Crisis

The American Medical Association (AMA), with more than 190 state and specialty medical societies, is the largest medical association in the United States. In 2022, it adopted a policy declaring that "climate change a public health crisis that threatens the health and well-being of all people"—a landmark achievement that enabled it and its affiliated societies to advocate for federal legislation to reduce GHG emissions and to support policies that promote a transition to clean energy and climate justice. This policy resulted from a multiyear effort by several AMA members.

By advocating that the AMA adopt this policy, they illustrated several principles of movement building: identifying available resources, building power to create systemic change from within, declaring a major problem (climate change) a public health crisis, collaborating with like-minded health professionals across the United States, developing a shared strategy, and using international science-based goals (those set by the 2015 Paris Agreement) to develop policy within the organization.

They also illustrated the importance of persistence. The proposal for a new AMA policy was introduced three times before it was considered for a vote. The failed attempts were utilized as opportunities for developing shared relationships and building support for the policy within the AMA community—which led to its ultimate adoption.[16]

MOVEMENTS ADDRESSING CLIMATE CHANGE

The Climate and Health Movement consists of a large number of professional organizations and nongovernmental organizations that are working to address climate change from a public health perspective. It recognizes that many policies and actions to address the climate crisis also have benefits that can improve the health of individuals and communities. Box 19-1 provides an extensive list of these organizations and their activities. Box 19-2 focuses on greening of the healthcare sector.

Innovative Initiatives

An increasing number of innovative initiatives are addressing climate change. Many of these initiatives rely on social media and the Internet as well as on the capabilities, commitments, and energy of young people. Box 19-3 provides three examples of these initiatives.

Box 19-1 The Climate and Health Movement

Linda Rudolph

The rapidly expanding Climate and Health Movement deploys the voice and expertise of health professionals to limit further global warming, protect people today and future generations from the adverse consequences of climate change, and promote strategies that create healthy, equitable, and sustainable communities. With their expertise, credibility, and public trust, health professionals are uniquely positioned to inform the public and policymakers that climate change is a health emergency and to advocate for healthy and equitable climate solutions.

More than 150 health organizations have endorsed the 2019 U.S. Call to Action on Climate, Health, and Equity: A Policy Action Agenda. Priorities for climate health action include

- Aggressive reduction of GHG emissions sufficient to limit global temperature increases to 1.5°C (2.7°F) above pre-industrial levels
- Rapid transition away from the production and use of coal, oil, and "natural gas" (methane) to clean and safe renewable energy and energy efficiency
- Emphasis on active transportation during the transition to zero-carbon transportation systems

- Promotion of healthful, sustainable, and resilient farms and food systems, forests, and natural lands
- Access to safe and affordable drinking water for everyone in the United States and a sustainable water supply
- Investments and policies that support a just transition for workers and communities adversely impacted by climate change and by the transition to a low-carbon economy
- Active health-sector engagement in climate action, including proactive assessment of the health and equity impacts of proposed climate action to optimize the health, equity, and environmental justice benefits of climate programs and policies
- Robust and meaningful community involvement in climate program and policy decision-making at all levels of government
- Building and maintaining healthy, equitable, and climate-resilient communities and redressing historical discrimination and its adverse impacts on low-income communities and communities of color
- Investments in climate and health to build health-sector capacity to address the climate crisis.

The following are some nonprofit health organizations, academic programs, government agencies, and other programs that are engaged in the Climate and Health Movement, listed below in alphabetical order:

- The Alliance of Nurses for Healthy Environments, which educates, organizes, and mobilizes nurses to address climate change (It coordinates the Nurses Climate Challenge, which includes more than 540 Nurse Climate Champions from 6 continents, 16 countries, and 42 U.S. states in partnership with Health Care Without Harm.)
- The American Lung Association Healthy Air Campaign, which advocates for the use of sound science in policy processes, strong federal regulation of air pollutants, and policies that support electrification of vehicles
- The American Public Health Association's Center for Climate, Health, and Equity, which advances healthy climate policy and galvanizes public health workers to address climate change from a health equity perspective
- The Climate MD program of the Center for Climate, Health, and the Global Environment at the Harvard Chan School of Public Health, which prepares community health centers to address climate-related impacts

on health and healthcare delivery and offers fellowship programs for health professionals

- The Columbia University Mailman School of Public Health Global Consortium on Climate and Health Education, which develops health professional standards for climate and health knowledge and practice, shares educational resources, and advocates for the inclusion of climate and health competencies on health professional licensure examinations
- EcoAmerica's Climate for Health, which is a national initiative led by a network of health-sector leaders that offers tools, resources, online training, and other communications to enable health-sector leaders to speak about and advocate for climate solutions
- Fossil Free for Health, which is a new initiative that aims to address the intertwined fossil fuels and climate change health crises through addressing the direct health harms throughout the fossil fuel life cycle and the role of the fossil fuels industry in perpetuating harm
- The Global Climate and Health Alliance, which works to integrate health into global and national policy responses to climate change, such as through strengthening country commitments to climate action, making health and equity central to national climate action plans, and collaborating with the World Health Organization
- Health Care Without Harm, which works to reduce the environmental footprint of healthcare and promotes environmental health and justice globally (Box 19-2.)
- The Health Professional Call for a Fossil Fuel Non-Proliferation Treaty, which has been signed by organizations representing 46 million health professionals and thousands of individual health professionals to urge governments to develop a legally binding global plan to end all exploration, production, and use of fossil fuels
- The Medical Society Consortium on Climate & Health (comprised of 46 national U.S. medical societies representing over 700,000 physicians), which organizes, empowers, amplifies, and mobilizes medical doctors to advocate for equitable and effective health-focused climate solutions and offers the Climate and Health Equity Fellowship, which seeks to empower physicians of color to lead on climate and health equity
- The National Academy of Medicine's Grand Challenge on Climate Change, Human Health, and Equity, which launched a multiyear global initiative to address climate change and the Action Collaborative on Decarbonizing the U.S. Health Sector, which aims to address the health sector's environmental impact and strengthen its sustainability and resilience

- The Office of Climate Change and Health Equity in the Department of Health and Human Services, which collaborates with other parts of the Department, other government agencies, and external partners to identify disproportionate climate risks and vulnerabilities, promote and translate climate and health research, support efforts to reduce health-sector emissions, and build capacity of the workforce addressing climate and health
- Physicians, Scientists, and Engineers for Healthy Energy, which is a non-profit research institute that studies health impacts of energy production and use
- Physicians for Social Responsibility, which, with chapters in 20 U.S. states, works to oppose fracking, replace methane gas stoves and heaters with electric appliances, reduce the risk of new technologies such as carbon capture, and mobilize health professionals to advocate for healthy and equitable climate solutions.

In addition, there is a rapidly expanding network of state-based climate and health organizations that are working to educate and mobilize health professionals for climate action and to advocate for healthy climate policies at the state and local levels.

Box 19-2 Greening of the Healthcare Sector

Gary Cohen

Healthcare has critical roles to play in reducing the health consequences of climate change and improving the resilience of the communities in which healthcare facilities are located. In many areas, healthcare is both a major energy consumer and a large employer. Hospitals are the second-most intensive energy user in the U.S. economy.[1] The healthcare sector is responsible for 8.5% of all U.S. GHG emissions.[2] Healthcare is also a leading user of toxic chemicals. And it produces more than 5 million tons of medical waste annually.[3]

GHG emissions are divided into three scopes (Figure 19-2). Scopes 1 and 2 relate to direct emissions from healthcare facilities and purchased energy from fossil fuel sources to operate these facilities. Scope 3 relates to indirect GHG emissions related to purchased goods, patient travel, waste, food, pharmaceuticals, and healthcare system investments in fossil fuels. In the healthcare sector, more than 70% of GHG emissions are in Scope 3.[4]

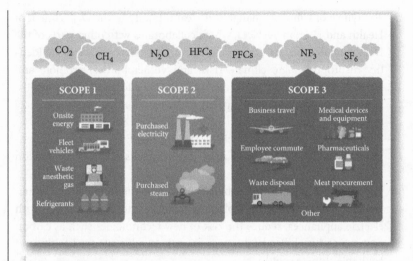

Figure 19-2 Common greenhouse gas emission sources in healthcare. CO_2: Carbon dioxide, CH_4: Methane, N_2O: Nitrous oxide, HFCs: Hydrofluorocarbons, PFCs: Perfluorocarbons, NF_3: Nitrogen trifluoride, SF_6: Sulfur hexafluoride. Other Scope 3 categories can be found at: https://ghg protocol.org. (*Source*: Practice Greenhealth. Courtesy of Gary Cohen.)

Leading healthcare organizations are becoming leaders in low-carbon development and climate resilience. Due to their significant economic clout—healthcare accounts for 18% of the U.S. economy—hospitals can exert both upstream influence on their supply chains and downstream influence on their employees and patients.

The healthcare sector is taking on three new roles and responsibilities:

1. Mitigating GHG emissions by adopting energy efficiency measures and transitioning to renewable energy sources, which reduce expenses and address the public health impacts of climate change. Over the past few years, healthcare systems have made major commitments to mitigating GHG emissions. In the United States, more than 870 hospitals, suppliers, and group purchasing organizations have signed the Department of Health and Human Services Climate Pledge, committing to reduce their GHG emissions by 50% by 2030 and 100% by 2050. For example, Mass General Brigham in Boston invested more than $61 million over a 3.7-year period in a strategic energy master plan and in energy efficiency measures for its entire 15-facility system. It installed onsite combined heat and power technology, invested in

renewable energy, and designed new facilities with greater natural light, energy-efficient materials, and smart design. These measures reduced energy use by 25%, yielding savings equal to $61 million.[5] In 25 countries, more than 14,000 hospitals have committed to the Race to Zero Initiative, a program facilitated by the UNFCCC and supported by Health Care Without Harm.

2. <u>Anchoring community resilience</u>, such as by preparing for the health impacts of extreme weather events and by financially investing in local sustainable development. For example, the Gundersen Health System in Wisconsin has made a variety of energy investments to achieve energy independence for its campuses and co-invested in community-wide energy projects, such as wind power and methane gas recovery, to increase the resilience of its host communities.

3. <u>Providing leadership</u> by advocating for policies to address climate change and other environmental issues. The Health Care Climate Council, facilitated by Health Care Without Harm and comprised of 20 leading healthcare systems, has been advocating for policy changes that would support the health sector decarbonizing its buildings and supply chain. For example, it advocated for a "direct pay" option to allow non-profit hospitals to receive financial incentives for investing in renewable energy to operate their facilities—a policy that was adopted in the 2022 Inflation Reduction Act.

Health Care Without Harm is advocating for the Center for Medicare and Medicaid Services (CMS) to require hospitals to develop climate resilience plans and commitments to decarbonize their systems as a condition of participation in CMS reimbursements. It is also advocating for the Joint Commission, which accredits hospitals and other healthcare organizations, to require hospitals to measure their GHG emissions and make significant commitments to reduce these emissions.

In 2021, the COP26 Health Program was launched, which asked countries to commit to low-carbon and climate-resilient health systems. Seventy-three countries have committed to this program (called the Alliance for Transformative Action on Climate and Health), which is administered by the WHO.

Collectively, these initiatives are reducing the carbon footprint of healthcare and helping the healthcare sector lead by example by implementing a low-carbon and resilient pathway within the context of its broader healing mission.

Box References

1. U.S. Department of Energy. Integrating health and energy efficiency in healthcare facilities. June 2021. Available at: https://www.energy.gov/sites/default/files/2021-06/integrating-health-ee-healthcare-facilities.pdf. Accessed December 28, 2022.

2. The University of Chicago Medicine. Health care accounts for eight percent of U.S. carbon footprint. November 10, 2009. Available at: https://www.uchicagomedicine.org/forefront/news/health-care-accounts-for-eight-percent-of-us-carbon-footprint. Accessed February 13, 2023.

3. Practice Greenhealth. Waste. Available at: https://practicegreenhealth.org/topics/waste. Accessed February 13, 2023.

4. Practice Greenhealth. Scope 3 GHG emissions accounting tool. 2022. Available at: https://practicegreenhealth.org/tools-and-resources/scope-3-ghg-emissions-accounting-tool. Accessed December 28, 2022.

5. Quinlan D. Health care and climate change: An opportunity for transformative leadership. April 2014. Available at: https://noharm-uscanada.org/sites/default/files/documents-files/2704/Health%20Care%20Climate%20Change%20-%20Opportunity%20Transformative%20Leadership_0.pdf. Accessed February 13, 2023.

Box 19-3 Three Innovative Climate Change Initiatives

Olivia M. Dyrbye-Wright

The following are three innovative climate change initiatives.

350.org

350.org, an international grassroots organization, established in 2007 by six students and environmental activist and writer Bill McKibben, advocates for a future without fossil fuels. Its name derives from James Hansen, a NASA climatologist who warned that the safe upper limit for the level of atmospheric carbon dioxide is 350 parts per million. The organization fights to protect the Earth's atmosphere by accelerating renewable energy sources and ending financial support of fossil fuels. 350.org calls upon governments to address climate change through global campaigns. The organization also provides resources for those wanting to become engaged in climate activism, such as online and self-led courses and training sessions on how to build successful movements.

350.org has transformed political advocacy by optimizing online tools, such as sending mass email messages to raise money and organize demonstrations. By using Facebook, Twitter (now X), Instagram, and other platforms, the organization maximizes outreach. The utilization of social media has increased participation and mobilized collective action with demonstrations, thereby providing an "online-to-offline" framework.[1]

Fridays for Future

Fridays for Future (FFF) is an international organization that facilitates and enhances the demands of young people for immediate responses to the climate crisis. It began in 2018, when Greta Thunberg, then 15, protested for 3 weeks in front of the Swedish parliament, demanding action to fulfill commitments in the 2015 Paris Agreement. By documenting her protests on social media, she inspired young activists around the world.

Without a traditional hierarchical structure of organization, FFF relies on an online framework for connecting activists, disseminating protest messages, and coordinating offline action. It has used Instagram, Twitter, and other social media to facilitate the collective action of young people globally.[2] For example, in 2019, FFF led the Global Climate Strike, in which about 7.6 million young people in more than 180 countries demanded action to address the climate crisis.[3]

FFF has utilized social identity theory to link pro-environmental social identification with pro-environmental action. By reinforcing shared pro-environmental values, such as protecting the most vulnerable people and ecological systems, FFF promotes unified climate action.[4]

The Sunrise Movement

Since 2017, the Sunrise Movement has been a major driver of environmental activism in the United States, most notably recognized for its advocacy for the Green New Deal, which proposed public policy to address climate change, create jobs, and reduce economic inequalities. In 2018, it staged a sit-in at House Speaker Nancy Pelosi's office, during which it called for a total transition to renewable energy, accompanied by climate justice, social justice, job creation, and economic benefits.

The Sunrise Movement is guided by 11 core principles which highlight community-led activism, social justice, intersectionality, nonviolence, mutual respect, and unity. It has recognized that environmental hazards disproportionately impact the health of people of color through institutional structures and policies, thereby expanding the collective identity of environmental activism.[5]

The movement's success is a result of mobilizing collective action by creating a shared narrative of environmental justice. It is based on value-added theory, which proposes that collective action is prompted by perceived injustices and serves to decrease societal strain. The theory provides an analytical framework for understanding youth mobilization through six determinants of collective action utilized by the Sunrise Movement:

- Structural ability: Physical and social environments permit collective behavior, such as free speech
- Structural strain: Related to both anxiety of youth due to inadequate attention to climate change and their indignation in the face of environmental racism and societal injustices
- Generalized beliefs: A common understanding of the climate emergency and a shared belief in the value of political action
- Precipitating factors: Events that initiate collective action and mobilization
- Mobilization: Collective action through demonstrations, protests, and unified offline initiatives
- Social control: Governments and other entities counter collective action by providing disincentives or alternatives or by satisfying participants by addressing the root causes of structural strain.[6]

The growth of social media has empowered young activists with a shared collective identity. 350.org, FFF, and the Sunrise Movement have successfully utilized online tactics, such as posting information and photos from strikes and protests, paired with offline demonstrations. Digital technology has enabled successful mobilization and collective action, with social media creating online environments for deepening collective narratives.

Box References

1. Hestres L. A case of online-to-offline activism. Civic Media Project, 2015.

2. Maier BM. "No Planet B": An analysis on the collective action framing of the social movement Fridays for Future [master's thesis], 2019. Available at: https://www.diva-portal.org/smash/get/diva2:1393821/FULLTEXT01.pdf. Accessed January 23, 2023.

3. Cologna V, Hoogendoorn G, Brick C. To strike or not to strike? An investigation of the determinants of strike participation at the Fridays for Future climate strikes in Switzerland. PLoS One 2021; 16: e0257296. https://doi.org/10.1371/journal.pone.0257296.

4. Barth M, Masson T, Fritsche I, et al. Collective responses to global challenges: The social psychology of pro-environmental action. Journal

of Environmental Psychology 2021; 74: 101562. https://doi.org/10.1016/j.jenvp.2021.101562.

5. Rapid Transition Alliance. The Sunrise Movement: How a US grassroots youth movement helped set the national climate agenda for rapid change. January 28, 2021. Available at: https://www.rapidtransition.org/stories/the-sunrise-movement-how-a-us-grassroots-youth-movement-helped-set-the-national-climate-agenda-for-rapid-change/. Accessed January 23, 2023.

6. Saffer AJ. Value-added theory. In RL Heath, W Johansen (eds.). The International Encyclopedia of Strategic Communication. Hoboken, NJ: John Wiley & Sons, 2018, pp. 1–10.

SUMMARY POINTS

- Effective social movements are necessary to address the climate crisis.
- Much can be learned from the experience of other social movements that have addressed environmental issues.
- Systematic movement-building frameworks can provide valuable guidance.
- Effective social movements engage in practical and systemic actions, and they build and maintain relationships.
- Policies and actions to address climate change are likely to succeed if they are associated with social movements that engage people, develop shared goals, pool their resources, and focus their energy.

References

1. Earthday.org. The history of Earth Day. Available at: https://www.earthday.org/history/. Accessed July 1, 2022.
2. Schlosberg D, Collins LB. From environment to climate justice: Climate change and the discourse of environmental justice. WIREs Climate Change, 2014, p. 2.
3. Planetary Heath Alliance. Planetary health. Available at: https://www.planetaryhealthalliance.org/planetary-health. Accessed July 9, 2022.
4. Britannica.com. Social movement. Available at: https://www.britannica.com/topic/social-movement. Accessed September 15, 2022.
5. Ganz M. Why David Sometimes Wins: Leadership, Organizing, and Strategy in the California Farm Movement. New York: Oxford University Press, 2009.
6. Leading Change Network, Ganz M, New Organizing Institute. Organizing: People, power, and change: The one on one 1:1 meeting. Available at: https://commonslibrary.org/organizing-people-power-and-change-the-one-on-one-meeting/. Accessed July 1, 2022.
7. Ganz M. Organizing: People, power, change. 2014. Available at: https://commonslibrary.org/wp-content/uploads/Organizers_Handbook.pdf. Accessed July 1, 2022.

8. Ganz M. The power of story in social movements. In the Proceedings of the Annual Meeting of the American Sociological Association, Anaheim, California, August 18–21, 2001. Available at: https://dash.harvard.edu/bitstream/handle/1/27306251/Power_of_Story-in-Social-Movements.pdf. Accessed September 8, 2022.

9. Pipher M. The Green Boat: Reviving Ourselves in our Capsized Culture. New York: Riverhead Books, 2013.

10. Halpin J, Cook M. Social movements and progressivism: Part three of the progressive tradition series. Center for American Progress, April 14, 2010. Available at: https://www.americanprogress.org/article/social-movements-and-progressivism/. Accessed July 1, 2022.

11. Gannon A. Working in groups: Start here. Available at: https://commonslibrary.org/working-in-groups-start-here/. Accessed July 1, 2022.

12. Kerr R, Nerbonne J, Potter T. Sparking a movement for a healthy climate through leadership development. Creative Nursing 2019; 25: 216–221. doi: 10.1891/1078-4535.25.3.216

13. Equity Research Institute, USC Dornsife. Available at: https://dornsife.usc.edu/eri. Accessed February 13, 2023.

14. Minnesota 350. Line 3 Report: A giant step backward. Available at: https://mn350.org/giant-step-backward/. Accessed September 8, 2022.

15. Health Professionals for a Healthy Climate. Health risks of the Enbridge Line 3 pipeline replacement project. Available at: https://drive.google.com/file/d/1OWH6ohWaL92Urry6n6i3MVOn1DegO2Lb/view. Accessed September 8, 2022.

16. American Medical Association. Declaring climate change a public health crisis D-135-966. Available at: https://policysearch.ama-assn.org/policyfinder/detail/climate?uri=%2FAMADoc%2Fdirectives.xml-D-135.966.xml. Accessed September 8, 2022.

Further Reading

Stern P, Dietz T, Abel T, et al. A value-belief-norm theory of support for social movements: The case of environmentalism. Human Ecology Review 1999; 6: 81–97. Available at: https://cedar.wwu.edu/hcop_facpubs/1. Accessed January 31, 2023.

The Value-Belief-Norm Theory asserts that successful movement building involves specific values, beliefs, and norms that are threatened and the obligation for individuals to preserve them.

Fritsche I, Barth M, Jugert P, et al. A social identity model of pro-environmental action (SIMPEA). Psychological Review 2018: 125: 245–269.

The Social Identity Theory focuses on how identification processes lead to social ingroups and outgroups. The Social Identity Model of Pro-Environmental Action, which derives from Social Identity Theory, identifies four processes for behavioral responses to environmental crises: shared motivation, shared identification, shared goals, and collective efficacy.

van Zomeren M, Postmes T, Spears R. Toward an integrative social identity model of collective action: A quantitative research synthesis of three socio-psychological perspectives. Psychological Bulletin 2008; 134: 504–535.

The Social Identity Model of Collective Action, which derives from the Social Identity Theory, is an analytical framework for understanding and predicting collective action through social and political identity, structural injustice, and efficacy.

Caren N. Political process theory. The Blackwell Encyclopedia of Sociology, 2007. https://doi.org/10.1002/9781405165518.wbeosp041.

The Political Process Theory, also known as the Political Opportunity Theory, frames social movements as political, rather than psychological. It asserts that the outcome of social movements is influenced mainly by political opportunities.

Resources

Cambridge Health Alliance. Climate Health Organizing Fellows Program. Available at: https://www.healthequity.challliance.org/climate-health-2023-24. Accessed November 1, 2023.

Climate Health Organizing Fellows participate in in-depth community organizing and public narrative workshops, developed in collaboration with Marshall Ganz. The 6-month program is offered to groups of three people for free.

The P3 Lab at Johns Hopkins University. Available at: https://www.p3researchlab.org/. Accessed February 14, 2023.

The P3 Lab conducts research on the science of movement building and social change. Its mission is to disseminate knowledge about organizing and to promote the vision of a "world where people's participation in public life across race and class is possible, probable, and powerful."

Leading Change Network. What is organizing? Available at: https://leadingchangenetwork.org/resource_center/what-is-organizing/. Accessed February 13, 2023.

The Leading Change Network includes movement leaders, educators, practitioners, and institutions from 75 nations. Its mission is to "To further the knowledge, capacity and leadership of community organizers by connecting ideas, building learning spaces and developing relationships to create organized people power." It provides excellent resources, coaching and support, education for leaders, and a community of practice. Its website provides additional information about Marshal Ganz's five steps of organizing (https://leadingchangenetwork.org/our-approach-to-change/).

The Commons Social Change Library. Guide to campaign and movement building. Available at: https://commonslibrary.org/guide-to-campaign-and-movement-building-by-the-leading-change-network/. Accessed February 13, 2023.

The Social Change Library offers comprehensive resources and toolkits for leading movements and organizing change. In addition to the five steps of movement building, the library provides information on theories of change, communication strategies, and well-being.

20

Promoting Climate Justice

Rohini J. Haar and Barry S. Levy

Climate justice recognizes the imbalances between responsibilities and harms. Climate justice promotes policies and interventions to correct these imbalances, support human rights, empower people, and facilitate community alliances and self-sufficiency in order to improve population health.[1] And climate justice addresses climate-related inequities and inequalities by ensuring the following:

- Those countries and other entities most responsible for climate change and its consequences assume primary responsibility for mitigation of greenhouse gas (GHG) emissions, adaptation to reduce these consequences, and reparation (compensation for loss and damages) to those countries and groups most severely affected
- All people have a voice in policy decisions about mitigation, adaptation, and reparation, especially marginalized populations that have been severely affected by climate change
- Policies and measures for mitigation, adaptation, and reparation are fair and equitable, create co-benefits for affected populations, increase the resilience and resources of populations that have been most severely affected by climate change, and do not worsen societal inequities and inequalities.

Advocates for climate justice call for eliminating the use of fossil fuels and for requiring high-income countries to assume the primary responsibility for compensation for loss and damages. And many advocates for climate justice perceive the climate crisis as an opportunity to address a wide range of related societal inequities.

At the global level, low- and middle-income countries (LMICs), which have contributed the least to causing climate change, are suffering the most from its adverse health consequences. In contrast, high-income countries, which have contributed the most to causing climate change, are experiencing less frequent and less serious related health consequences (Figure 1-4 in Chapter 1). And high-income countries have considerably more resources for protection against climate change.

Rohini J. Haar and Barry S. Levy, *Promoting Climate Justice* In: *Climate Change and Public Health.*
Edited by: Barry S. Levy and Jonathan A. Patz, Oxford University Press. © Oxford University Press 2024.
DOI: 10.1093/oso/9780197683293.003.0020

At the local level, low-income people, people of color, and Indigenous Peoples, who have contributed the least to causing climate change, are suffering the most from increased temperatures, extreme weather events, and other manifestations of climate change. And high-income people, who are suffering less from the consequences of climate change, frequently have the resources to alleviate its most severe consequences.[1] (Disparities and inequities associated with climate change are described in more detail in Box 1-1 in Chapter 1.)

ROOTS OF THE CLIMATE JUSTICE MOVEMENT

The roots of the Climate Justice Movement include the Human Rights Movement, responses to historical injustices, public awareness of environmental disasters, the Environmental Justice Movement, and reactions to misinformation and disinformation about climate change and its consequences.

The Human Rights Movement

The U.S. Declaration of Independence (1776) and France's Declaration of the Rights of Man and of the Citizen (1789) asserted that all people are created equal and have inalienable rights to life, liberty, and the pursuit of happiness. While these documents heralded the modern human rights era, the concepts on which they were based antedated these documents by centuries.

Human rights, many of which are threatened by the consequences of climate change, have been codified in international law. The right to health was articulated in 1946 in the Constitution of the World Health Organization, which stated that "the enjoyment of the highest attainable standard of health is one of the fundamental rights of every human being without distinction of race, religion, political belief, economic or social condition."[2] The United Nations General Assembly, in 1948, adopted the Universal Declaration of Human Rights (UDHR), which enshrined the right to security and the right to a standard of living adequate for health and well-being, including food, clothing, housing, medical care, and necessary social services.[3] Human rights were also enshrined in (a) the International Covenant of Economic, Social, and Cultural Rights and the International Covenant on Civil and Political Rights in 1966; (b) in other human rights treaties, such as the Convention on the Elimination of All Forms of Discrimination Against Women in 1981, the Convention on the Rights of the Child in 1990, and the Convention on the Rights of Persons with Disabilities in 2008; and (c) in the laws of many countries.

Because the UDHR protects a standard of living adequate for health and well-being, one can infer that it protects the right to a healthy environment. However, access to a "clean, healthy and sustainable environment" was formally recognized as a distinct and universal human right by the UN Human Rights Council in 2021 and the UN General Assembly in 2022.[4]

In general, international human rights law provides a legal framework for advocacy for bringing claims to a jurisdictional body. But this framework has significant limitations, such as when addressing allocation of limited resources. There are also significant legal challenges to a rights-based approach. International human rights treaties have limited domestic influence when they have not been ratified by a country or when it neglects its obligations. And enforcement within a judicial system, even for the most egregious violations, has rarely occurred.

Responses to Historical Injustices

The profound long-term socioeconomic, political, and health consequences of imperialism, slavery, and colonization have contributed to climate-related inequities. Growing awareness of these consequences has influenced the evolution of the concept of climate justice and the Climate Justice Movement.

Exploitation by Western European countries in Africa, Latin America, Asia, and Australia, which began in the early 1500s, greatly expanded in the mid-1800s, especially growing in reaction to their increased need for natural resources and labor during the Industrial Revolution. The forced migration and enslavement of millions of Black people during the transatlantic slave trade between 1501 and 1867 resulted in legacies of racism, marginalization, and heightened vulnerability to environmental and health problems related to climate change. Colonial rule left many affected countries with neither systems for internal growth and infrastructure nor natural resources to protect them from environmental disasters—and left them less resilient to adapt to climate change.

Public Awareness of Environmental Disasters

In the decades before climate change was widely recognized, public awareness of environmental hazards increased as a result of major environmental disasters and recognition of associated sociopolitical inequities. These disasters included about 3,000 cases of methylmercury poisoning due to discharges from a chemical plant in Minamata, Japan, in the 1950s; the deaths of at least 15,000 people from an explosion of a pesticide plant in Bhopal, India, in 1984; the disaster at the nuclear power plant in Chernobyl, Ukraine, in 1986; and the *Exxon Valdez* oil spill in Alaska in 1989. More recent environmental disasters, such as widespread lead contamination of drinking water in Flint, Michigan, in 2014, have further heightened awareness of environmental hazards and their disproportionate impacts on people of color, low-income people, and other marginalized populations. Concurrently, public awareness of occupational health and safety hazards grew with recognition of the disproportionate impact of these hazards on marginalized working populations, such as coal miners, migrant farmworkers, and sweatshop workers.

The Environmental Justice Movement

The term *environmental justice* originated with protests by Black communities in the late 1970s against toxic waste dumping in Warren County, North Carolina. The protests did not stop the dumping, but they focused attention on waste dumping in Black communities. And they led to the establishment of the Environmental Justice Movement, which drew lessons from the Civil Rights Movement and environmental activism. The Environmental Justice Movement initially focused on environmental toxins and resultant harm to racial minorities and other marginalized groups within the United States and other high-income countries.[5,6] In 1991, its focus broadened when the First National People of Color Environmental Leadership Summit adopted 17 core principles of environmental justice, including participation and self-determination in environmental discussions, control of toxic substances, and the right of victims to compensation.

Since the 1980s, the Environmental Justice Movement has grown substantially in the United States and globally, raising awareness of the linkages between race, poverty, and environmental and health inequities and the need to protect and empower poor and marginalized people. This global movement now addresses a wide range of concerns, including resource extraction by multinational corporations and the export of hazards from high-income countries to LMICs. It has empowered marginalized people and their organizations to oppose siting of hazardous facilities in their communities. It has inspired governments and nongovernmental organizations in LMICs to resist the import of hazardous chemicals and waste. It has contributed to research that has documented environmental injustice and to the development of principles, laws, regulations, and guidelines aimed at eliminating environmental injustice. However, although participants in the global Environmental Justice Movement share some goals, frames, and forms of mobilization, there is no united organization in charge; much of its work is local and targeted at specific local grievances and inequities.[7]

Reactions to Misinformation and Disinformation about Climate Change and Its Consequences

The Climate Justice Movement has reacted to *misinformation* (incorrect statements) and *disinformation* (purposely misleading statements) about climate change and its consequences generated by the fossil fuel industry, politicians, and others. Reactions to misinformation and disinformation campaigns have helped to energize the Climate Justice Movement and improve public awareness about the reality of climate change and its consequences. (See section on "Misinformation and Disinformation" in Chapter 18.)

JUSTICE

Justice can be broadly defined as the fair treatment of people based on the principles of law and ethics. *Social justice* is based on fairness and equity, including the fair distribution of wealth and opportunities. *Procedural justice* is based on ethical, inclusive decision-making procedures that engage all stakeholders equitably. *Distributive justice* gives all members of a society a "fair share" or a "fair chance"—based on *equality* (everyone gets the same resources), *equity* (one's resources are based on what one needs), or other frameworks.

Within countries, a climate justice approach supports engagement and empowerment of populations that have previously been marginalized. For example, it supports the inclusion of Indigenous Peoples in the needs assessment, planning, implementation, enforcement, and evaluation of climate solutions.

Globally, countries that have long suffered from exploitation of their land and resources, violations of human dignity, genocide, enslavement, indentured labor, and oppression are suffering the most as a result of climate change. Poverty as well as inadequate resources and infrastructure protections have resulted in their heightened vulnerability to the consequences of climate change.

The Climate Justice Movement aims to empower countries and populations that have been historically, racially, culturally, or politically marginalized or mistreated. It aims to reframe the climate crisis not only in scientific and health terms, but also in the context of historical and ongoing disparities. It works to ensure that policies and programs are designed to (a) mitigate further harm and (b) adapt to climate change with an inclusive approach, one that shifts the power balance toward those who have been marginalized or are most vulnerable.

EVOLUTION OF THE CLIMATE JUSTICE MOVEMENT

As climate science was crystalizing, the Rio de Janeiro Earth Summit, in 1992, adopted the United Nations Framework Convention on Climate Change

(UNFCCC), the first international environmental treaty to address "dangerous human interference with the climate system" by limiting emission of GHGs.[8] The UNFCCC mandated regular meetings of the signatory countries, known as Conference of the Parties meetings, or COPs. Although the UNFCCC did not acknowledge the differential impacts of climate change explicitly, it noted that (a) signatory countries (parties) should act to protect the climate system on the basis of "common but differentiated responsibilities and respective capabilities," and (b) industrialized country parties should take the lead in addressing climate change.[8] (See Box 1-2 in Chapter 1.)

The Kyoto Protocol, which was signed at COP3 in Kyoto, Japan, in 1997, established legally binding obligations for 36 industrialized countries to reduce their GHG emissions by an average of 5%. While the United States and many other countries did not ratify the Protocol initially, the countries that did ratify it subsequently met their individualized, targets.[9]

In the years since then, the conception of climate change has evolved from a scientific and environmental problem to a climate justice challenge. It has been increasingly seen as "[a] human rights issue that is interwoven with issues of fundamental justice, impacts some populations and regions more than others, exacerbates existing social inequalities, and in which loss and damage compensation to countries with lower adaptive capacity must be provided."[10]

In 2000, the first Climate Justice Summit took place in The Hague in parallel with COP6. This conference, which was organized as a radical alternative to the official talks, affirmed climate justice as a human rights issue, highlighting that the real victims of climate change were people in less developed countries who had not been given a voice in the process.

In 2002, international environmental groups met at the Earth Summit in Johannesburg, where they adopted the Bali Principles of Climate Justice, which aimed to redefine climate change as a human right and social justice issue. The Bali Principles stated, in part, that "communities have the right to be free from climate change, its related impacts and other forms of ecological destruction." The Principles recognized "a principle of ecological debt that industrialized governments and transnational corporations owe the rest of the world as a result of their appropriation of the planet's capacity to absorb greenhouse gases." They supported "the rights of victims of climate change and associated injustices to receive full compensation, restoration, and reparation for the loss of land, livelihood and other damages." And they affirmed "the right of all people, including the poor, women, rural and indigenous peoples, to have access to affordable and sustainable energy." (A complete list of the Bali Principles of Climate Justice can be found at https://www.corpwatch.org/article/bali-principles-climate-justice.)

In 2007, the Climate Justice Now! network was established at COP13 in Bali. This network popularized the concept of climate justice. It acknowledged that different groups and networks used various strategies and tactics to address not only climate-related injustices but also "local struggles against dispossession, exploitation, contamination or industrial expansion in local areas." Leaders asserted that climate injustice resulted from current and historical social relations and that, to address the politicization of climate change, fundamental changes in economic and political systems were necessary.

In 2009, the Climate Action Network was founded with the goal of driving collective and sustainable action to fight the climate crisis and achieve social and racial justice. Today, with more than 1,900 civil-society organizations in more than 130 countries, it convenes and coordinates civil society at COP meetings and other international forums. It focuses on giving voice to those most affected by the impacts of climate change, ending the use of fossil fuels, holding governments accountable to the promises that they made for keeping global warming to less than 1.5°C (2.7°F) above preindustrial levels, and advocating for political outcomes that ensure a climate-safe, just, and sustainable future.[11]

In 2013, an initiative of the Mary Robinson Foundation and the World Resources Institute led to the Declaration on Climate Justice, which called on world leaders to take bold action on climate change to create an equitable future for all people. The Declaration identified the following priority pathways to achieve climate justice:

- Giving voice to people and communities affected by climate change
- Transforming the economic system to one based on low-carbon production and consumption
- Achieving a just transition to a low-carbon economy
- Promoting access to sustainable development for all
- Creating a new investment model that builds capacity and resilience while lowering emissions
- Ensuring that governments commit to bold action informed by science and delivering on their commitments to reduce emissions and provide climate finance
- Ensuring strong legal frameworks to ensure transparency, longevity, credibility, and effective enforcement of climate-related policies.[12]

In 2015, at COP21 in Paris, an ambitious global agreement to reduce GHG emissions was reached by 196 countries. This agreement, known as the Paris Agreement, became the first treaty under the UNFCCC to include "climate justice" as a concept—although it depended on individual countries' commitments. The Paris Agreement aimed to limit global warming by the year 2100 to less than 2.0°C (3.6°F)—and possibly to 1.5°C (2.7°F)—above preindustrial levels through a system of nationally determined contributions (NDCs) and reviews of these pledges. Because the system permits countries that exceed their NDCs to trade excess reductions to other countries, there has been significant opportunity for abuse of the system and coercion of less powerful countries.

In 2022, at COP27 in Egypt, parties to the UNFCCC agreed on an explicit plan to acknowledge climate justice and develop a "loss and damage fund" to assist developing countries that have suffered the impacts of historical injustices and are especially vulnerable to the adverse effects of climate change. The involvement of youth groups and additional commitments on just energy transitions and collaborative actions demonstrated additional progress.

APPROACHES TO ACHIEVING CLIMATE JUSTICE

Climate justice attempts to ensure climate policies that are fair and just through various strategies, including financing LMICs, adopting sustainable development practices, and decolonizing the climate change narrative by involving LMICs in policies to address climate change.[13] As actions to address climate change increase in scale and urgency, it is critically important to make sure that respect for human rights is included in the design and implementation of these actions.[14]

Climate Justice Goals

Climate justice goals focus on (a) an economic transition to renewable energy sources; (b) investment in "frontline communities," which have been historically marginalized and face disproportionate risk of harm from climate change; and (c) creation of climate-friendly jobs. which do not harm the environment or make climate change worse. These goals, which are described below, are designed not only to reduce disproportionate damages resulting from climate change, but also to produce equitable health and other benefits resulting from mitigation and adaptation measures.[15]

An economic transition to renewable energy sources: The Program on Climate Change Communication of Yale and George Mason University has found that 70% of Americans support a clean energy transition.[15] (Climate justice policies for a renewable energy transition are described in Box 20-1.)

Investment in frontline communities: Most (68%) of Americans support increasing funding to communities that have been disproportionately harmed by pollution. The homes in these communities are much more likely to have low energy efficiency; therefore, their residents pay higher

Box 20-1 Steps for a Just Energy Transition

Barry S. Levy

During the renewable energy transition, measures must be taken to ensure climate justice, including the following:

- Ensuring that development and placement of new energy technologies do not repeat the environmental injustices brought about by old energy technologies
- Prioritizing removal of polluting facilities from communities that have been socially, economically, and politically marginalized
- Equitably placing greenspace and other environmental assets to improve the health and well-being of individuals and communities
- Offering training to workers in old-technology facilities, such as coal mines and coal-fired power plants, to enable them to work in new-technology facilities, such as plants that produce electric cars or wind turbines
- Providing governmental income support and social services for communities impoverished by the shutdown of old technology.

During the transition to renewable energy, measures must also be taken to ensure that disadvantaged and marginalized communities do not bear disproportionate health risks from renewable energy technologies, have equal access to the benefits of these technologies, and are equitably represented in policymaking and programmatic discussions and decisions. This means that they are not only given "a seat at the table," but that they are also provided access to the news media and social media, allowed to publicly protest, given unchallenged access to legislative and executive branch government decision-makers, and provided ample opportunities to challenge governmental decisions in courts. Climate justice also means considering reparations (payment for loss and damage) to historically disadvantaged and marginalized communities to compensate for previous inequities and resultant damages.

On a global scale, climate justice means implementing measures to ensure that low- and middle-income countries have opportunities equal to those of high-income countries in reaping the benefits of renewable energy. It also means that high-income countries compensate low- and middle-income countries for previous loss and damages.

energy bills and have less protection from extreme temperatures. Most (79%) Americans support provision of federal government funding to make residencies in low-income communities more energy efficient.[15]

Climate-friendly job creation: Most (61%) of Americans think increasing production of clean energy will produce more jobs than will increasing fossil fuel production. Most Americans also support climate-friendly policies that will create jobs:

- 83% support reestablishing the Civil Conservation Corps, which would employ people to protect natural ecosystems, plant trees, and restore soil on farmland.
- 81% support creating a jobs program to hire unemployed oil and gas workers to safely close down abandoned oil and gas wells.
- 81% support creating a jobs program that would hire unemployed coal workers to safely close down old mines and restore the natural landscape.
- 79% support measures to increase energy efficiency in buildings in low-income communities.[15]

Need for Development of Political Will

Support for policies to achieve the three climate justice goals described above does not automatically translate into political action. Although there is a large and diverse base of support for these goals, it is necessary to develop the public and political will to implement them. Doing so will require increasing public awareness of climate change and its consequences (Chapter 18), increasing advocacy for specific policies and programs, and public demonstrations and other expressions of support (Figure 20-1).

Further research on public support for climate justice can help identify public and political support for these goals—such as support from underrepresented groups, labor unions, and grassroots organizations that can build the public and political will to enact laws and implement other policies. Research can also be used to stimulate individuals and communities most affected by climate change to organize and participate in developing and implementing adaptation measures and other climate solutions.

Work of Cities in Implementing Climate Justice

Many large U.S. cities have incorporated climate justice principles into their climate action plans, such as distributive justice and procedural justice. Some

Figure 20-1 Greta Thunberg and Luisa Neubauer demonstrate with Fridays for Future, Berlin, September 2021. (Wikimedia. Photograph by Stefan Müller.)

cities have developed tools to implement climate justice policies, such as the following:

- Justice partnerships, which promote participation by historically underrepresented communities and promote engagement of vulnerable groups
- Equity advisory boards, which represent and engage vulnerable populations, propose justice-centered policy objectives and actions, and review policies and programs to ensure that they are aligned with justice goals
- Equity tools, which are decision-making frameworks that guide city governments in recognizing and systematically incorporating justice and equity concerns throughout the policy process
- Justice indicators, which are comprehensive metrics to monitor and evaluate the justice and equity impacts of climate plans and policies.[16]

Addressing Climate-Induced Migration

Climate change is forcing more people to flee their communities—and often their countries—because of higher temperatures, extreme weather events, sea-level rise, and resultant consequences, such as reduced agricultural yields and shortages of food and water (Box 10-1 in Chapter 10). Climate justice can address the inequities and inequalities caused by climate-induced migration by implementing measures such as

- Reducing the need to migrate by developing and implementing adaptation measures, such as improved agricultural practices and development of drought-resistant seeds
- Providing short-term humanitarian assistance and economic support to climate migrants, especially for those who are internally displaced within their own countries
- Facilitating resettlement of climate migrants by providing access to jobs, housing, education, and healthcare and assisting them in integrating into host communities
- Providing economic assistance to host communities and host countries to facilitate the resettlement of climate migrants
- Respecting, protecting, and fulfilling the human rights of climate migrants
- Ensuring that climate migrants are protected by the laws of their host countries and by international humanitarian law
- Facilitating the return of climate migrants to their native communities and countries if they choose to return.[17]

Proposed International Commission to Assess Policy Options

Andy Haines, Former Director and Professor of Environmental Change and Public Health at the London School of Tropical Hygiene and Medicine, has proposed an international commission of independent experts and others to assess policy options that "optimize the climate, development, and social and health equity outcomes of GHG mitigation actions in different socioeconomic settings."[18] The proposed commission could make recommendations on how to estimate the benefits and tradeoffs of various policy options, including their equity impacts, and how to monitor implementation of policies. And it could support GHG mitigation measures to reduce inequities in health and wealth, such as by increasing access to clean energy for cooking and heating, expanding access to health and social services, improving affordability and accessibility of healthy and sustainable diets, and building infrastructure.[18]

Reparations

During the COP26 meeting in Glasgow in 2021, the parties to the UNFCCC formally acknowledged for the first time what had been suggested for many years—that high-income countries provide reparations for climate-related loss and damage suffered by low-income countries. They called for a dialogue between parties, relevant organizations, and stakeholders to discuss how to fund reparations for climate-related damages.[19]

Audrey R. Chapman, Professor of Medical Ethics and Humanities, and A. Karim Ahmed, Adjunct Professor of Occupational and Environmental Medicine, both at the University of Connecticut School of Medicine, have proposed establishing the Global Climate Reparations Fund, to be overseen by the United Nations Human Rights Council and multilateral agencies. The proposed Fund would administer the raising of money and the distribution of reparations for climate-related harms—linking reparations to human rights standards, including transparency, equity, and accountability. Adapting the principles of philosopher Henry Shue to climate justice, they have asserted that climate reparations are justified by principles of equity and fairness, including the following:

- Countries that have benefited the most from industrialization and extraction of fossil fuels should finance reparations to assist low-income countries, which have been most adversely affected by climate change.
- High-income countries have a greater responsibility because they are able to pay for adaptation and reparations; therefore, major corporations in the fossil fuel industry bear some of the responsibility for reparations.[20]

Need for Transformational Change

Many analysts believe that climate injustice reflects underlying social injustice and inequity issues in our global society—issues that cannot be resolved only by mitigation, adaptation, and reparation. They believe that achieving climate justice will require a major reorganization of the global economy.[21,22]

THE FUTURE

The Climate Justice Movement will likely grow and further influence public awareness of climate change and actions to address climate change. It will benefit from the support of a wide spectrum of global society, including academic institutions, nongovernmental organizations, and governmental agencies. And it will likely continue to exert a major influence on policies and actions for climate change mitigation, adaptation, and reparation.

SUMMARY POINTS

- Climate justice addresses inequities between those who have been responsible for causing climate change and those who are suffering most from the consequences of climate change.
- Climate justice asserts that those most responsible for climate change assume primary responsibility for mitigation of GHG emissions, adaptation to climate change, and reparation for those who are most severely affected.
- Climate justice ensures that all people, especially marginalized communities, have a voice in public policy decisions, mitigation, adaptation, and reparation.
- Climate justice ensures that mitigation, adaptation, and reparation measures create co-benefits and do not worsen existing inequities.
- And climate justice means addressing a wider range of societal inequities.

References

1. Mari-Dell'Olmo M, Oliveras L, Barón-Miras L, et al. Climate change and health in urban areas with a Mediterranean climate: A conceptual framework with a social and climate justice approach. International Journal of Environmental Research and Public Health 2022; 19: 12764. https://doi.org/10.3390/ijerph191912764.
2. World Health Organization. Constitution. Available at: https://www.who.int/about/governance/constitution. Accessed February 21, 2023.
3. United Nations. Universal Declaration of Human Rights. Available at: https://www.un.org/en/about-us/universal-declaration-of-human-rights. Accessed February 21, 2023.
4. United Nations Human Rights Council. The human right to a clean, healthy and sustainable environment. Resolution adopted by the Human Rights Council on 8 October 2021 (A/HRC/RES/48/13). Available at: https://documents-dds-ny.un.org/doc/UNDOC/GEN/G21/289/50/PDF/G2128950.pdf?OpenElement. Accessed March 13, 2023.
5. Bullard RD. Dumping in Dixie: Race, Class, and Environmental Quality. Boulder, CO: Westview Press, 1990.
6. Brulle RJ, Pellow DN. Environmental justice: Human health and environmental inequalities. Annual Review of Public Health 2006; 27: 103–124.
7. Martinez-Alier J, Temper L, Del Bene D, Scheidel A. Is there a global environmental justice movement? The Journal of Peasant Studies 2016; http://dx.doi.org/10.1080/030661150.2016.1141198.
8. United Nations. United Nations Framework Convention on Climate Change. Available at: https://treaties.un.org/doc/source/recenttexts/unfccc_eng.pdf. Accessed August 21, 2023.
9. United Nations. Kyoto Protocol to the United Nations Framework Convention on Climate Change. Available at: https://unfccc.int/resource/docs/convkp/kpeng.pdf. Accessed March 19, 2023.

10. Gach E. Normative shifts in the global conception of climate change: The growth of climate justice. Social Science 2019; 8: 24. https://doi.org/10.3390/socsci8010024.
11. Climate Action Network International. Available at: https://climatenetwork.org/. Accessed January 25, 2023.
12. Mary Robinson Foundation - Climate Justice. Climate justice dialogue. Available at: https://www.mrfcj.org/our-work/climate-justice-dialogue/#:~:text=The%20Decl aration%20on%20Climate%20Justice,that%20is%20fair%20for%20all.   Accessed January 25, 2023.
13. Hasan M. Climate justice and COP26: A new perspective on the climate crisis. Medicine, Conflict, and Survival 2022; 38: 242–250.
14. Robinson M, Shine T. Achieving a climate justice pathway to 1.5°C. Nature Climate Change 2018; 8: 564–569.
15. Carmen J, Lu DL, Low J, et al. Exploring Support for Climate Justice Policies in the United States. New Haven, CT: Yale Program on Climate Change Communication, August 4, 2022. Available at: https://climatecommunication.yale.edu/publications/ exploring-support-for-climate-justice-policies-in-the-united-states/.   Accessed January 20, 2023.
16. Diezmartinez CV, Short Gianotti AG. US cities increasingly integrate justice into climate planning and create policy tools for climate justice. Nature Communications 2022. doi.org/10.1038/s41467-022-33392-9.
17. Dwyer J. Environmental migrants, structural injustice, and moral responsibility. Bioethics 2020; 34: 562–569.
18. Haines A. Use the remaining carbon budget wisely for health equity and climate justice (Comment). Lancet 2022; 400: 477–479.
19. United Nations. Climate Change. Glasgow Dialogue, June 2022. Available at: https:// unfccc.int/event/glasgow-dialogue#:~:text=COP%2026%20established%20the%20 Glasgow,adverse%20impacts%20of%20climate%20change.   Accessed   January 30, 2023.
20. Chapman AR, Ahmed AK. Climate justice, human rights, and the case for reparations. Health and Human Rights 2021; 23: 81–94.
21. Chomsky A. Is Science Enough? Forty Critical Questions about Climate Justice. Boston: Beacon Press, 2022.
22. Cripps E. What Climate Justice Means and Why We Should Care. London: Bloomsbury Continuum, 2022.

Further Reading

Levy BS, Patz JA. Climate change, human rights, and social justice. Annals of Global Health 2015; 81: 310–322.

This article describes the multiple interrelationships among climate change, human rights, and social justice.

Gutierrez KS, LeProvost CE. Climate justice in rural southeastern United States. A review of climate change impacts and effects on human health. International Journal of Environmental Research and Public Health 2016; 13:189. doi: 10.3390/ijerph13020189.

This review article focuses on the southeastern United States, which is especially susceptible to climate change because of its geography and the vulnerabilities of resident populations. It also addresses local and regional mitigation and adaptation strategies to combat human health effects related to climate change.

Hernández D. Climate justice starts at home: Building resilient housing to reduce disparate impacts from climate change in residential settings. American Journal of Public Health 2022; 112: 66–68.

People of color, low-income individuals, older people, and those who are medically vulnerable face not only disproportionate climate risks, but also various forms of housing insecurity. Resilient housing can uphold the principle of climate justice by ensuring that all populations have structurally and socially sound homes to withstand the impacts of climate change.

Smith GS, Anjum E, Francis C, et al. Climate change, environmental disasters, and health inequities: The underlying role of structural inequalities. Current Environmental Health Reports 2022; 9: 80–89.

This article provides a conceptual framework that maps how long-standing structural inequalities in policy, practice, and funding shape the vulnerability of low-income people and racially and ethnically marginalized individuals.

Rudolph L, Gould S. Climate change and health inequities: A framework for action. Annals of Global Health 2015; 81: 432–444.

This article provides a conceptual framework that depicts the complex interrelationships among climate change, health, and equity, and the many opportunities for greater public health engagement concerning these issues.

Climate Justice Alliance. Available at: https://climatejusticealliance.org/. Accessed September 13, 2023.

An alliance of 89 communities, organizations, and supporting networks in the Climate Justice Movement.

Climate and Economic Justice Screening Tool. Available at: https://screeningtool.geoplatform.gov/en/#3/33.47/-97.5. Accessed September 13, 2023.

This tool is used by federal agencies to help identify disadvantaged communities that will benefit from investments in programs to address the impacts of climate change.

Index

11th Street Bridge Park, 378

350.org, 416, 417

acclimatization, 55, 59
Action Limit, 55
active transportation (active travel), 23, 59, 273, 274, 342, 410
acute renal failure, 73
acute stress disorder, 187
adaptation, 19, 23, 198, 321, 324, 329–331, 364, 422, 433
adaptive management, 364
adjustment disorders, 189
Administrative Procedures Act (APA), 229, 231
"Advanced Clean Cars," 277
Advanced Notice of Proposed Rulemaking (ANPR), 231
adverse birth outcomes, 75
Aedes, 109, 113, 123, 326
aeroallergen exposure, 88
Africa, 211, 212
Agassiz Glacier, 38
"Agenda 21," 20
agricultural
   extension programs, 172
   lands, 369
   management practices, 145
   Model Intercomparison and Improvement Project, 162
   practices, 433
   subsidies, 303, 304
   waste, 137
Agriculture Improvement Act of 2018 (the "Farm Bill"), 302–305, 309
agriculture policy, 22, 292–311
agri-food systems, recommendations for action, 172
Ahmed, A. Karim, 434
air conditioning, 52, 64
air filtration systems, 89
air pollution, 98, 240, 253, 294, 299
air quality, 252, 323, 365, 366
Alaska, 183–185
algae, 136
algal blooms, 142

allergic rhinitis, 86
Alliance for Transformative Action on Climate and Health, 332, 415
American Housing Authority (AHA), 350
American Medical Association (AMA), 409
American Planning Association, 375
Antarctic Oscillation, 30
anthroponoses, 111. See also *zoonoses*
anxiety and worry, 187, 188
apathy, 190, 191
Arctic, 37, 41
Arctic amplification, 43, 44, 155
Arctic Oscillation, 30
Arctic sea ice, 43
Argentina, 73, 115
armed conflict, 208–218
assessment of risk, 321
asthma, 75, 86, 87, 92, 93, 321
Athens, 72
Atlanta, 364, 365
Atlantic Multidecadal Oscillation (AMO), 31, 42
attribution of extreme weather events, 39
Australia, 4, 77, 186, 207

Bali Principles of Climate Justice, 427
Bangladesh, 214
beef, 305
benzene, 92
Berkeley, California, 349
beryllium, 237
bicycling, 274, 282, 352
Biden, Joe, 260
   administration, 277
Big Oil vs the World, 398
Billion Tree Campaign, 371
biodiesel, 298
biodiversity, 298
biofuel, 298–302, 308, 309
biomass, 99, 100
black carbon, 101
Black Lung Benefits Program, 236
blood pressure, 274
Bloodborne Pathogen Standard, 237
blue carbon, 370

Bonn Challenge, 371
Boston, 15, 76, 226
Boston University School of Public Health
    (BUSPH), 238, 239
BRACE program, 329
Brazil, 94
bronchitis, 93
built environment, 323, 339–355
Burlington, Vermont, 378
Bush, George W., administration, 276
bushfires, 186. See also wildfires

C40 Cities Climate Leadership Group, 21
California, 70, 72, 77, 79, 114, 115, 121, 272,
    273, 276–278, 285, 346, 352
California Occupational Safety and Health
    Administration (CalOSHA), 51
California's Air Resources Board (ARB),
    277, 285
California's SB 1000, 352, 353
Cambodia, 3, 324
Campylobacter, 136, 137, 141
Canada, 3, 73, 115, 117, 330
cancer, 274, 285, 306
Cancun Adaptation Framework, 326
cap and trade, 266, 267
capital improvement plans (CIPs), 350
carbon debt, 300
carbon dioxide, 6, 7, 32, 89, 251, 259, 277, 292,
    294, 300, 368
    removal, 368
carbon emissions, 351
carbon monoxide, 92, 299
carbon offsets, 266, 267
carbon sequestration, 293, 369
carbon sinks, 292, 344
carbon tax, 266, 267
cardiac dysrhythmia, 73
cardiovascular diseases, 74, 92, 274, 285
caring-for-country approach, 197
Center for Climate Change
    Communication, 387
Center for Community Health and
    Development at the University of
    Kansas, 242
Center for Medicare and Medicaid Services, 415
Centers for Disease Control and Prevention
    (CDC), 71, 134
cerebrovascular diseases, 74
Chapman, Audrey R., 434
C-HEAT Project, 238, 239
Chelsea, 238
Chicago, 72, 77

chikungunya, 109
children, 16, 91, 135, 169
China, 3, 63, 77, 91, 94, 209, 262, 266, 267, 371
China's Air Clean Plan, 252
cholera, 133, 144
chronic diseases, 76. See also specific diseases
chronic kidney disease, 56
chronic obstructive pulmonary disease
    (COPD), 86
cities, 345, 360, 363, 364, 431, 432
citizen advisory committees, 236
citrus canker disease, 158
Civil Conservation Corps, 431
civil-society organizations, 330, 332
civil war, 212
Clark, Helen, 207
Clean Air Act of 1970, 237, 265, 273
Clean Electricity Standards (CESs), 261, 265
clean energy, 409
Clean Power Plan, 265
Climate Action Network, 428
climate action plan (CAP), 350–352
Climate Adaptation Policy Statement, 226
Climate, Aggression, and Self-Control in
    Humans (CLASH) model, 216
Climate Change and Health Program, 332
climate-friendly job creation, 431
Climate and Health Movement, 410–413
climate justice, 409, 422–435
Climate Justice Movement, 20, 404, 424,
    426, 434
Climate Justice Now! network, 427
Climate Justice Summit, 427
climate migrants, 214, 215, 433
climate modeling, 32, 34
climate-proofing of healthcare facilities, 331
climate science, 30–45
climate services, 173
    for health, 328
climate shocks, 168
climate-smart agriculture, 309
climate variability, 19
clothing, 60
coastal (blue carbon) ecosystems, 370
coastal flooding, 56
co-benefits, 100, 101, 196, 197, 364, 366, 369,
    372, 393. See also health co-benefits
coccidioidomycosis, 94
Code of Federal Regulations, 231
coffee rust disease, 159, 160
cold extremes, 9
collective violence, 208–218
Colombia, 366

Colorado Transportation Commission, 280
combined sewage outflow events, 139
commodity prices, 302
communicating, 387–400
communitarian approach, 235
communities LEAP, 241
community
    centers, 344
    groups, 236
    input and engagement, 265, 322
    involvement, 411
    populations, 70–79
    toolbox, 242
    violence, 207, 208
commuting by walking or bicycling, 282, 352
"complete streets" policies, 280
complex models, 34
comprehensive design guidelines, 345
Comprehensive Plans, 352
computation, 35, 36
concentrated animal feeding operations
    (CAFOs), 137, 306, 307
Conference of the Parties (COP) meetings,
    19, 20, 332. *See also specific COP meetings
    listed by number*
congenital malformations, 56
congestion pricing, 281
Congress, U.S, 227–229
Congressional Review Act (CRA), 231
connectivity, 341
consent decrees, 376
conservation agriculture, 308, 370
Conservation International, 37
Conservation Reserve Program (CRP), 300,
    301, 308
conservation and restoration of diverse
    ecosystems, 368
contributions to warming, 33
cooking fuels, 100, 101
cookstoves, 99, 172
cooling center, 78
COP3 in Kyoto, Japan, 427
COP6 in The Hague, 427
COP13 in Bali, 427
COP21 in Paris, 428
COP26 in Glasgow, Scotland, 332, 391, 433
COP26 Health Program, 415
COP27 in Sharm el-Sheikh, Egypt, 332, 429
Copenhagen, 281
coping, 197
coral reefs, 373
corn, 295–297, 299, 300, 302, 304, 305
corn-based ethanol, 278, 302, 308

coronary artery disease, 305, 306
corporate average fuel economy (CAFE), 272
cost of energy, 257
Costa Rica, 261
cotton, 304
Coupled Model Intercomparison Project
    Phase 6, 34
COVID- 19, 59, 207, 231, 344, 345
crime rate, 354
crop workers, 56
crop yield, 155, 167
croplands, 370
Cryptosporidium, 136, 137, 143
Culex, 113, 114, 121
Cyanobacteria, 136, 142
cyanobacterial blooms, 143
cyclones, 11

Debunking Handbook, 398
decarbonizing electricity sources, 263, 265
Declaration on Climate Justice, 428
Declaration of Independence, 423
Declaration of the Rights of Man and of the
    Citizen, 423
deforestation, 374
dehydration, 70, 73
democracy, 242
dengue, 109, 326
denial, 190, 191
Denver, Colorado, 353
    Housing Authority, 353
Department of Agriculture (USDA), 227, 295
Department of Energy (DOE), 241
Department of Health and Human Services
    Climate Pledge, 414
Department of Housing and Urban
    Development (HUD), 283
Department of Transportation (DOT), 272, 283
depression, 183, 187
    and grief, 188
diabetes, 74, 92, 275, 282, 305, 306
diarrhea, 134, 375
diesel particulate matter, 285
diet, 166, 167, 306
dietary guidelines, 305
diptera, 110
disasters, 182
    resilience, 373
    response, 197
    response plans, 323
disinformation, 190, 398, 399, 426
disparities, 15–17, 423
displacement and relocation, 192

distributive justice, 426
Doha, Qatar, 58
domestic violence, 187, 207
downscaling, 35
drinking water, 11, 143, 144
drought, 3, 9, 12, 93, 142, 157, 161, 162, 168, 169, 193, 212
　resilience, 373
drug and alcohol abuse, 187
dynamical regional-scale model, 35

"E8 gasoline," 295
early-warning and response systems, 79, 173, 321, 323
Earth Day Movement, 404
Earth Summit, 426
earth-system models of intermediate complexity, 34
East Africa, 118, 210
East Harlem, 362
eastern equine encephalomyelitis virus (EEEV), 117
Ebola, 374
ecodriving, 283
ecological grief, 188
economic assessment, 56
economic assistance, 433
economic impacts, 62
economic transition, 429
ecosystem, 192, 367–369
effects on physical and biological systems, 37
El Niño, 8, 161
El Niño-Southern Oscillation (ENSO), 31, 42, 109, 118, 161, 210
electric power, 259
electric vehicles, 277, 278
electricity generation, 255
emergency department visits, 74, 75, 99
Emergency Planning and Community Right to Know Act, 237
emergency temporary standards (EFSs), 231
emissions caps, 278, 279
emissions trading program, 267
energy efficiency, 263
energy policy, 22, 251–268
Energy Policy Act of 2005, 295
energy sector, 254
enteroviruses, 136, 137
environmental disasters, 424, 425
environmental justice, 352, 353, 425
Environmental Justice in Local Land Use Planning" (SB 1000), 352
Environmental Justice Movement, 404, 425

environmental migrants, 215
environmental pathogens, 135, 138
Environmental Protection Agency (EPA), 88, 91, 226, 227, 231, 236, 265, 272, 277, 283, 284
epidemiologic evidence, 274
equity, 324, 325, 377
Equity Research Institute, 406
Escherichia coli, 136, 137, 141
ethanol, 295, 299, 309
Ethiopia, 171, 328, 371
Europe, 41, 72, 73, 117
European Union, 266, 267
evaporation rate, 59
export of hazards, 425
extreme event attribution, 39
extreme precipitation, 141, 168
extreme rainfall, 3, 41, 143
extreme weather events, 4, 11, 19, 37–39, 41, 89, 157, 158, 161, 166, 168, 214, 321, 323

farm output, 157
farmworkers, 52, 166
Federal Coal Mine Safety and Health Act, 236
federal courts, 229, 231
Federal Energy Regulatory Commission, 266
Federal Register, 229, 230
federalism, 232
feedback processes, 41, 42
fetal and infant growth, 93
fine particulate matter ($PM_{2.5}$), 92, 100, 252, 294, 330
First National People of Color Environmental Leadership Summit 425
flood, 3, 11, 12, 15, 16, 56, 88, 134, 135, 139, 141, 157, 162, 168, 324
　resilience, 373
flood plains, 12
Florida, 117, 280
fluorinated gases, 6, 251, 266
food
　access, 164
　availability, 163
　groups, 306
　insecurity, 25, 153–174
　prices, 303
　price shocks, 301, 302, 308
　production, 293
　riots, 301–303
　security, 153, 302, 309, 321
　stability, 165
　systems, 293, 410
　utilization, 165

Food Insecurity Nutrition Incentive (FINI), 309
foodborne disease, 165, 170
forests, 369, 370
   ecosystems, 374
formaldehyde, 92
frames, 13, 391–393
France, 58, 77
Fraser Health Climate Change and Health Adaptation Framework, 325
freshwater, 308
Fridays for Future, 417, 432
frontline communities, 429
fruit, 305, 306
fuel-economy standards, 277–279
fuel taxes, 284
Fukushima Daiichi nuclear disaster, 268
fungi, 86, 95

Ganz, Marshall, 405–406
gasoline, 298
gastrointestinal diseases, 74
gender-based violence, 207
General Plans, 352
George Mason University, 387
geothermal power, 255, 258
Germany, 77, 363, 371
gestational hypertension, 14
Giardia, 136, 137, 143
Glacier National Park, 38
glaciers, 43
global climate models (GCMs), 34, 42
Global Climate Reparations Fund, 434
Global Climate Strike, 417
Global Heat Health Information Network, 58
global heating, 36
global warming
   hiatus, 42
   level, 9, 10
   potential, 292
Global Warming's Six Americas, 387–391
Grand Challenge on Climate Change, Human Health and Equity, 21
grasslands, 369
Great Green Wall, 371
Greece, 373
Green Belt Movement, 371
green building standards, 349
green facades, 363
green gentrification, 377
green homes, 348
green infrastructure, 360, 365, 376

Green New Deal, 417
green roofs, 363
green spaces, 344
green stormwater infrastructure (GSI), 364
green transportation, 348
greenhouse effect, 5, 32
greenhouse gases (GHGs), 5, 6, 18, 31, 32, 52, 63, 97, 100, 154, 171, 208, 225, 227, 240, 251, 253, 256, 258–261, 264–267, 272, 273, 276, 278, 281, 284, 292, 294, 295, 298, 300, 303–308, 310, 326, 331, 339, 346, 349, 352, 365, 366, 409, 410, 413, 414, 422
   mitigation, 329
   reductions, 277
greening communities, 321
Greenland Ice Sheet, 31, 37, 43
GreenRoots, 238
greenspace, 25, 430
   inequity, 366
groundwater, 141
growth management policies, 350
Gundersen Health System, 414

Haines, Andy, 433
Haiti, 133
handwashing, 144
Hansen, James, 416
harmful algae blooms (HABs), 138, 297
Harvard Kennedy School of Government, 405
health adaptation, 321–334
Health Care Climate Council, 415
Health Care Without Harm, 414, 415
health co-benefits, 251–253, 274, 285, 347, 349, 362, 366, 368, 387
health equity, 322, 325
health frame, 393
Health Impact Assessment (HIA), 353
health national adaptation plans (HNAPs), 327, 329
health in all policies, 329
Health Professionals for a Healthy Climate (HPHC), 409
health relevance of climate change, 387–400
health systems, 327
healthcare sector, 413–415
heat, 90, 94, 193
heat acclimatization programs, 55
heat action plans, 323
heat exhaustion, 53
heat exposure, 15, 53
heat hazard, 55
Heat Index, 71
heat-level rise, 52

heat rash, 53
heat-related disorders, 14, 51–64, 70–79, 323
heat-related deaths, 56, 58, 323
heat-response plans, 79
"HeatRisk," 71
heat strain, 53
heat stress, 53, 54, 59
heat stroke, 53
heat syncope, 53
heat-warning systems, 62, 64, 70
heat wave, 3, 9, 13, 71–73, 142, 157, 162, 187
heat-wave days, 25
heat-wave mortality, 25
heavy precipitation events, 9, 140
heavy-duty truck travel, 284
Hendra virus, 374
hepatitis A, 136, 137
hepatitis E, 136, 137
high-income countries, 327
high-performance computer systems, 36
Highland Park, 239–241
Hindu-Kush Himalaya (HKH) mountain
    system, 155
historical injustices, 424
home energy use, 348
home rule states, 232
Homer-Dixon, Thomas, 209
homes, 342, 343
homicide, 208
Hong Kong, 53
Horn of Africa, 3, 161
hospitalizations, 75
hospitals, 413, 415
host-associated pathogens, 135, 137, 138
hottest temperature ever recorded, 41
House of Representatives, U.S., 227–229
household air pollution, 100, 101
housing, 16, 354
Houston area, 61
Hsiang, Solomon, 209
human behaviors, 144
human rights, 332, 423, 433
Human Rights Movements, 423, 424
humanitarian assistance, 433
human-scale design of facilities, 342, 345
humidity, 59
Hunter Point South Park, 364
Hurricane Katrina, 12, 139, 207
Hurricane Rita, 139
Hurricanes, 11
hydraulic fracturing (fracking), 259
hydrocarbons, 299
hydroelectric power, 255, 258

hydrofluorocarbons, 6, 227
hydrological cycle, 308
hypercholesterolemia, 305
hyperthermia, 70

ice-albedo feedback mechanism, 41
ice core data, 7
ice sheets in Greenland and West
    Antarctica, 37, 43
impaired cognitive function, 75
impaired lung function, 75
income support, 430
incubation period, 112
India, 41, 56, 59, 60, 73, 91, 155, 211, 252,
    262, 371
Indigenous Peoples, 16, 17, 192, 197, 260, 265,
    321, 423, 426
indirect psychosocial impacts, 191, 192
Indonesia, 37
indoor air quality, 89, 98
indoor moisture, 16
indoor work, 58
inequities, 14, 366, 423
infant mortality, 93
infectious diseases, 14, 19, 169, 170, 369,
    374, 375
Inflation Reduction Act (IRA), 227, 260,
    261, 342
innovative initiatives, 416–418
insurgency, 212
Intergovernmental Panel on Climate Change
    (IPCC), 19, 20, 34, 43, 70, 155, 162, 205,
    259, 366, 376
intergroup conflict, 191, 192
international humanitarian law, 433
International Union for Conservation of
    Nature, 371
interpersonal violence, 206
investments, 264
Iraq, 161
irrigation, 172
ischemic heart disease, 73. See also coronary
    artery disease
ischemic stroke, 73, 306
Islamabad, 331

Jakarta, 37
Japan, 58, 268
jet stream, 44
just energy transition, 430
just transition, 410
justice, 422–435
justice-based mechanisms, 264, 265

Kathmandu, 362
Kenya, 207, 212
kidney disease, 58
kidney disorders, 74
Kyoto Protocol, 20, 427

La Niña, 161
labor productivity, 62, 329
Lake Michigan, 133, 141
Lancet Countdown, 23, 25
land clearing, 300
land subsidence (land sinking), 37
landslides, 12
land use, 12, 340, 374
Leadership in Energy & Environmental Design
    (LEED), 349
legal obligations, 218
Legionella, 138
Leptospira, 136
Let Communities Choose report, 240, 241
Line 3/93, 308
liver disorders, 74
livestock, 292, 307, 308
London, 72
Los Angeles, 362
lost labor productivity, 56
lost school days, 75
low-birthweight, 169
low-carbon energy sources, 254
low-carbon fuel standard (LCFS), 278
low-income people, 260, 265, 423
low- and middle-income countries (LMICs),
    15, 16, 60, 153, 216, 327, 361, 422, 425,
    429, 430
lung cancer, 92
Lyme disease, 4, 120, 326

maize, 293. *See also* corn
malaria, 25, 124, 374
Mali, 212
malnutrition, 14, 153–174, 294
Mangrove forests, 373
manual labor, 60
Mariposa Health Living Initiative, 353, 354
Mariposa Healthy Living Tool, 354
market-based mechanisms, 266, 267
Mary Robinson Foundation, 428
masks, 89
Mass General Brigham, 414
McKibben, Bill, 416
meat, 307, 308, 310
meat consumption, 305
meat production, 304, 305

mediators and moderators of psychological
    impacts, 192, 193
Medical Society Consortium on Climate &
    Health, 390
Mekong River, 3
Melbourne, 96
mental health impacts, 4, 14, 53, 75, 180–199,
    321, 323
mental models, 391–393
Merchants of Doubt, 398
meta-analyses, 213
metabolic rate, 55
methane, 6, 7, 32, 227, 251, 268, 292
Mexico, 371
microorganisms, 19, 135
migration, 215, 432, 433
mileage-based user fee (VMT fee), 283
Millennium Ecosystem Assessment, 367
Milwaukee, 133, 141
misinformation, 398, 399, 426
mitigation, 19, 22, 23, 331, 422
modes, 30
mold, 16, 96, 97
monoculture row crops, 300
monsoon, 3, 162
Moscow, 72
mosquitoes, 111, 117, 121. *See also specific
    mosquito species*
movements, 404–419
Mycobacterium, 138

National Academy of Medicine, 21
National Academy of Sciences, 39
National Adaptation Plans (NAPs), 326, 327
National Ambient Air Quality Standards, 237
National Archives, 229
National Highway Traffic Safety Administration
    (NHTSA), 277
National Renewable Energy Laboratory, 284
National Weather Service (NWS), 71, 79
Nationally determined contributions (NDCs),
    20, 170, 260, 428
natural disasters, 182
natural gas, 268, 349. *See also* methane
natural gas stoves, 253
nature-based climate solutions, 360–378
nature's contributions to people, 367, 369
negative emissions, 368
Nelson, Gaylord, 404
Nepal, 362
net-zero emissions, 251, 261, 263
    vehicle program, 277
net-zero targets, 259, 267, 268

New Delhi, 16
new energy technologies, 430
New York City, 99, 213, 345–347, 362, 364, 407
New Social Movement Theory (NSMT), 405
NIMBYism (for "Not in My Back Yard"), 265
nitrate, 298
nitrites, 298
nitrogen, 297
nitrogen dioxide, 86
nitrogen oxides, 90, 251, 299
nitrogen trifluoride, 6
nitrous oxide, 6, 251, 292
noncommunicable diseases, exacerbation,
    56, 321
nongovernmental organizations, 225. *See also*
    specific nongovernmental organizations
noroviruses, 133, 136, 137
North Carolina, 58, 137
Northern Annular Mode, 30
Northwestern United States, 3, 73
Norway, 4, 261
nose, eye, and throat irritation, 93
nuclear power, 254, 258, 268
nutrition, 294. *See also* malnutrition

Obama, Barack, 276
    administration, 260, 265, 283
obesity, 166, 275, 282, 352
occupational asthma, 87
occupational illnesses and injuries, 321
occupational respiratory disorders, 87–89
Occupational Safety and Health Act, 231
Occupational Safety and Health Administration
    (OSHA), 231, 237
ocean acidification, 13
ocean surface temperatures, 12
older people, 58, 76
open space, 354
oral rehydration therapy, 144
Oregon, 278, 351
organizational responses to climate change, 19–21
Our Common Future, 20
outdoor laborers, 14
overconsumption, 304
overproduction, 304
overweight, 166
overwintering, 113
ozone, 3, 75, 86, 88–92, 94, 96, 98, 321

Pacific Institute, 211
Pacific-North American Pattern, 31
Pacific Northwest, 41. *See also* Northwestern
    United States

Pakistan, 3, 41, 56, 73, 134, 135
Paris Agreement, 20, 23, 25, 170, 260, 368, 409,
    417, 428
parks, 362
particulate matter, 86, 92–94, 98, 251, 299
Partnership for Sustainable Communities, 283
Partnerships for Climate-Smart Commodities,
    227
Pastor, Manuel, 406–408
pathological eco-anxiety, 188
pay-as-you drive insurance, 281
people of color, 15, 260, 265, 423
People's Climate March, 407
perfluorocarbons, 6
personal protective equipment, 59
pesticides, 304
Pew Research Center, 390
physical inactivity, 352
planck feedback, 42
Planetary Health Movement, 404, 405
planetary health reference diet, 306
Planning for Urban Heat Resilience, 375
plant diseases, 158–160
Plant-for-the-Plant Foundation, 371
plant pathogens, 158–160
plant-based diets, 308, 310
pocket parks, 344
poikilotherms, 110, 111
polar vortex, 44
political will, 431
pollen, 89, 94–96, 98, 99
pollen levels, 88
polyfluoroalkyl substances (PFAS), 237
population density, 340
population growth, 22
Portland, Oregon, 351
positive feedback process, 41
positive radiative forcing, 32
posttraumatic stress disorder (PTSD), 183, 187
poverty, 15
precipitation, 8, 294. *See also* rainfall
predictive capability, 42, 43
preeclampsia, 14
pregnancy, 75
pregnant women, 14
premature mortality, 92
preparedness, 62, 64, 70, 78
preterm birth, 92
preventive measures
    heat-related disorders, 62, 78, 79
    respiratory disorders, 97–99
primary energy resources, 254
principles of climate change communication, 396

procedural justice, 426
Project on Climate Change
    Communication, 387
Pseudomonas aeruginosa, 138
psychiatric, 180
psychological, 180
    framework, 194
    impacts, classes of, 181
    well-being, 274
public health agencies, 322, 330
public health crisis, 409
public infrastructure, 347
public participation, 232–237, 242, 243
public policy, 225
public policymaking, 225–243
public and political support, 25

Race to Zero Initiative, 414
radiation, 75
ragweed, 95
rainfall, 11, 12, 139, 141, 155, 157, 210–213
    decreased, 12
recreational water, 145
Reducing Emissions from Deforestation
    and Forest Degradation in Developing
    Countries (REDD+), 21, 376
regenerative agriculture, 309, 368, 370
renewable energy, 254, 255, 262, 410
renewable fuel standard (RFS), 278, 295,
    296, 299
Renewable Portfolio Standards (RPSs), 261, 265
reparations, 433, 434
Republic of Korea, 267
resilience, 170, 171, 195, 196, 285, 325, 327, 332,
    408, 414
    of health systems, 332
Resource Mobilization Theory (RMT), 405
respiratory disorders, 3, 14, 74, 75, 86–103,
    323, 366
response planning, 78
rice production, 292, 293, 294
Rift Valley fever (RVF), 118, 119
"Rio Earth Summit," 20, 426
risk factors, heat, 75–77
Rodney Cook Sr. Park, 364, 365
rotaviruses, 136, 137
Russia, 72, 88

Safe Drinking Water Act, 144
safely managed sanitation services, 143
Safer Affordable Fuel-Efficient ("SAFE")
    Vehicles Rule, 277
Salmonella, 133, 136, 137

San Francisco, 280
sand and dust storms, 93
scarcity, 209
Scotland, 261, 332
sea ice, 37
sea-level rise, 13, 14, 36, 37, 142, 157, 214
Seattle, 280
self-inflicted violence, 206. *See also* suicide
Senate, 227–229
septic systems, 139
sequestering carbon, 368–370, 372, 373
sequestering greenhouse gas emissions,
    365, 368
serum cholesterol, 274
sewer systems, 364
Shared Socioeconomic Pathway (SSP)
    scenarios, 34, 35
Shue, Henry, 434
simple models, 34
Singapore, 59
siting and design of healthcare facilities, 321
Small Business Regulatory Enforcement and
    Fairness Act, 231
Small Island Developing States (SIDS), 13, 16,
    21, 157, 164
"smart growth," 279, 280
social capital, 344, 365
social cohesion, 354, 365
social justice, 193, 426
social movement, 405
social protection, 171, 173
    services, 172
social services, 430
soil, 172, 292
solar photovoltaic (PV) devices, 256
solar power, 252, 255, 257
solar-thermal systems, 256
solastalgia, 192
solid fuels, 252
Somalia, 3, 168, 210
somatic disorders, 187
soulardarity, 240
South Africa, 266, 373
South America, 3
Southern Annular Mode, 30
soybean, 293, 298, 304, 305
Spain, 77
special districts, 232
speed limits, 283
Sri Lanka, 124
St. Louis, 208
staple crops, 156
states' rights, 232

Stewart's wilt, 159
Stop Line 3 Movement, 408, 409
storm surges, 12
stormwater, 344, 346, 363, 364
stress, 75
stunting, 167, 168
Sub-Saharan Africa, 210, 252
subsidies, 264, 304, 305
suburb, 350
suburban style of development, 342
suicide, 53, 187, 206
sulfates, 101
sulfur dioxide, 251
sulfur hexafluoride, 6
Sunrise Movement, The, 417, 418
Superstorm Sandy, 285
Supplemental Nutrition Assistance Program
    (SNAP), 171
Supreme Court, U.S., 227–229
surveillance, 145, 321, 322, 330
sustainable development, 20
Sustainable Development Goals (SDGs), 100,
    366, 369
sweating, 59, 75
synoptic analysis, 79
Syrian Civil War, 212
systems approach to climate services, 328

technological disasters, 182
telecommuting, 283
teleconnections, 30
temperature, 8, 11, 55, 91, 142, 155, 157, 294.
    See also heat
  hottest ever recorded, 41
  warmer, 213
Thailand, 58, 63
The Hague, 427
"third places," 343, 344
threshold heat index, 71
Threshold Limit Value (TLV), 55
thunderstorms, 88, 96
tickborne diseases, 110
tickborne encephalitis, 120
Toronto, 361
traffic, 90
traffic policy, 284
training to workers in old-technology
    facilities, 430
transferrable development rights, 351
transit-oriented development (TOD), 280,
    343, 348
transition to renewable energy, 265, 430
transpiration, 76

transport, 340–342
transportation, 340, 348, 354
transportation policy, 22, 272–286
tree canopy, 362
tree planting, 96, 98, 363
trees, 362, 364, 366, 371, 372
Trees in Cities Challenge, 363
Trillion Tree Campaign, 371
tropical countries, 56
tropical cyclones, 9
tropical storms, 3, 157
trucks, 284, 285
Trump, Donald, administration, 260, 277
tuberculosis, 94
Turkey, 371

Uganda, 211
undernutrition, 167–169, 323
Union of Concerned Scientists, 240
United Kingdom, 15
United Nations, 20, 70, 367
United Nations Conference on Environment
    and Development, 20
United Nations Development Program, 207
United Nations Environment Program, 20,
    327, 371
United Nations Framework Convention on
    Climate Change (UNFCCC), 19, 20, 259,
    326, 327, 376, 414, 426, 427, 429, 433
United Nations General Assembly, 332,
    423, 424
United Nations Human Rights Council,
    424, 434
United Nations Intergovernmental Panel on
    Climate Change (IPCC), 19, 20, 34, 43, 70,
    155, 162, 205, 259, 366, 376
United Nations World Commission on
    Environment and Development
    (WCED), 20
United States
  Congress, 227–229
  Department of Agriculture (USDA), 227, 295
  Department of Housing and Urban
    Development (HUD), 283
  Department of Transportation (DOT),
    272, 283
  Environmental Protection Agency (EPA),
    88, 91, 226, 227, 231, 236, 265, 272, 277,
    283, 284
  Global Change Research Program, 71
  House of Representatives, 227–229
  Senate, 227–229
  Supreme Court, 227, 229

Universal Declaration of Human Rights
    (UDHR), 423, 424
Universal Healthy Reference Diet, 23
Universal Thermal Climate Index, 60
University of Southern California, 406
urban centers, 360
urban design, 63
urban forests, 365, 366
urban growth boundary (UGB), 351
urban heat island, 13, 41, 76, 77, 109, 321, 346,
    347, 362
urban structure, 346, 349
urban tree cover, 362
U.S. Call to Action on Climate, Health, and
    Equity, 410
utilitarian approach, 235

Vancouver, 143
Vancouver Coastal Health, 325
vasodilation, 75
vector management, 323
vectorborne diseases, 14, 109–125, 323
vectorial capacity equation, 112
vegetables, 305, 306
vegetation on building roofs and walls, 363
vehicle GHG standards, 272, 273
vehicle miles traveled, 279
vehicle technology and efficiency, 273,
    276–283
ventilation, 89
Vibrio species, 136, 138, 142
vicarious impacts, 189, 190
Vietnam, 41
violence, 4, 14, 75, 205–218
    prevention, 214, 216, 217
vitamins, 294
volatile organic compounds (VOCs), 90, 92, 96
vulnerability, 331
vulnerability and adaptation (V&A) assessment,
    23, 183, 321, 328, 329, 332, 333
vulnerable populations, 14
    heat, 75–77
    respiratory disorders, 97

walking, 274, 282, 352
warm/hot extremes, 9
warning systems, 98

Washington, DC, 378
Washington, State of, 278
wastewater, 143
wasting, 167
Water Conflict Chronology, 211
water, sanitation, and hygiene systems, 323
water supply, 410
water vapor feedback, 41
waterborne diseases, 11, 14, 133–146, 323
water-related conflicts, 211
weather, 4
weather-related disasters, 373
West Africa, 155, 212
West Antarctic ice sheet, 37, 43
West Nile virus (WNV), 110, 114–117, 326
western equine encephalomyelitis virus
    (WEEV), 120, 121
Wet Bulb Globe Temperature (WBGT), 53,
    55–57, 60, 61
wet surface evaporation approach, 63
wetland ecosystems, 369
wheat, 293, 294, 304
wheat stem rust fungus, 158
wildfires, 12, 78, 88, 92, 98, 183–185, 330
wind power, 252, 255, 256
wind and solar power, 252, 255. See also
    solar power
wood as fuel, 101
work hours, 60
work intensity level, 55
workers, 14, 51–64
workplaces, 343
World Health Organization (WHO), 21, 53,
    143, 145, 329, 330, 332, 415, 423
World Meteorological Organization (WMO),
    20, 53, 71
World Resources Institute (WRI), 428
World Weather Attribution (WWA), 39, 41

Yale, 387
yield, 156

zero-emissions, 268
    vehicle, 284
Zika, 109
zoning, 350
zoonoses, 111, 137